Essentials of Nanobiotechnology

Essentials of Nanobiotechnology

Clive Jarvis

STATES
ACADEMIC PRESS
www.statesacademicpress.com

States Academic Press,
109 South 5th Street,
Brooklyn, NY 11249, USA

Visit us on the World Wide Web at:
www.statesacademicpress.com

ISBN: 978-1-63989-187-0 (Hardback)

Cataloging-in-Publication Data

Essentials of nanobiotechnology / Clive Jarvis.
p. cm.
Includes bibliographical references and index.
ISBN 978-1-63989-187-0
1. Nanobiotechnology. 2. Nanotechnology. 3. Biotechnology. I. Jarvis, Clive.
TP248.25.N35 N36 2022
660.6--dc23

Table of Contents

Preface

Nanobiotechnology is a sub-discipline of nanotechnology and biology. It is an inter-disciplinary field which draws on the principles of mechanics, electronics, optics, biology and nanotechnology. The key objective of nanobiotechnology is the application of nanotools to the biological and medical problems. Some of the key applications of this discipline are nanomedicine, stem cell treatments, protein folding, etc. Nanomedicine is a field of medical science that focuses on biological machines and nanorobots. Nanobiotechnology is also applied in the fields of nanodevices, nanoparticles and various other nanoscale phenomena. Nanobiotechnology is an upcoming field of science that has undergone rapid development over the past few decades. Different approaches, evaluations and methodologies on nanobiotechnology have been included herein. Coherent flow of topics, student-friendly language and extensive use of examples make this book an invaluable source of knowledge.

To facilitate a deeper understanding of the contents of this book a short introduction of every chapter is written below:

Chapter 1- The discipline wherein tools from nanotechnology are used and developed to the study of biological phenomena is termed as nanobiotechnology. It is involved in the development of nanodevices. This is an introductory chapter which will introduce briefly all the significant aspects of nanobiotechnology as well as the role of nanobiotechnology.

Chapter 2- The particles which range in size from one nanometer to hundred nanometers are known as nanoparticles. The structures whose at least one dimension lies in the nanometer scale range is termed as a nanostructe. This chapter discusses in detail the major aspects of nanoparticles and nanostructures such as carbon based nanomaterials and nucleic acid nanostructure.

Chapter 3- Nanotechnology has a wide range of applications in the field of medicine. These fall under the umbrella of medical nanobiotechnology. It is used for the development of nanodiagnostic tools as well as drug discovery and development. The topics elaborated in this chapter will help in gaining a better perspective about the different aspects of medical nanobiotechnology.

Chapter 4- Nanotechnology can be used within the food industry for various purposes. A few of them are using nanobiosensors for increasing food safety, nanoformulations and using nanotechnology for food packaging. This chapter closely examines the different applications of food nanobiotechnology to provide an extensive understanding of the subject.

Chapter 5- Biogenic nanoparticles are the particles which are synthesized through the use of biological organisms are known as biogenic nanoparticles. Nanoparticles which are able to avoid recognition by the immune system is known as a stealth particle. This chapter has been carefully written to provide an easy understanding of the varied facets of biogenic and stealth nanoparticles.

I owe the completion of this book to the never-ending support of my family, who supported me throughout the project.

Clive Jarvis

Chapter 1

Nanobiotechnology: An Introduction

The discipline wherein tools from nanotechnology are used and developed to the study of biological phenomena is termed as nanobiotechnology. It is involved in the development of nanodevices. This is an introductory chapter which will introduce briefly all the significant aspects of nanobiotechnology as well as the role of nanobiotechnology.

Bionanotechnology is a science that sits at the convergence of nanotechnology and biology. Nanobiology and Nano biotechnology are other names that are used interchangeably with bionanotechnology. The field applies the tools of nanotechnology to biological problems, creating specialized applications.

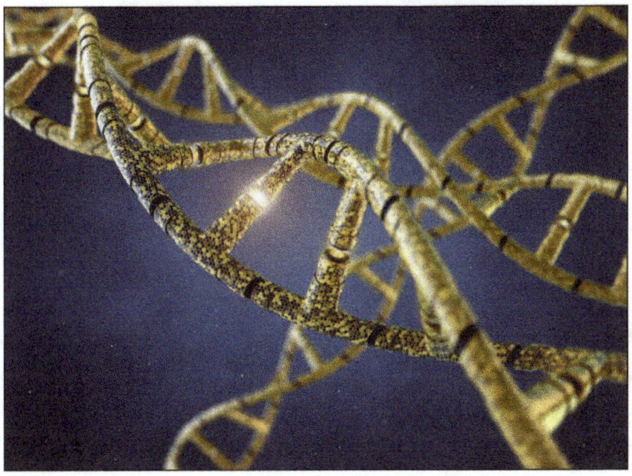

Nanotechnology is usually defined as the manipulation of materials that range from the nanometer (nm) to the micrometer (um) scale. For comparison, the cells of living organisms are typically around 10 um across. That puts the machinery and components of living cells within the nanoscale size range, making them ideal for interactions with functional nanoparticles and nanomachines.

Nanoscale materials have unusual properties distinct from bulk materials. Some of these enhanced properties include surface area, cation exchange capacity, ion adsorption, and complexation. A high proportion of the atoms in a nanoparticle are present on its surface, meaning that compared to macro-scale materials, nanoparticles have different surface compositions, different reactivity, and different types of surface interaction sites.

Role of Nanotechnology

Nanotechnology is enormously being investigated in the field of drug delivery with the aim of maximizing therapeutic index of bioactives. Growth in the field of nanomedicine has made remarkable progress in the recent years, yet the availability of nanotechnology based products require further investigation since the variation in physiological barriers, biological environment, and physico-chemical properties of new reagents are far more demanding than in the beginning might have been expected. Nanotechnology based nanocarriers including nanoparticles, liposomes, polymeric micelles, nanostructured lipid carriers (NLCs), and polymeric drug conjugates with a size less than 200 nm, have shown hopeful efficiency in the field of medicine. Please refer to Figure where different nanotechnology based carriers have been shown along with their unique rewards.

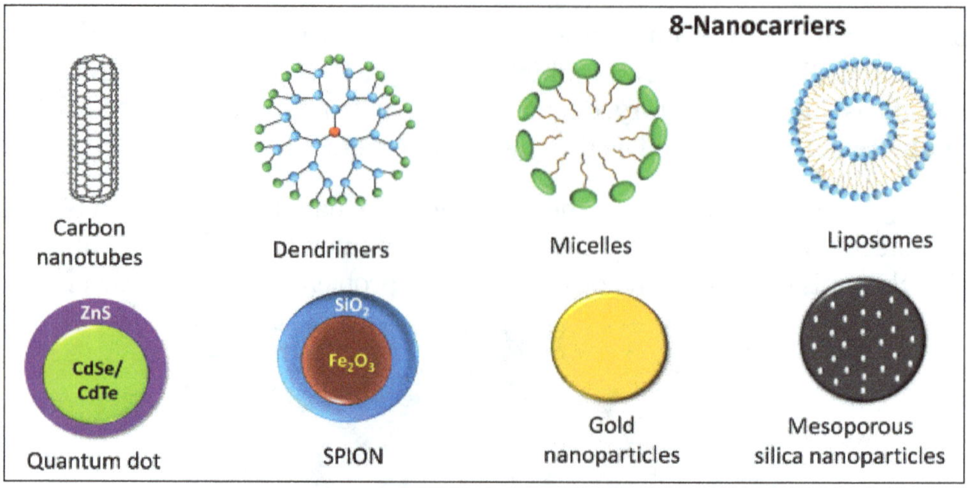

Nanocarriers

Nanomedicines

Nanotechnology is showing an important part in the field of medicine and drug delivery. Healthcare system is currently witnessing a lot of progress attributable to nanotechnology. Nanomaterials used in medicine are studied under the science "nanomedicine" and the aim of nanomedicine is to increase the therapeutic benefit of medicinal substances with simultaneous reduction in the toxic illustrations to increase the therapeutic index of medicine. Apart from medicine, nanotechnology is also playing important role in various industrial applications.

Significant progress has been made in last decades in the meadow of nanotechnology which rendered the possibility of characterization, fabrication, and alteration of useful properties of nanoparticles for biomedical applications. Nanobiotechnology links the physical and biological sciences, with uses of nanophase and nanostructures in various fields of science, especially in biomedicine, in which these substances are of immense interest. Nanoparticles are swaying the field of medicine from biosensors, drug delivery, nanobiotechnology, microfluidics, and microarrays to tissue micro-engineering. Metal, polymeric and organic nanoparticles and liposomes are frequently being investigated and designed for the site-specific delivery of drugs.

Applications of Nanobiotechnology

Nanobiotechnology is being employed in different branches of science and industry for unique and fascinating benefits of nanomaterials. In today's world nanoscale materials are touching most of the items of routine use like cosmetic products, musical instruments, sports accessories and medicines etc. It has been forecasted for analysts that selling of nanotechnology based products may reach up to \$3.1 trillion by 2015. Some important industrial applications of nanotechnology are depicted in figure. Figure shows how nanotechnology is being investigated in textile industries. In Figure you can see nanotechnology is found to be useful in sports industry to increase durability, strength and reduce weight of sports items. Figure is showing various industrial applications of nanotechnology. Figure is illustrating particularly biological applications of nanotechnology.

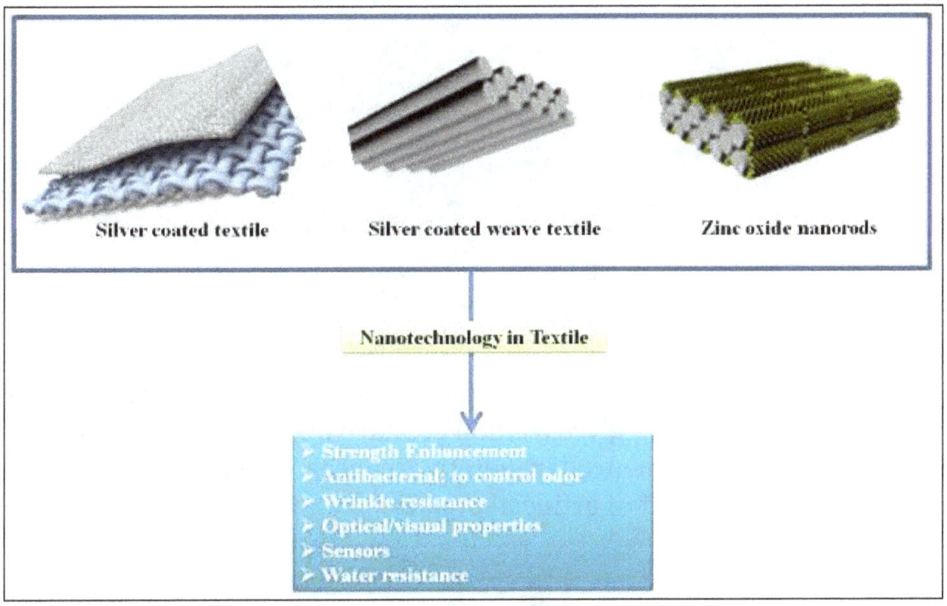

Applications of nanotechnology in textile industry.

Some of the important applications of nanobiotechnology are discussed in the following points:

- Nanobiotechnology is being developed as potential tool to increase global production of agricultural products to meet the day to day demands by increasing population in the form of nanofertilizers that can increase yield of crops, combat with harsh environmental conditions and plant-growth-enhancing nanomaterials that can increase agronomic yield of crops etc.
- Nanotechnology is speculating new hope in drug delivery to ease targeted and excellently tuned treatment of different diseases ailments even at molecular level to enlarge efficiency of treatment strategy.
- Biocompatible nanomaterials like nanoparticles, solid lipid nanoparticles, nanostructured lipid carriers, nanocrystals, lipid drug conjugates, nanoliposomes, nanoshells, dendrimers, nanoemulsion, quantum dots and nanotubes etc. are extensively being explored for delivery of therapeutic agents including smart bioactives like nucleic acids, proteins, artificial RBCs and so on.
- Nanomedicines are showing enormous potential in drug delivery because of nanoscale size-derived chemical, physical, and biological properties.
- One of the most important features of nanomaterials is that they can be easily manipulated to incorporate desirable drug delivery properties to improve therapeutic index and patient compliance; the ultimate aim of any treatment strategy.

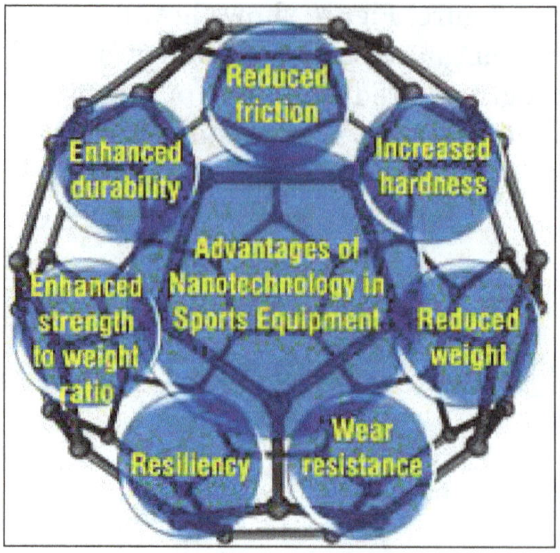

Applications of nanotechnology in sports.

- Surface engineered nanocarriers with different strategies like ligand, antibody conjugation, PEGylation, acetylation, glycosylation etc have been explored in treatment of various serious disorders including neurodegenerative, cancer, genetic, HIV and autoimmune diseases as well as development of artificial organs etc.

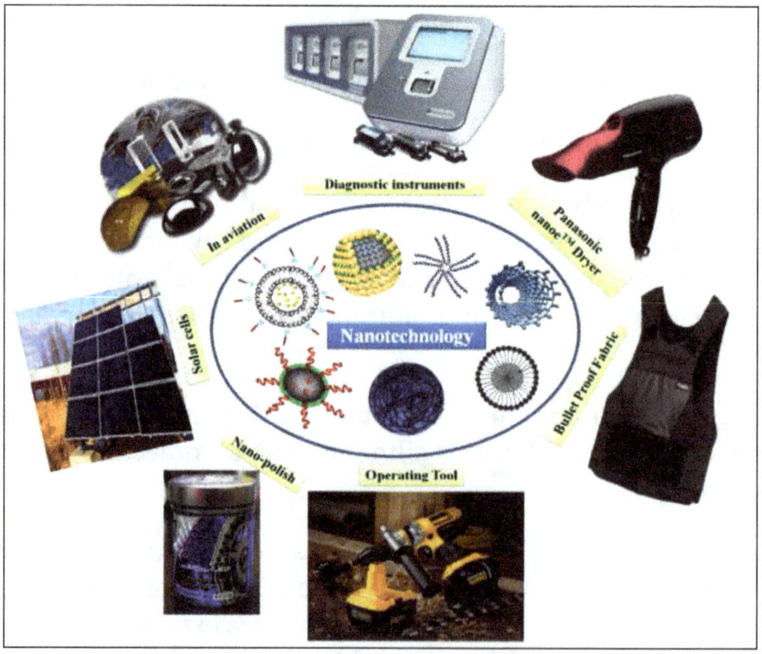

Various industrial applications of nanotechnology.

- Vesicular nanocarriers like liposomes, ethosomes, transfersomes, invasomes etc have been investigated by scientists for skin delivery and were found to increase skin permeability of hydrophilic drugs. Figure shows graphically how these vesicular carriers like ethosomes can modulate the skin permeability. Transdermal delivery of bioactives endures the advantages like bypass first-pass metabolism, maintains constant plasma drug concentration for a long period of time, and delivers the drug to the local site of action with improved patient compliance.

The success of nanotechnology in the field of nanomedicine is evinced by the presence of marketed formulations based on nanomaterials. Various potential applications of nanotechnology particularly relevant to field of medicine are summarized in figure.

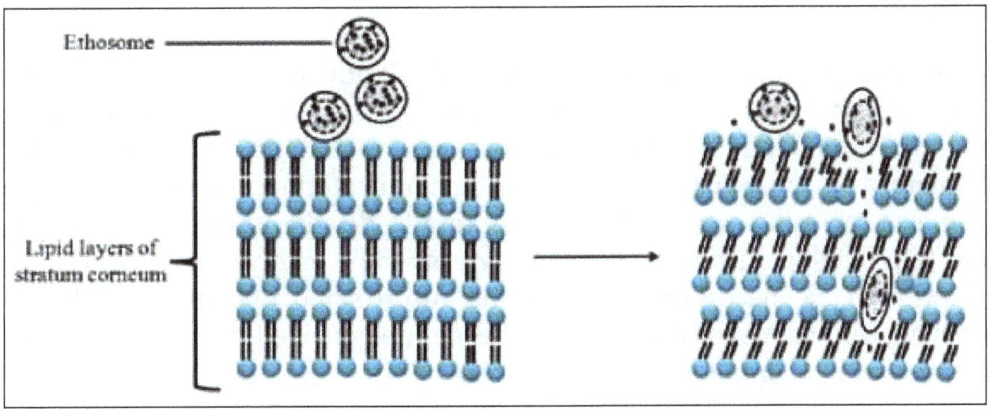

Modulation of skin permeability with vesicular nanocarriers like ethosomes.

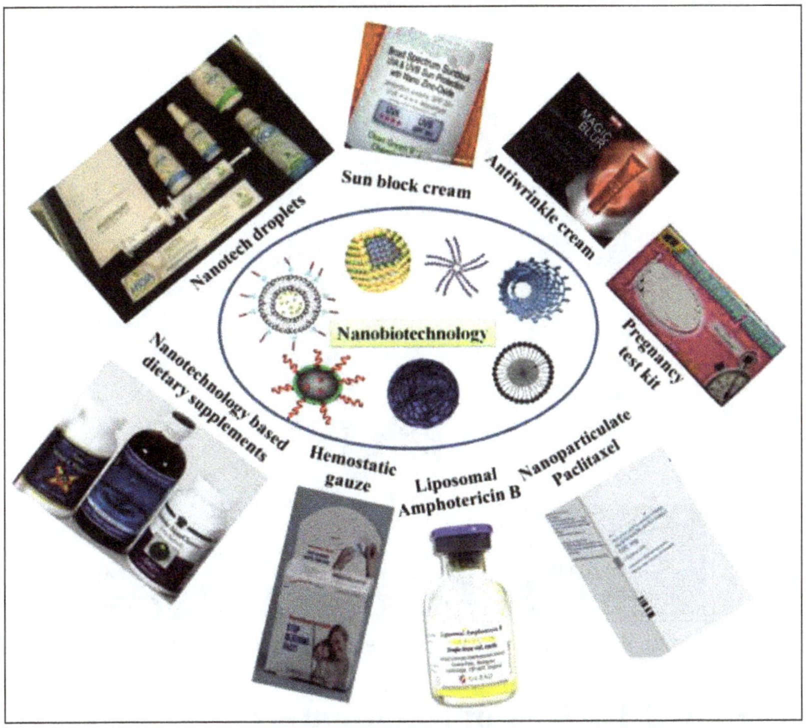

Biotechnological products based on nanotechnology.

Nanobiotechnology and Healthcare Industry

The potentials of nanomedicines are fascinating due to their ability to provide nanomedicines having improved therapeutic efficacy, reduced toxicity, targeted delivery, sensitive and less expensive diagnostic tools.

Nanostructured materials shows exclusive medical benefits attributed to their size and structure. Scientists are focusing on drugs that are difficult to formulate when using conventional techniques due to their poor water solubility, and hence to improve bioavailability. Development of engineered nanocarriers having ability to deliver drugs preferentially to diseased target site is an important implication immense potential.

Nanotoxicology

Although, very small size of nanomaterials, arbitrarily 1-100 nm or less, renders them unique physical, chemical and biological properties, which results in some potential benefits, yet their non-specific interaction with biological components may result into some toxicities like oxidative stress, inflammatory responses or even some drastic manifestations. To ensure safety and efficacy of nanotechnology, scientific committees are working to develop systematic report on their applications and to provide guidance for safety programs to be involved with nanotechnology. It is therefore necessary to take necessary steps to assure future safety of mankind and environment as well as other living beings. A combined effort is desperately needed to ensure environmental fate, hazards and risk of nanotechnology.

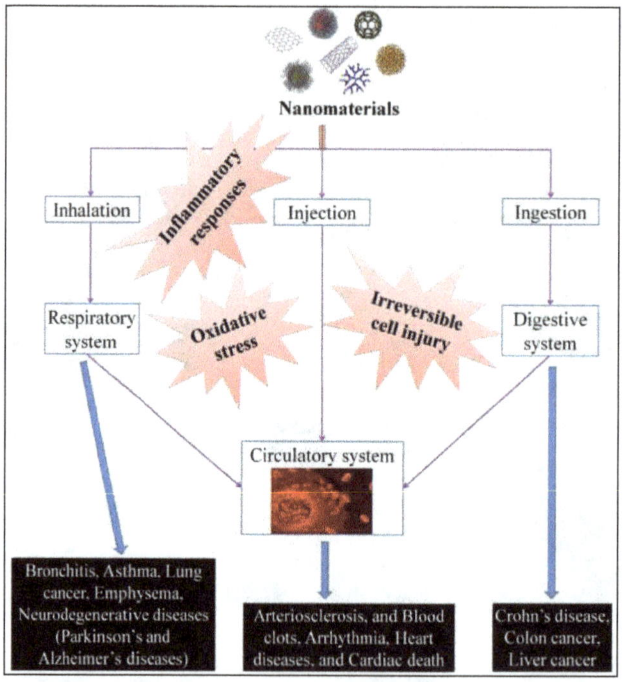

Schematic illustration of toxicity mechanism of nanotechnology.

Reasons for Anticipated Toxicity of Nanomaterial

- Attributed to their extraordinarily small size and high surface area per unit of mass, nanomaterials could interact with cellular and sub-cellular components leading to toxicological consequences.
- Although nanotechnology has shown promising potential in drug delivery applications yet it believed that these nanomaterials may sometimes elicit toxic manifestations which could be speculated to be fatal.
- It has been observed that one of the nanomaterials, multi walled carbon nanotubes (MWCNTs), can elicit upshots like a human carcinogen, chrysotile asbestos. Fortunately

till date no direct case of nanomaterials linked toxicity has been reported in human even when around 2000 nanomaterials are currently in use or being developed for use.

- Owing to extremely fine size these nanoparticles may enter through three portals of entry i.e. lung via inhalation, gastrointestinal tract by ingestion, and systemic circulation via injection. After entry into biological environment these particles could reach and react with vital organs including lung, lymph node, heart, kidney, bone marrow, spleen and central nervous system and may lead to potential toxicity.
- In various in vitro studies these nanoparticles have shown toxicity including inflammation, cardiovascular and respiratory disorders depending on chemical nature and dose.

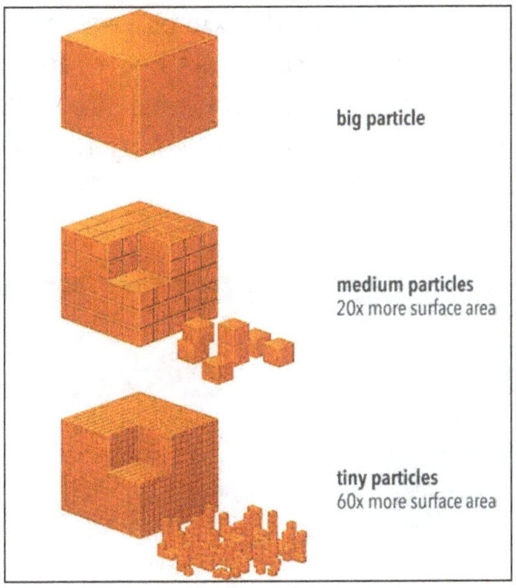

Diagrammatic representation of enormous increase in surface area of nanomaterials in comparison to big particles.

The high reactivity of nanoparticles in comparison to larger particles is due to their high surface/ mass ratio, which is the result of their nanoscale size since particle surface is the principle site of chemical reactivity and biological interactions.

Mechanism of Toxicity

Researchers are continuously investigating as to how nanomaterials interact with cells, animals, humans and environment as well as the influence of chemical, physical, and other important properties of nanomaterials on these interactions. Rather the interaction of nanomaterials with environment is a complex process yet a prospective mechanism of toxicity of nanomaterials lies in their continuous contact with gastro-intestinal tract, human skin, lungs and the environment. The primary protective system of body i.e. skin and mucus membrane can prevent the interaction of foreign particles. But the other two portals of entry for foreign particles gastrointestinal tract and lungs are rather susceptible to entry of natural or anthropogenic nanomaterials. After entry through these portals nanomaterials may translocate in to systemic and lymphatic circulation, which then transport these nanoparticles to different organs, tissues and cells. Interaction of some of these nanoparticles may cause permanent damage to cells by injuring the cell organelles or oxidative stress.

Not all nanomaterials are toxic or produce adverse effect to health; various factors such as size, composition, aspect ratio, surface groups, charges, physical and aggregation state of nanomaterials and genetic constituents of an individual etc. affect toxicity of nanoparticles to living organisms. Toxicity of nanomaterials is different and may be more severe in comparison to normal toxicants owing to their nanometric size range, which ascribes them larger surface area and ability to translocate within intracellular compartment which is not the case with normal toxicants.

Toxicity of nanomaterials is not a new concept, as we are aware of toxicity of natural asbestos fibers and synthetic biopersistence vitreous fibers, which can increase the risk of pulmonary fibrosis and one of the most fatal diseases, cancer. Carbon nanotubes (CNTs) which have dimensions in nanometers and fibrous structure are also found toxic in few reports owing their high aspect ratio (length to diameter ratio) of greater than 100. It has been reported that CNTs could affect adversely biological system depending upon the interactions site and may impose potential hazards. Toxicity of some of the nanomaterials have investigated by few scientists in "in vitro" conditions and "in vivo" preclinical studies as shown in figures.

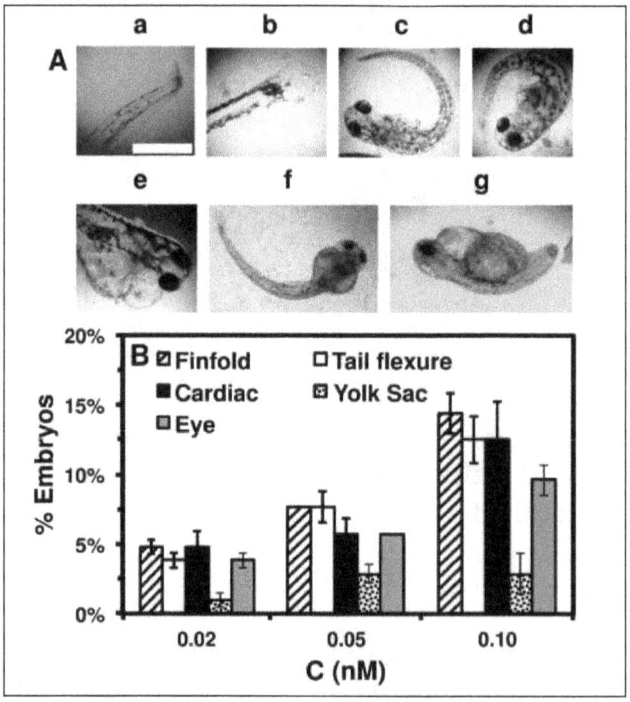

Figure shows dependence of types of deformed zebrafish on the silver nanoparticles concentration. (A) Optical images of deformed zebrafish show: (a–b) finfold abnormalities; (c–d) tail/spinal cord flexure; (e) cardiac malformation and yolk sac edema; and (f–g) eye abnormality. (B) Histograms of distributions of embryos that developed to deformed zebrafish with five distinctive types of deformities at nanoparticle concentration of 0.02, 0.05 and 0.10 nM. A total of 36 embryos were studied for each nanoparticle concentration and control. The percentage of embryos that developed to any given type of deformed zebrafish at any given concentration was calculated by dividing its number with the total number of deformed zebrafish observed in all concentrations. The means and standard deviations (error bars) of each given percentage at each given concentration from each of 3 replicates are presented. Scale bar is 500 μm for all images in (A).

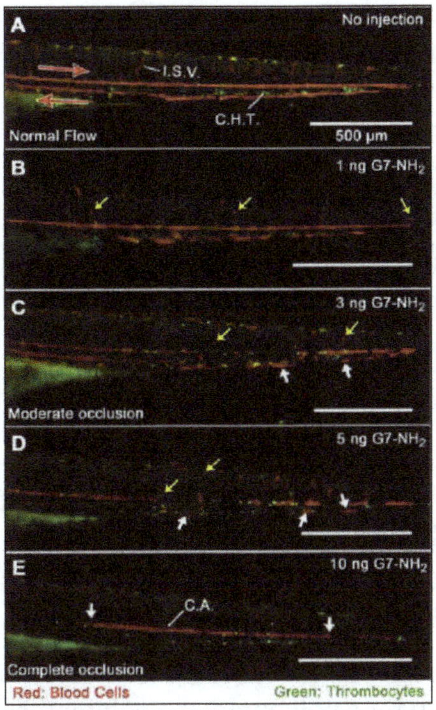

Dose-dependent G7-NH2 occlusion in zebrafish embryos one minute post-injection; (A) Unperturbed caudal blood flow as streaks of red and green light is visible in the major caudal artery (CA, left to right), vein (right to left), and intersegmental vessels (ISV). Punctate green and red spots in between the major caudal artery and vein indicate the location of the caudal hematopoietic tissue (CHT). Subsequent images (proceeding downward (B-E), with increasing G7-NH2 dose) depict increased cellular adhesion and vascular occlusion resulting from dendrimer injection, culminating in no visible blood flow following an injection of 10 ng G7-NH2. The yellow arrows depict individual red blood cells trapped inside the vessel while; the large white arrows show significant blood clots inside the vessels. Notice in figure, the caudal artery is completely occluded with no observable flow in the major caudal vein.

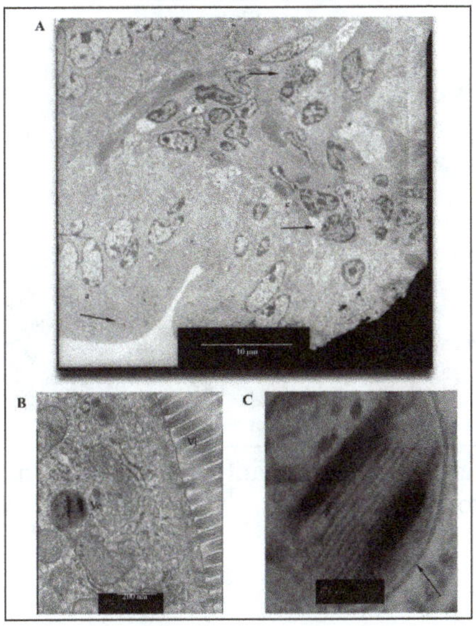

Figure shows the absorption of SWCNTs through intestinal columnar epithelial cells. (A) SWCNTs (arrows) found in the intestinal mucous membrane. (B) Magnification of the cell indicated by the left arrow in (A). Ve, transportion vehicles; Vi, villus of the columnar cells. (C) Magnification of the Ve, which has membrane with double lipid layers (arrow).

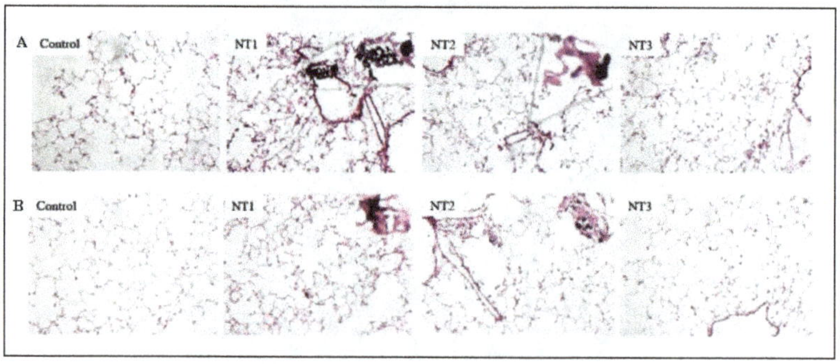

Figure shows the effects of MWCNTs in vivo in mice - lung histology; (a) Lung histology 1 day after a single intratracheal instillation of vehicle or NT1, NT2 or NT3 (100μg/mouse, magnification×20), (b) Lung histology 30 days after a single intratracheal instillation of vehicle, NT1, NT2, or NT3 (100μg/mouse, magnification ×20).

NT1: Uncoated multi-walled carbon nanotubes (MWCNTs), NT2: MWCNTs coated with carboxylic polyacid polymer, and NT3: polystyrene polybutadiene polymethacrylate polymer coated MWCNTs.

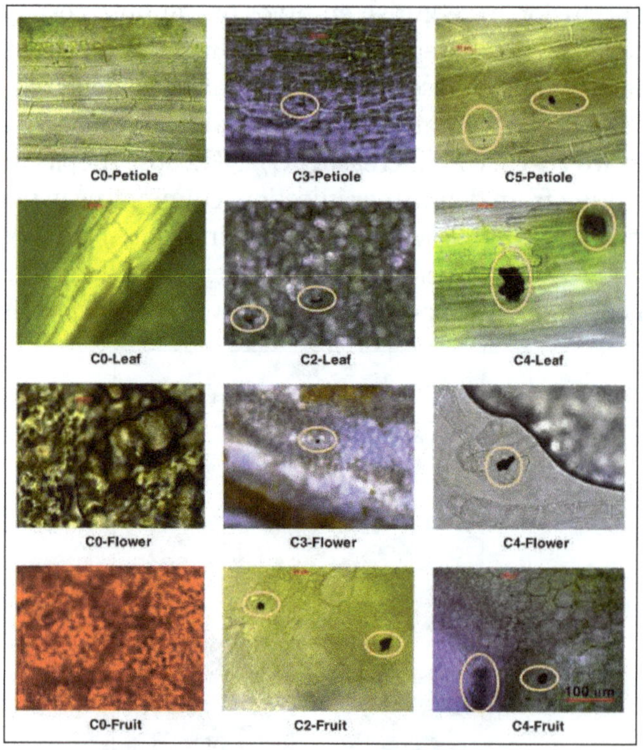

Accumulation of fullerols into different organs of plant, bitter melon. The circles highlight black aggregates and C0-C5 indicates concentration of fullerol (0.0, 0.943, 4.72, 9.43, 10.88, and 47.2 nM, respectively).

Regulatory Aspects

Nanotechnology is offering extraordinary breaks in the treatment of diseases by providing opportunity for development of vaccines, new treatment strategies, diagnostic and theranostic tools, gene and immune-therapy and a new direction to personalized medicine system.

- The most important obstacles in the clinical translation and regulatory approval of nanomedicines are the associated ambiguous toxicity and unwanted health hazards.
- As nanotechnology is progressing it is also necessary to frame stipulated policy, laws and ethics to ensure their safety and efficacy as we know that some nanoparticles might elicit toxicity that can cause potential hazards to human beings and environment.
- Address of regulatory, ethical and social viewpoints of nanotechnology is important to minimize their toxic effects and maximize their benefit.
- This could be done only after performing systematic and cumulative experiments to evaluate the toxicity of nanomaterials and management of risk associated with nanotechnology.
- The first question in this regard is that 'can regulatory system check the potential risks associated with nanotechnology'. To answer this question it is necessary for various federal agencies to present stipulated legislative guidelines to render a proactive regulatory approach to develop nanotechnology.

One another important parameter to check safety of nanomedicine is education of society members about the pros and cons of nanotechnology that will maintain the public support for government to maintain the quality of nanomaterials.

Drug Targeting

Conventional administration of the drug leads to non-selective drug distribution and peripheral drug loss. This results into sub-therapeutic dose at desired site. Consequently drug resistance may occur. Although oral or parenteral administration of conventional drug delivery systems often results in a fast response but it is frequently associated with the problems like:

- Decreased efficacy due to degradation of bioactive.
- Low therapeutic index.
- Side effects.
- High cost of medication due to excess dosing required to give therapeutic concentration at diseased site.
- Administration pain with parenteral route.

Hence, to get rid of all these drawbacks, explorations are going on throughout the globe to deliver the drug effectively to the diseased site. This involved development of drug carriers like liposomes, ethosome, nanoparticles, nanocomposites, CNTs, dendrimers etc. which are smart enough to deliver the drug in intended concentration/rate at site of action known as targeted drug delivery.

The objectives of targeting include:

- Desired therapeutic response.
- Reduced or no side effects.

- High therapeutic index i.e. high therapeutic efficacy with reduced side effects.

The very first report on drug targeting was given by Paul Ehrlich as his theory of "the magic bullet" as an approach to treat diseases. Ehrlich speculated that development of a moiety with propensity to preferentially target the diseased site or the organism causing the disease. A drug that can kill these disease causing organisms could be delivered along with this moiety having selectivity towards the diseased site or disease causing organism. Hence, a "magic bullet" can be designed that will selectively kill targeted organism. The concept of a "magic bullet" could be seem to came in reality with the development of monoclonal antibodies which have high affinity and specificity for their corresponding antigens.

Components of Drug Targeting

Since the early 1960s, many scientists have dedicated their research to advancement of drug targeting strategies for the treatment of various diseases. Usually the objective of targeted therapies is to augment the effectiveness and decrease the toxicity of drugs. A drug targeting system usually has three components namely; targeting moiety, carrier and therapeutic drug molecule. Role of targeting moiety is to recognize and bind the target. Carriers carry the drug molecule and therapeutic drug molecule provides a therapeutic action to the specific/desired/target site.

A targeting system could be briefly defined as a nanocarrier that carries attached drug to the site of action, with its path monitored by attached targeting moiety which can recognize and bind a particular target site.

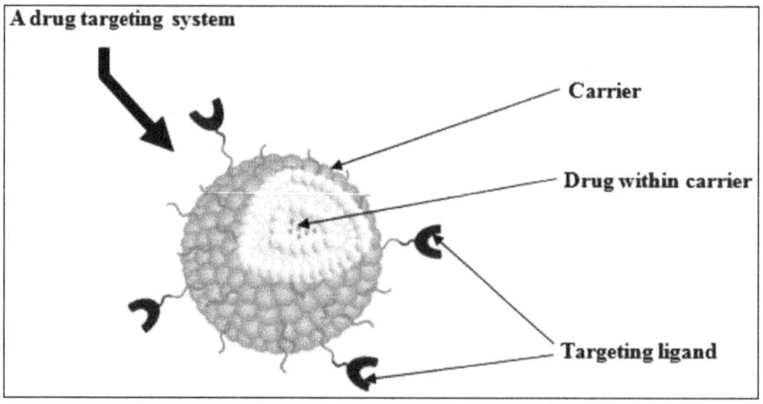

Schematic representation of a typical drug targeting system.

Technical Advantages of Targeting

- Higher efficacy resulting from selectively targeting and/or killing diseased cells or disease causing moiety.
- Reduced toxicity and lower side-effects.
- Potentially more cost-effective.
- Prospect for shorter treatments times.

Therefore with the aid of targeted drug delivery medication could be delivered preferentially to the right place in a body, at perfect time, in a right dose for the right period of time with minimum reach to the places where drug related side-effects may be elicited.

Levels of Targeting

Depending on the part being targeted, drug targeting could be divided in following levels/orders:

- First order: First order targeting encompasses restricted distribution to the capillary bed of a predestined target site which could be any tissue or organ. This order is also known as organ compartmentalization. Ex. targeting to liver, brain and cancer tissues etc.
- Second order: Second order targeting includes selective delivery of drug molecule to a specific cell type. Ex. Targeting of drug to cancer cells, alveolar cells.
- Third order: In third order targeting drug is specifically delivered to intracellular site of target cells is known as intracellular or third order targeting. Ex. Targeting to mitochondria, lysosome, nucleus.

Barriers to Drug Targeting

As discussed earlier to attain a needed pharmacological response at a selected site with minimal toxic side effects is the most important objective of targeted drug delivery. This objective is being difficult to achieve due to certain barriers to targeted drug delivery including physiological barriers, biochemical and pharmaceutical challenges. Biochemical challenges present obstacle in identification and validation of molecular targets and pharmaceutical challenges presents difficulty in design of appropriate techniques of conjugating targeting ligands to the drug delivery system. Further after targeting, retaining the drug at a specific site for the desired duration and its uniform distribution also constitute another major challenge. In order to tackle these challenges different approaches for drug targeting have been explored here.

Approaches of Drug Targeting

Major approaches for drug targeting could be classified in the following targeting approaches:

- Active targeting,
- Passive targeting,
- Inverse targeting,
- Double targeting,
- Dual targeting,
- Combination targeting.

1. Passive Targeting: Passive targeting is achieved by incorporating therapeutic agent into a macromolecular or nanoparticulate carrier that passively reaches the target organ. Due to size in the colloidal range, drugs loaded within nanoparticles or conjugated to macromolecules could target cancer cells passively via enhanced permeation and retention (EPR) effect. Drugs encapsulated in

nanoparticles or drugs coupled to macromolecules can passively target tumours through the EPR effects. Accumulation of drug/drug carrier system at a particular site due to physicochemical or pharmacological factors or by making use of pathophysiological and anatomical opportunities is known as passive targeting. Such pathophysiological and anatomical factors are:

- Pathophysiological factors:
  - Inflammation/infection,
  - EPR effect.
- Physicochemical factors:
  - Size,
  - Molecular weight.
- Anatomical opportunities:
  - Catheterization,
  - Direct injection.
- Chemical approaches:
  - Prodrugs,
  - Chemical delivery systems.

2. Active Targeting: In drug targeting, mainly two types of strategies can be distinguished: passive targeting and active targeting. In the case of passive targeting, the carrier–drug complex is often delivered to macrophages and other cells of the monocyte-phagocytic system. This leads to gradual degradation of the carrier and (slow) release of the liberated drug from the cells either into the blood circulation or into the tissue environment. By size exclusion, extravasation of the carrier drug complex can be limited, thereby preventing the drug from being distributed to non-target sites. As a consequence, toxicity can be reduced.

Active targeting should lead to a higher therapeutic concentration of the drug at the site of action. This can be accomplished by cell specific delivery of the drug. In the ideal case, the dose of the drug can be reduced and side-effects can be diminished. The majority, if not all, active drug targeting strategies exploit receptor-based drug targeting principles, in which receptor-specific ligands attached to the carrier–drug complex or directly to the drug itself deliver the drug to the target cell of choice. Depending on the subsequent routing of the receptor complex, the drug will arrive in a specific compartment in the target cell.

Specific modification of drug/drug carrier systems with 'active' agents having selective affinity for recognizing and interacting with a specific cell, tissue or organ in the body constitutes active targeted drug delivery.

For receptor mediated active targeting different receptors have been explored like:

- Folic acid receptors.

- Low density lipoprotein (LDL) receptors.
- Peptide receptors.
- Lipid components of the cell membrane.
- Surface antigens/proteins.

The moieties that could be recognized by these receptors have been used as Targeting Ligands such as:

- Folic acid.
- Sugars (Galactose, fucose, mannose etc).
- Lectins (Wheat germ agglutinin etc).
- Modified albumins (Cationized albumin, sugar modified albumins etc).
- Peptides (VEGF, LHRH, somatostatin etc).
- Antibodies (MAbs).

(* VEGF: Vascular endothelial growth factor; LHRH: Luteinizing hormone releasing hormone; MABs: Monoclonal antibody).

3. Inverse targeting: In inverse targeting approach attempts are mad to circumvent and avoid passive uptake of colloidal carriers by reticuloendothelial system RES system by reversing biodistribution trend of the carrier.

4. Dual Targeting: Carrier molecules which have their own intrinsic antiviral effect thus synergies the anti-viral effect of the loaded active drug.

5. Double Targeting: Combined spatial and temporal control of drug delivery or combined active / passive drug targeting to control drug release and achieve:

- Sustained release,
- Stimuli responsive release,
- Self-regulating release.

6. Combination Targeting: The targeting systems are equipped with carriers, polymers and homing devices of molecular specificity that could provide a direct approach to target site. (For proteins and peptides).

Emerging Trends in Targeting

- Vascular targeting,
- Intracellular targeting,
- Promoter/transcriptional targeting,
- Physical targeting,

- Pre-targeting or sandwich targeting,
- Chemical strategies.

1. Vascular Targeting: This approach is gaining importance due to (a) distinct biochemical differences and (b) easy accessibility by all drugs absorbed or injected in the blood stream. Ex. Targeting to tumor vasculature via cell adhesion receptor, integrins (αvβ3), annexin A1 etc.

2. Intracellular Targeting: For drugs -whose site of action is located in the intracellular compartments or which undergo extensive efflux from the cell by the efflux transporters such as multidrug resistance proteins (MRP) and P-glycoproteins (Pgp). Ex. Glucocorticoids (e.g., dexamethasone), most anticancer drugs (DNA intercalaters), Proteins & plasmid DNA.

3. Promoter/transcriptional Targeting: It could be exemplified by targeted gene therapy which involves use of specific tissue, tumour or induced promoters that can limit gene expression to target cells which express a specific transcription factor. This targeting approach:

- Restricts the transgene expression in target tissue.
- Allows flexibility to regulate the duration and level of expression exogenously.

Ex. Promoter for telomerase gene – tumor specific α-fetoprotein gene

4. Physical Targeting: Selective drug delivery programmed and monitored at the external level with the help of physical means like ultrasound, magnetic field etc is known as physical targeting:

- Ultrasound: Triggers the release at the target site and enhance the cellular uptake of the drug.
- Proposed mechanisms:
  - Ultrasonic triggered extravasation of polymeric micelles.
  - Ultrasonic activated drug release from micelles.
  - Ultrasonic enhanced uptake into the cells.

5. Magnetic field: The use of localized magnetic field gradients to attract the particles to a chosen site, and to hold them there until the therapy is complete and then to remove them. Effective targeting of magnetic nanoparticles via systemic administration would require a targeting magnetic force that offset the force due to linear blood flow rates of .0.05 cm/s in capillaries to .10 cm/s in arteries and .50 cm/s in the aorta – higher magnetic moments/ magnets are required.

6. Pre-targeting or Sandwich Targeting: Pre-targeting of specific cells with a ligand followed by treatment with a drug linked to another molecule having high affinity for the ligand used e.g., avidin-biotin, streptavidinbiotin. Ex. Angiogenesis in tumors, targeted immunotherapy for treatment of cancer.

7. Chemical Strategies: Chemical strategies of targeting include prodrug approaches like:

- Antibody directed enzyme prodrug therapy (ADEPT).

- Gene directed enzyme prodrug therapy (GDEPT).
- Chemical delivery systems (CDS).

Targeting is attained by means of chemical design of the system wherein the active drug is converted into a biologically inert form by attaching targetor and a modifier moiety to it, via bioremovable links.

- Targetor – Targeting & providing site specificity to the system.
- Modifier – Lipophilizer & prevents unwanted metabolism of the system.
- CDS undergo sequential enzymatic transformations targeting and releasing the active drug in a controlled manner at the desired site e.g., targeting neuropeptides to brain.

Drug Targeting Systems

For targeting drugs to specific sites, and maintaining pharmacologically relevant drug levels at the site for a period required for desired therapeutic action.

- Soluble carriers (MAbs, Dextrans, Soluble synthetic polymers etc).
- Vesicular/Particulate carriers (Liposome, Micro-and nanoparticles, Microspheres etc).
- Target specific recognition moieties (Mabs, Carbohydrates, Lectins etc).

Smart Technologies:

- Antibody-directed enzyme/prodrug therapy (ADEPT).
- Virus-directed prodrug/enzyme therapy (VDEPT).
- Chemical drug delivery systems.

Liposomes

- Small vesicles composed of unilamellar or multilamellar concentric phospholipid bilayers surrounding one or several aqueous compartments.
- Variation in charge, lipid composition and size – Variation in in vivo behavior.
- Use of PEG prevents liposome recognition by phagocytic cells – Stealth liposomes (longer circulation time, increased distribution to the peripheral tissues).
- Targeting device/homing ligand can be included at the external surface.

Majorly investigated for following purposes in drug targeting:

- Macrophage specific delivery of drugs.
- Passive drug targeting.
- Non-viral vector mediated gene therapy (cationic liposomes & lipoplexes).

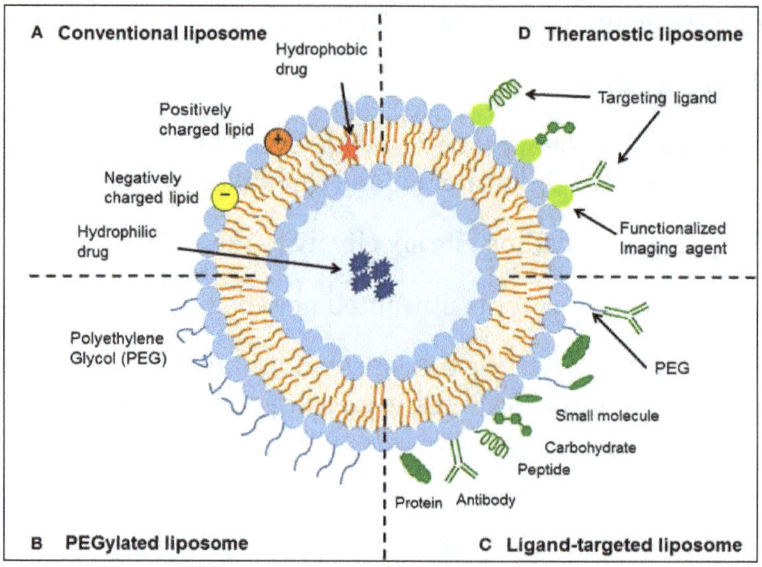

Diagrammatic representation of conventional and modified liposomes.

Monoclonal Antibodies (MAbs) and Fragments

Developed by Kohler and Milstein, MAbs are use in targeting as:

- Targeted effecter molecules.
- Carriers for selective delivery of drugs, toxins, enzymes, radioisotopes and adenoviral vectors.

Antigen recognition by antibodies has been explored in Cancer therapy: Some examples of MAbs used in therapy include anti-TNF α antibodies for treatment of chronic inflammatory diseases; anti-VEGF antibodies inhibit new blood vessel formation or angiogenesis.

Lipoproteins

These are the endogenous lipid particles (LDL i.e. low density lipoproteins and HDL i.e. high density lipoproteins) containing a lipid and apoprotein moiety could work as natural targeted liposomes.

- Lipophilic drugs/prodrugs have been incorporated into lipid cores.
- Apolipoprotein moiety could be glycosylated or modified with other (receptor) targeting ligands.
- Size and charge of particles determines in vivo behaviour.

Exploration with lipoproteins in targeting:

- For hepatocyte targeting of antiviral & antimalarial drugs using lactosylated lipoproteins.
- Oxidized & acetylated LDL has been explored for kupffer cells and sinusoidal endothelial cells targeting.

Microspheres and Nanoparticles

Microspheres are generally made up of biocompatible polymers such as poly(N-(2- hydroxypropyl)methacrylamide) (HPMA), dextrans, sepharose, poly-l-lysine, alginate, PLG etc and belong to particulate carriers. Backbone can be modified with sugar molecules or antibody fragments to introduce cellular specificity.

Nanoparticles are submicron sized polymeric colloidal particles with a therapeutic agent encapsulated within their polymeric matrix or adsorbed/conjugated onto the surface.

- Intracellular targeting by passive or ligand mediated targeting approaches.
- Rapidly opsonized and cleared by macrophages of RES useful for targeting drugs to macrophages and for chemotherapy of RES localized tumours.

Polymeric Micelles

Polymeric micelles could be defined as di-block structure with a hydrophilic shell and a hydrophobic core in size range of 10-100 nm size. Drugs can be incorporated by chemical conjugation or physical entrapment. Hydrophobic core consists of biodegradable, Non-or poorly-biodegradable polymers and serve as reservoir for an insoluble drug.

Functional groups at their surface could be conjugated with a targeting moiety such as attachment of antibodies or sugars, or introduction of a polymer sensitive to temperature or pH could also be done to increase drug delivery specificity. Ex. Polymeric micelles have been explored in passive targeting of anticancer drugs to tumours.

Challenges Appeared/Anticipated in Drug Targeting in Clinical Research

Several problems have been faced/could be anticipated in development of a clinically successful targeted delivery as:

- Immunogenicity after repeated administration.
- Counterproductive liver clearance.
- Production yield.
- When target cell is of non-hepatic origin, extensive uptake by liver is not desirable.
- Validation of each drug targeting system under going development is required via in vivo animal models of diseases in order to ensure therapeutic effectiveness and absence of toxicity even after treatment for a prolonged period of time in terms of following requirements:
    - Proof of targeting potential is required.
    - Pharmacokinetics can be established comparatively with non-targeted therapy.
    - Pharmacological activity of drug in target cell.
    - Kinetics of release of drug.

- To understand structural requirements for targeting important in designing of delivery system.
- Facilitate transdermal administration or programmable infusion system to circumvent the barriers for long term parenteral administration of drug.
- Integration of drug targeting technology in the earliest phases of drug development as many drugs, active at low concentration, are withdrawn from R&D pipeline due to toxicity in non-target cells.

References

- Bionanotechnology-Applications, life-sciences: news-medical.net, Retrieved 15, January 2020
- Nanobiotechnology: inflibnet.ac.in, Retrieved 23, April 2020

Chapter 2

Nanoparticles and Nanostructure

The particles which range in size from one nanometer to hundred nanometers are known as nanoparticles. The structures whose at least one dimension lies in the nanometer scale range is termed as a nanostructe. This chapter discusses in detail the major aspects of nanoparticles and nanostructures such as carbon based nanomaterials and nucleic acid nanostructure.

Nanoparticle

Nowadays production of different nanoparticles (NPs) with plausible biomedical benefits is tremendously increasing. NPs are of great interest in drug delivery systems, drug formulation, medical diagnostic, and biosensor production. Aside from the importance of NPs in medicine, their negative side effects including potential cytotoxicity, inflammatory response induction, and drug interruption should be carefully considered. Several molecular and physicochemical mechanisms are involved in toxicity induction of NPs. Finding the negative effects of NPs on human tissues and investigation of their mechanism of action are a way for preventing the happening of unpleasant event.

Nanotechnology is referring to the technology of production, characterization, and application of materials in nanoscale. After the definition of this term by Norio, nanoparticle (NP) production and application in several different fields gain much attention. The small size and high surface area of NPs are the causes of their tunable physicochemical properties such as improved thermal conductivity, light absorbance, significant chemical stability, and high catalytic activity. Furthermore the surface layer of NPs can be functionalized using chemical and biological agents like small molecules, surfactants, and polymers for enhancing their activity and specificity. It is revealed that each NP, regarding its size, surface charge, shape, surface groups, and type of ions, shows unique biological and physicochemical properties. Owing to such diversity, these materials got immense biomedical applications such as drug delivery, radionuclide therapy, biosensors, cancer therapy, diagnostics, magnetic resonance imaging (MRI), and biological molecules purification.

Although a wide variety of NPs with diverse ions and surface modifications are produced and preclinically tested, only a limited number of them gain approval for clinical uses. The long-term stability, general cytotoxicity and inflammatory response induction, and lack of guideline for relevant biological testing are the main reasons for low-approved NPs. Due to the fast development of NPs, it is necessary to identify correlation between the physical and chemical attributes of NPs and their corresponding biological effects. For instance, it is shown that although the positive charge of NPs enhances the efficacy of gene delivery, and imaging, it also enhances the cytotoxicity of corresponding constructs.

NP Production Methods

Physicochemical Methods

NPs can be produced through two main approaches including top-down approach which is the production of NPs by making smaller and smaller structures by etching the bulk agents and bottom-up approaches which is the building up of NPs from atoms. Physical, chemical, biological, and in some cases hybrid technique are the main ways of NP production. The physical methods of NP production include methods like laser ablation, high-energy irradiation, spray pyrolysis, and ion implantation, and the chemical one includes chemical reduction technique, sonochemical method, solgel process, microemulsion method, and electrochemistry. The biological method which is also called green NP biosynthesis involves application of plants extracts, microorganisms, enzymes, and even some agricultural wastes for NP production.

Although the physical and chemical methods resulted in bulk amount of NPs a few times, application of chemical agents during the NP production in coordination with production of environmentally dangerous compounds simultaneously with NP production limited their applications. For instance, thermolysis which is a chemical method for dissociation of organometallic precursors is performed at high temperatures by using organic solvents. Also in some cases, surfactant is added to the reaction medium for reducing coalescence of particles. Chemical reduction technique is an adopted chemical method which used a wide range of reducing agents such as sodium borohydride, hydroxylamine, and N, N-dimethylformamide for production of zerovalent ions. Wave-assisted chemical method used ultrasonic waves in coordination with surfactant or reducing agent for production of NPs. The formation of micro cavities with high temperatures upon ultrasonic induction can start chemical reduction of substrates.

The physical NP production methods are mainly energy intensive and need special devices. For instance, milling process is a way by which metallic microparticles are crush using high-energy ball mills. The gas-phase process or aerosol process which is divided into four main types (including flame reactor, plasma reactor, laser reactor and hot wall reactor, and chemical gas-phase deposition) is a particular way for the production of NPs like fullerenes and carbon nanotubes. All types of these methods need special devices and are mainly high energy consuming. NP production by wet chemical synthesis takes place at low temperature and is one of the most employed methods for NP production. Limitations in increasing batch reactor because of limited mixing and low heat transfer are mentioned as the main disadvantage of wet chemical synthesis method. The main advantages and disadvantages of NP production by physicochemical methods are summarized in table.

Table: The main physicochemical methods of NP production and their corresponding advantages and disadvantages.

Method	Advantages	Disadvantages
Chemical vapor deposition and chemical vapor condensation (CVD and CVC)	High pure NPs production.	High temperature of procedure (above 300°C) Uses of chemical agents.
Gas condensation	Production of ultrafine nano-crystalline metals and alloys.	Need for special devices Extremely slow.
Laser ablation	High-purity NP production.	Need special devices Difficult to control size, agglomeration, and crystal structures.

Solgel	Simple method Production of large range of materials Uses low temperature.	Using chemical agents Undesirable agents production. The cost of materials may be high.
Chemical reduction	Cost-effective Good production rate.	Application of toxic agents Hazardous by product formation.

NP Biomedical Applications

With respect to special properties of NPs discussed before, they have various applications. Here we investigate some of these applications with special look at their uses in biomedical fields.

Drug Delivery

NPs are of great interest for being used as a device for site-specific drug delivery with optimum dosage drug release. Current NP-based drug delivery approaches focused mainly on enhancing drug shelf life though improving drug uptake efficiency. NP-based drug carriers are able to cross the blood-brain barrier and tight junctions of the skin epithelial tissue. Also they improve hydrophobic molecule solubility and increase stability of biological therapeutic agents.

NPs enabled us to deliver drugs by various routes including nasal mucosa and oral administration, aerosol method, and topical vaccination. The aerosol technology is used for respiratory disorder drug delivery. Target drug delivery approaches using magnetic NPs are widely being used for cancer therapy, gene therapy, MRI, and cell sorting. For instance, Fe_3O_4, γ-Fe_2O_3, and super magnetic iron oxide NPs (SPIONs) are the main NPs used for site-specific drug delivery. The surface properties and particular shape of fullerenes and carbon nanotubes make them attractive for drug delivery. These particles are such small that can pass through cell membrane and deliver agents like DNA and protein into the cells.

Antibacterial Agent

The prevalence of antibiotic-resistant bacteria species becomes a threat for human health. NPs with significant antibacterial properties and no bacterial resistance are the best alternative for common antibiotics. Ag NPs are the leading NP-based antibacterial agents with significant bactericidal effects on both Gram-negative and Gram-positive bacteria. Every day various NPs with different physicochemical properties and bactericidal activities have been developed, and their mechanism of action and potential side effects are under investigations. Also application of common antibiotics such as ampicillin, chloramphenicol, and kanamycin in the presence of NPs demonstrated the positive effects of this combination. Previous studies showed that NPs can be used as a vehicle for antibiotic delivery. The attachment of NPs to the bacterial surface and induction of damages are reported as the main mechanism bacterial death with NPs. Interaction of NPs with bacterial cell membrane and disruption of its normal function are the most common way of NP bacterial killing. NPs are also able to hindrance bacterial biofilm formation. Furthermore, NPs are able to produce different types of ROS species. For example, Mg NPs are able to produce O^{2-}, and ZnO NPs produce H_2O_2 and OH. These ROS species interact with bacterial cells and cause acute stress reactions and finally lead to acute microbial death.

Biosensor

The optical and electronic properties of NPs make them suitable for biosensor application. The size, type of ion, and shape of NPs are critical parameters affecting SPR peaks and line widths of sensor. The noble metals like Au, Ag, and Pt NPs showed special physicochemical features which make them the most popular components of NP-based biosensors. NPs have different roles in any types of biosensors. For instance, electrochemical biosensor is performed by fixing the potential at a suitable value and determining the current changes versus time. The role of NPs in this type of biosensor is to improve sensitivity and signal detection. In optical biosensors, the free electron oscillation in conduction bands of some metals (Ag, Au, Cu) interacts with light photons and produces a polariton. Size tuning of plasmonic metals is a way for enhancing surface plasmon resonance and making the device suitable for biomedical applications. Using NPs leads to reaching to highest detection sensitivity. Au NPs because of their easy functionalization and showing different colors based on their size and shape are good choices for colorimetric biosensor, plasmonic sensing, immune sensors, and electrochemistry.

The Au NPs showed unique stability compared to other metals when used for bio-conjugation production and have valuable sensitive plasmon change which lead to their wide use in classic immunoassays. The stronger Raman and fluorescence enhancement of Ag NPs than Au NPs resulted to their broad uses in optical applications. Also they can easily be oxidized and be used in electrochemical sensors. Ag NPs with ability to detect proteins have been used for cancer detection. Furthermore they were also used for detecting glucose, DNA, dopamine, ascorbic acid, and several other biological molecules. Magnetic NPs are used in sensors through three main approaches including pre-concentration of analyte, magnetic tags, and integration into transducer materials.

Diagnostic Agent

The special features of NPs such as fluorescence properties, optical scattering and electromagnetic field enhancement, and even transferring light energy to heat resulted in a wide application of these compounds in medical diagnostic field. Furthermore, NPs are excellent carriers for delivery of active biomedical agents. Biomedical imaging is one of the useful tools for human disease diagnosis. The NPs with special optical, magnetic, and radioactive properties can enhance the quality of imaging. It is possible to functionalize NPs with multiple modals and by this way minimize the interface between each modality and provide multimodal agent for better imaging. The optical nano-probes can be designed for being used in linear optical imaging with high-emission quantum energy yield and expanded optical capacity.

In the case of fluorescent imaging, the degradation of organic dyes (photo bleaching) and metal complexes under light exposure is alleviated with fluorophore-doped silica NPS. These NPs have been used for untargeted imaging of human epithelial cells of the cervix and targeted imaging of cells A549, HeLa, and HepG2. Phosphorescence imaging using NPs produced images with lower background autofluorescence and scattered excitation light in the spectral range. Magnetic resonance imaging (MRI) used contrast agents for detection of small tumor and lesions in a normal tissue. The NPs with magnetic functionality are used in MRI, and the ones with larger magnetic moment are preferred.

Catalytic Agent

NPs have been developed for various catalytic applications. The NP catalytic reactions have several advantages including low reaction temperature, light transparency, and easily immobilization on solid supports, for instance, the catalytic activity of Au NPs in degrading methylene blue demonstrated by Researcher. Also the effect of geometrical parameters of supported Au NPs on its carbon monoxide oxidation has been evaluated. The NPs with an average diameter of 2 nm and height of six atomic monolayers showed optimum catalytic activity. The Au NPs on amorphous silica support produced by researcher were able to catalyze hydrogenation of cyclohexene. The Ag NPs produced by lychee (L. chinensis) extract showed significant photocatalytic activity even after three times of reusing.

Wound Healing Activity

NPs are a suitable wound dressing agent because of their valuable antibacterial, anti-inflammatory effects and ability to accelerate skin reepithelialization. Reports of healing effects of Ag NPs indicated that these nanoscale materials decrease local matrix metalloproteinase and neutrophil apoptosis. Also they showed inhibitory effects on pro-inflammatory cytokines interferon gamma and tumor necrosis factor alpha. The combination of Ag NPs and collagen results in the formation of component with suitable antibacterial activity. Au NPs do not have any antibacterial effects alone, but their combination with biological agents like collagen and gelatin improve their biocompatibility and biodegradability and make them suitable for wound dressing. Au NP antibacterial properties resulted from their interaction with cell membrane and inhibiting ATP synthase which consequently lead to ROS-independent cell death. The reports indicated that the combination of Au NPs, gallate, and epigallocatechin has positive effects in healing of mouse skin wounds through regulation of angiogenesis and anti-inflammatory effects. Pd, Pt, Se, and ZnO are other promising NPs for regenerative medicine and wound healing. The PAPLAL® solution (Toyokose Pharmaceuticals, Japan) which is a mixture of Pd NPs and Pt NPs showed protective effects against aging-related skin pathologies and normalized the gene expression levels of Mmp2, Has2, TNF-α, and IL-6 in the skin.

Zn NPs have valuable antibacterial effects, and its topical application leads to reduction of inflammation and improvement of skin reepithelialization. TiO_2 NP wound treatment enhances body fluid coagulation by making interaction with blood proteins. The formation of adherent crust of a nanocomposite improved healing of wound and inhibited infection and inflammation.

Also nanotechnology can be used for delivery of active agents with antimicrobial, anti-inflammatory, and healing effects. Curcumin treatment of diabetic wounds leads to significant enhancement in reepithelialization and an increase in fibroblast proliferation of injured tissue. Curcumin NPs not only have higher lifetime than curcumin but also showed valuable antibacterial effects against methicillin-resistant Staphylococcus aureus. With respect to molecular chemistry and self-assembly approaches, it is possible to develop peptide NPs with a variety of medical applications.

Polymeric NPs using both biological and synthetic polymers are of great interest for the development of wound dressing compound. Polymeric NPs are able to stimulate cell proliferation through enhancing angiogenesis and reepithelialization. They are able to stimulate the infiltration of inflammatory cells in the initial phase of healing. Furthermore, they are suitable carriers for therapeutic agents including cytokines, growth factors, and antibiotics which make them suitable for being used in treatment of both normal and delayed infectious wounds.

NP Toxicological Consideration

Regarding the extensive uses of NPs in foods, paper, drug delivery, biosensor, cancer therapy, and imaging, looking for possible toxicity and long-term exposure side effects and finding the mechanism underlying the adverse effects of NPs seem necessary.

Any toxicity induction of NPs is strongly related to NP base material, shape, size, and functional groups coated at their surface. The smaller NPs have a larger specific surface area which in turn leads to higher interaction cell components including nucleic acid, proteins, and carbohydrates. Also the smaller NPs can penetrate better into cells and interact higher with cells. Surface charge of NPs has strong correlation with their interaction with cells and absorbance. It is showed that the NPs with higher positive charge have higher cytotoxicity effects. Also NPs with positive charge are more toxic than negatively charged NPs. The shape of NPs is the other critical parameter which largely affects their cytotoxicity and antiproliferative effects. For instance, the amorphous TiO_2 NPs produced higher levels of oxidative stress and cell surface defects than anatase TiO_2 NPs. Also the spherical Fe_2O_3 NPs had lower cytotoxic effects than rod-shaped ones. Also cytotoxicity is strongly dependent on the type of cells. For instance, although citrate-capped gold NPs were nontoxic to human liver carcinoma and hamster kidney cells, they were severely toxic to human carcinoma lung cells.

NPs can easily penetrate into the cells and interact with cells' normal functions. ROS formation and consequently oxidative stress induction are the common side effects of metal NPs. The produced ROS disrupt normal cell function through attacking essential biological molecules including DNA, enzymes, and lipids. Peroxidation of membrane lipids; enhancing calcium entrance; release of calcium from intracellular stores, protein kinase C, and mitogen-activated protein kinase activation; and DNA damages are some of the main changes that lead to cell death after interaction with NPs. Furthermore the risk of early apoptosis upon exposure to some NPs such as ZnO and TiO_2 has been demonstrated. Also CuO, NiO, TiO_2, Fe_3O_4, ZnO, and Al_2O_3 NPs can arrest cell cycle and induce apoptosis. It is demonstrated that the phase of cell arrest depends on the type of cell and NPs. G2/M phase arrest is the most common type of cell arrest induced by metallic NPs. The induction of P53 pathway in NCM460 cells and cyclin-dependent kinase 1 downregulation in HaCaT cells after exposure to ZnO and cyclin B1 downregulation in A549 cells by TiO_2 have been reported as the main causes of cell proliferation disruption. Many researches have been done on different cell lines and animal models for finding the mechanism of NP toxicity and physiological changes. The high absorption of gold NPs and their aggregation inside cells are probably the main cause of gold NP toxicity.

Argyria is a condition of the skin and other organs' blue-gray discoloration as a result of long-time exposure to high levels of Ag NPs. Irritation, stomach pain, allergic reactions, and inflammation are reported as the main side effects of body exposure to high levels of Ag NPs. TiO_2 and ZnO NPs are widely being used in cream and lotions as sunscreen or materials for water- or stain-repellent properties. The cytotoxicity induction of TiO2 NPs through increasing reactive oxygen species and lactate dehydrogenase has been demonstrated. It revealed that UV and visible light irradiation enhanced ZnO NP cytotoxicity power. Also the Zn NPs were produced by microwave-assisted method, and its in vivo cytotoxicity and levels of distribution in different tissues have been evaluated. According to the obtained results, the produced Zn NPs were classified as nontoxic agents with highest distribution in the testis, liver, and brain.

Several assays have been developed for the determination of NP toxicity both in vitro and in vivo. Proliferation assay which measures the active cell metabolism is the most popular method for determining the antiproliferative potency of NPs. 3-(4,5-Dimethylthiazol-2-yl)-2,5-diphenyltetrazolium bromide (MTT), XTT, thymidine incorporation, alamar blue, and clonogenic assay are the most popular methods for determining cell proliferation rate. DNA damage and apoptosis induction of NPs which are mainly due to generation of free radicals can be determined by methods like annexin V, comet assay, DNA laddering, and TdT-mediated dUTP-biotin nick end labeling (TUNEL). NPs are able to interact with cell membrane and lead to cell integrity destruction and cell death. This phenomenon which is known as cell necrosis is mainly measured by neutral red and trypan blue exclusion assays. The in vivo assays including biochemical tests, histopathological analysis, hematology, and NP bio-distribution are also used for finding the effect of NPs on normal function of cells and tissues.

Quantum Dots

Quantum dots (QDs) are attracting scientists for various biotechnological applications due to their imaging potentials as well as ability to be explored for targeted drug delivery. Quantum dots are being investigated particularly to improve diagnosis of cancer to give a strategy for early detection of cancer. Recently QDs are particularly being explored as an alternative to fluorescent probe to study cell biology.

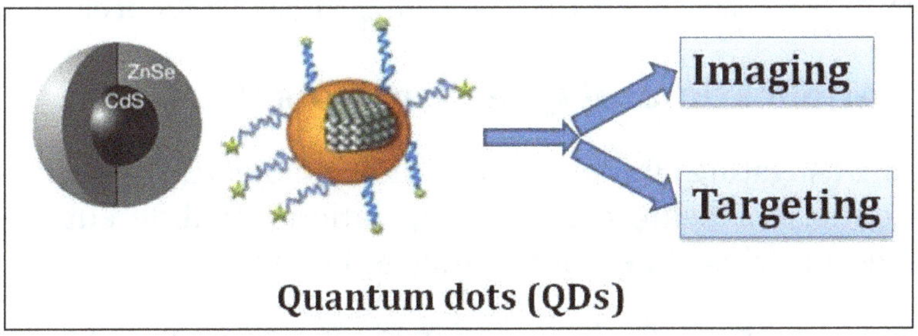

Schematic presentation of simultaneous imaging and targeting (i.e. theranostic) applications of quantum dots.

Properties of Quantum Dots

Various important properties of QDs are summarized in the following points. Quantum dots are:

- Semiconductor nanoparticles belonging to Group III-V and Group II-VI.
- Size is in nanoscale range varying from 1-10 nm.
- Have fluorescent properties which could be observed under the laser light source.
- Unique nanoscale size range and highly squashed structure gives quantum dots incomparable physical and chemical properties.
- QDs have been explored for various imaging applications including whole-animal imaging, live-cell imaging as well as imaging and treatment of cancer. Various QDs could be designed by changing their size to give emission at different wavelengths.

- Conductivity of QDs semiconductors ranges between conductivity of bulk semiconductors and distinct molecules.
- As the properties of semiconductors could be altered by limitiing properties of the electron and the hole, in the same way conductive properties of QDs could be changed by varying size and shape of individual crystals.
- Band gap in QDs increases with decrease in crystal size i.e. with small crystal size difference in energy between the highest valence band and the lowest conduction band will be larger resulting in larger band gap and higher energy requirement to excite the dot with release of more energy when dot returns to ground energy state.
- Colour of emitted light shifts from red to blue with decrease in crystal size.

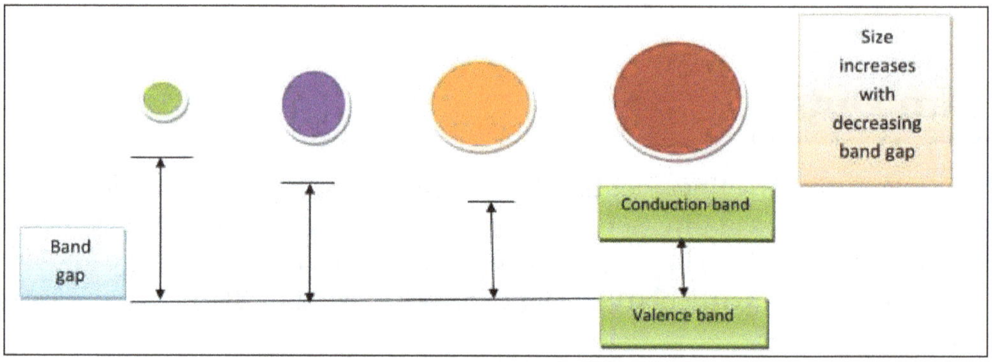

Diagramatic representation of relation between band gap and size of quantum dots.

Biotechnological Applications of Quantum Dots (QDs)

- Scientists are investigating QDs for various biotechnological applications like imaging, biosensing, targeting, biolabeling etc after conjugation with different biomolecules including oligonucleotides, proteins, different imaging and targeting agents.
- New generation of probes have been designed by modifying the surface of QDs with different functional groups for various biotechnological applications including drug delivery, targeted delivery of nucleic acids and biolabeling.
- QDs are of great interest to fundamental studies but also have potential applications as:
    - Biological probes.
    - Fluorescent biosensors.
    - Biological imaging and labeling probes because of their unique optical properties, including broad absorption with narrow photoluminescence spectra, high QY, low photobleaching, and resistance to chemical degradation, in comparison with other classical fluorophores for light emitting diodes (LEDs) and solar cells.

Characterization of Quantum Dots (QDs)

Quantum dots (QDs) are routinely characterized for size, structure, optical properties and composition. Various analytical techniques are used for the characterization of QDs for these properties as enlisted in following tables.

Characterization parameter	Characterization technique
Size, shape and structure	• Microscopic techniques: ◦ Scanning electron microscopy (SEM). ◦ Transmission electron microscopy (TEM). ◦ Atomic force microscopy (AFM). ◦ Scanning transmission electron microscopy (STEM). • Scattering techniques: ◦ Dynamic light scattering (DLS). ◦ Small angle neutron scattering (SANS). ◦ Laser light scattering. • Spectroscopic techniques: ◦ X-ray fluorescence. ◦ X-ray diffraction (XRD).
Optical properties	• UV–Visible spectroscopy. • Photoluminescence spectroscopy.
Composition	• Photoluminescence excitation. • Raman scattering spectroscopy. • Nuclear Magnetic resonance (NMR) spectroscopy. • Infra-red spectroscopy. • Mass spectrometry.
Electrical properties	• Electrochemistry. • Electrophoresis.
Rheological and physical properties	• Dielectric spectroscopy. • Differential scanning calorimetry.

Pharmacokinetics of Quantum Dots (QDs)

- Route of delivery: Parenteral.
- Absorption: Receptor mediated endocytic mechanism (at cellular level).

- QDs with targeting functional groups: Shown the ability to accumulate in selected target tissues.
- Only a little or even no information is available about blood–QDs interactions.
- Interaction between QDs and plasma protein is not clear till now or not investigated systematically.
- The QD core does not appear to be involved in extensive enzymatic metabolism, but the shell and coating appear to degrade under photolytic and oxidative conditions and hence release toxic cadmium cores.
- Only a very little information is available on the degradation products or biological effects of QDs, which may regulate the release of toxic cores.

Nanobiomedicinal Applications of Quantum Dots (QDs)

Quantum dots have been explored primarily for following biomedical applications in the healthcare systems:

- Cell imaging,
- Biosensor,
- Bio-analytical assay,
- Ex-vivo cell imaging,
- In-vivo animal targeting.

Due to their unique optical features, QDs are used for diagnostic purposes. In vivo imaging applications of QDs are critically influenced by the type and structure of organic/bioorganic shells of QDs, since it determines:

- Biocompatibility of QDs,
- Colloidal stability,
- Solubility in physiological fluids,
- Basic physiological parameters,
- Cytotoxicity.

Biocompatible water-soluble QD micelles demonstrate uptake and intracellular dispersion in cultured neurons. QDs–ligand interaction is used in detection of defects in DNA other biomolecular and protein detection, and cellular labeling. In vitro imaging applications of quantum dots include biomolecular tracking in cells, cellular imaging, and tissue staining. In vivo imaging applications of QDs include biodistribution, vascular imaging, QDs tracking, and tumor imaging.

QDs can concurrently tag multiple inter and intracellular components of live cells for time periods ranging from seconds to months. Different colors of QDs can label different cell components that can be easily visualized with fluorescent microscopy. CdSe QDs bind typically to cellulose and lignin in the cell wall and hence give a fluorescent image of plant cells; Biotinylated Cholera toxin B (CTxB) with QD–avidin conjugates has been investigated for labeling of ganglion.

QDs are also being explored as multifunctional nanomedicines for delivery of anticancer chemotherapeutic agents. In the last few decades, QDs have received considerable interest in the diagnosis, imaging and treatment of cancer and other diseases. QDs are the optical semiconducting nanomaterial. QDs inherently possess numerous advantages over traditional fluorescent dyes such as increased photostability, higher brightness, and narrow fluorescence spectra.

Till date various researchers and scientists have developed fluorescent nanoprobes and evaluated them on various animal models and represent one of the fastest moving and exciting interfaces of nanotechnology. QDs have unique optical properties, such as a broad absorption spectrum, narrow, size-tunable emission spectrum, high photostability, quantum efficiency, and strong non-linear response, from quantum confinement effects. The toxicity of QDs could be easily minimized by coating of PEG, polymers and other biomaterials, which make them more biocompatible and better candidate in cancer theranostic applications.

Ovarian cancer represents one of the most common gynecological malignancies in industrialized nations. Savla and co-workers designed tumor-targeted pH-responsive QDsmucin 1 aptamer-DOX (QD-MUC1-DOX) conjugate for chemotherapy of ovarian cancer, wherein DOX was attached to QDs via pH-sensitive hydrazone bond to provide greater stability in systemic circulation. The QDs-MUC1-DOX conjugate shows high potential in treatment of multidrug resistant ovarian cancer.

Carbon Based Nanomaterials

Numerous carbon based nanomaterials including Single walled CNTs (SWCNTs), Multi walled CNTs (MWCNTs), Fullerenes, Cup-stacked CNTs, Carbon nanohorns, Carbon nanotorous, Carbon nanopeapods, and Carbon nanobuds etc. are currently being investigated for various biotechnological applications along with other industrial applications. Among these carbon nanomaterials, carbon nanotubes (CNTs) are long, hollow molecules with sp2 hybridization and consist Fullerenes, of carbon atoms set in a sequence of condensed benzene rings rolled-up into a tubular structure. CNTs belong to the family of fullerenes, the allotropic form of carbon.

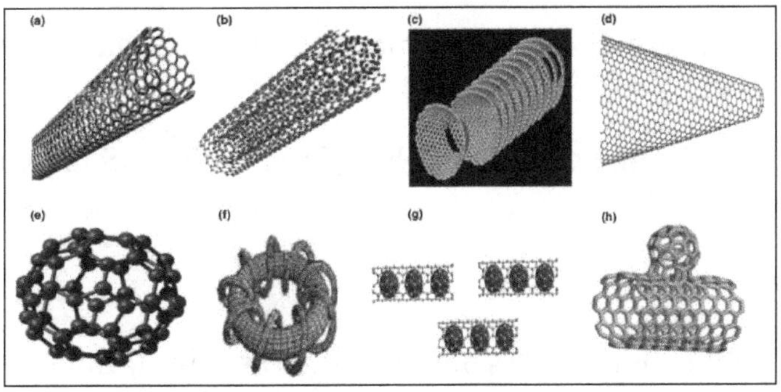

(a) Single walled CNTs (SWCNTs), (b) Multi walled CNTs (MWCNTs), (c) Cupstacked CNTs, (d) Carbon nanohorns, (e) Fullerenes, (f) Carbon nanotorous, (g) Carbon nanopeapods, and (h) Carbon nanobuds (Where CNTs – Carbon nanotubes).

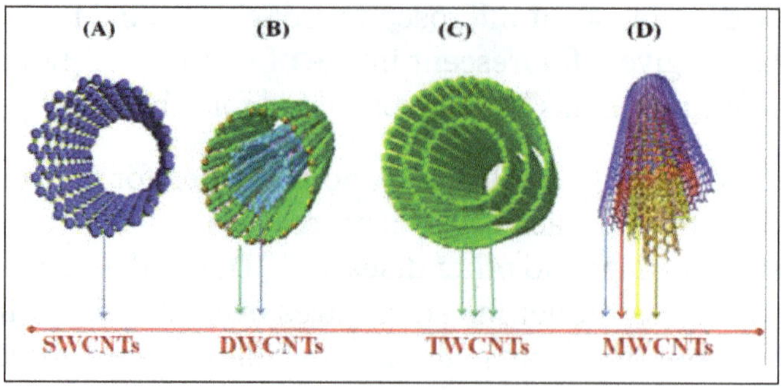

Different types of carbon nanotubes (CNTs).

Advantages of CNTs

CNTs have shown enormous advantages in regard of various industrial applications as given below:

- Highly uniform ordered structure with high aspect ratio.
- Ultra-light weight and easy availability.
- Photoluminescence property.
- Biocompatibility and non-immunogenicity.
- Highly elastic nature.
- Cellular internalization via endocytosis and passive diffusion mechanisms.
- Biodegradable.
- Exhibit minimum cytotoxicity based on the in vitro - in vivo studies.
- Excretion through biliary pathway (urine 96% and remaining 4 % by feces).
- Cell membrane penetration due to its tiny nanoneedle tubular structure.
- Open ends on both sides of CNTs make the inner surface accessible and further incorporation of bioactives within the tubes.
- Longer inner volume for endohedral filling relative to diameter.
- High mechanical strength, thermal conductivity, superconductivity metallic or semimetallic behavior.
- Rare retention of well functionalized CNTs (f-CNTs) in reticuloendothelial system (RES) due to small size (<1μm).

Limitations of CNTs

In regard of biomedical applications, CNTs have shown some limitations as listed below. Scientists are continuously working on various strategies to overcome these limitations for successful application of CNTs in healthcare and biotechnology industry.

- Low aqueous solubility in organic or non-organic solvents.
- Pristine CNTs exhibit some toxicity, hence un-suitable for drug delivery.
- Non-biodegradable.
- Bundling/aggregation phenomena.
- Accumulation in liver.

Functionalization of CNTs

CNTs suffers with some serious limitations. Functionalization of CNTs is one of the strategies to overcome these limitations including improvement in the targeting ability of CNTs, reduction in side effects, reduced accumulation in liver, improved dispersibility in water etc. Figure shows schematic representation of various functionalization strategies for CNTs.

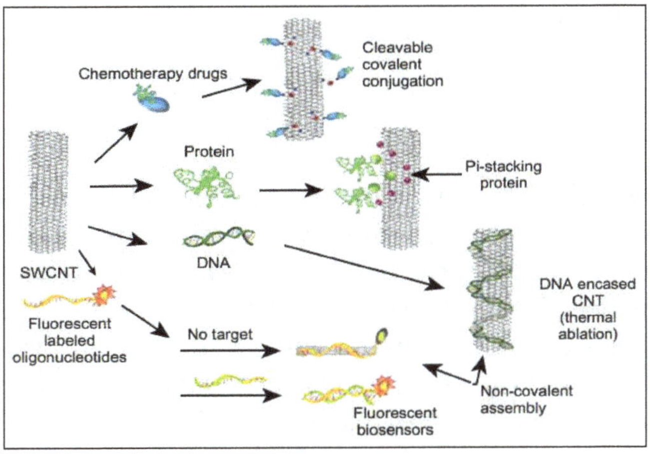

Functionalization of CNTs.

Characterization of CNTs

CNTs are basically characterized for their dispersibility, size, shape, aspect ratio, surface groups, and surface properties etc. Functionalized CNTs are also characterized for these parameters as well as for presence of groups/ligands/molecules used for functionalization. Various methods which are used for characterization of CNTs are enlisted below in table.

Table: Characterization techniques commonly employed for characterization of CNTs.

Characterization techniques		Characterization parameter
Microscopy	Scanning electron microscopy (SEM)	Size, shape, functionalization, surface characteristics.
	Transmission electron microscopy (TEM)	
	Atomic force microscopy (AFM)	
	Scanning transmission electron microscopy (STEM)	
Scattering techniques	Dynamic light scattering (DLS)	
	Small angle neutron scattering (SANS)	
	Laser light scattering	

Spectroscopy	UV-visible spectroscopy	Composition, dispersibility, functionalization, surface groups, bonding, chemical structure, crystallinity and crystal habit, surface morphology, hydrophilicity and tensile mechanical properties.
	Raman scattering spectroscopy	
	Nuclear Magnetic resonance (NMR) spectroscopy	
	Infra-red spectroscopy	
	X-ray diffraction (XRD)	
	X-ray photoelectron spectroscopy (XPS)	
Mass spectrometry (MS)	Gas chromatography – mass spectrometry (GC-MS)	Mass, Functionalization.
	Inductively coupled plasma mass spectrometry (ICP-MS)	
	Liquid Chromatography– mass spectrometry (LC-MS)	
	Matrix Assisted Laser Desorption Ionisation (MALDI)-Time of flight (TOF) {MALDITOF}	
High Performance liquid chromatography (HPLC)		Drug loading, functionalization.
Thermal methods	Differential scanning calorimetry (DSC)	Thermal stability, crystalline or amorphous nature.
	Differential thermal analysis (DTA)	
	Thermal gravimetric analysis	
Elemental analysis		Composition.

Impact of CNTs

In vitro and in vivo studies with CNTs have shown that CNTs could be used successfully in future for treatment of various deadly and infectious diseases like cancer, malaria, leishmaniasis, metabolic disorders etc. Particularly in the treatment of cancer, CNTs have shown promising results in preclinical human-tumour models. Similarly in case of other diseases also CNTs have shown successful outcomes in preclinical trials as indicated by prolonged survival. At present regulatory authority from any corner of world has posed any restrictions/recommendation/ruling on the therapeutic use of CNTs. Attributed to possible toxicity issues related to CNTs, a strict guidelines and comprehensive studies are needed for future medical use of CNTs. In the present conditions scientists are trying to develop chemically functionalized CNTs and physically modified CNTs to make them biocompatible and safe for use in biological environment and biotechnological applications.

Preclinical and Clinical Trials

Several CNTs based products reached the stage of clinical trials and many products are currently in market for various applications including industrial and cosmeceutical applications. Various research institutes and scientists have observed promising results with CNTs in preclinical trials particularly in treatment of cancer.

Nucleic Acid Nanostructure

Building with DNA

The Watson-Crick proposal for the structure of DNA is now over 50 years old. This proposal provided the chemical basis for understanding genetics; indeed, the last half-century of biology has been devoted to the exploitation of the implications of this model. The impact of DNA on society is just beginning to be felt, with the culmination of the Human Genome Project, and in the forensic and medical applications of DNA analysis. Regardless of its central importance in biology, the applications of DNA are not limited to the biological sciences. DNA is a molecule, and it functions successfully as genetic material because of its chemical properties. These properties include the affinity of complementary sequences, a well-stacked antiparallel double helical backbone that is largely regular regardless of sequence and a persistence length (a measure of stiffness) around 50 nm under standard conditions. It is logical to examine whether these properties can be exploited outside of biology. Structural DNA nanotechnology aims to use the properties of DNA to produce highly structured and well-ordered materials from DNA. This effort that has been underway since the early 1980's.

The structure of the DNA double helix is illustrated in figure. This classical DNA structure is known as B-DNA. On the left is the right-handed double helical molecule; it consists of a pair of backbones whose units are alternating sugars and phosphates that are connected by hydrogen bonding between flat molecular units, called bases. The base-sugar-phosphate unit is called a 'nucleotide'. Note that the two backbones have a directionality and that they are rotated upside down from each other; hence, there is a dyad axis of symmetry relating the backbones that is perpendicular to the helix axis. There is such a dyad going through every base pair and also half-way in between them. The phosphates are ionized at physiological pH, so the backbones are polyanions requiring neutralization by cations. The right side of figure shows the Watson-Crick base pairs between adenine (A) and thymine (T), and between guanine (G) and cytosine (C). Although other forms of base pairing are possible [ns-natco], these pairs are most favored within the double helical context. The double helical DNA molecule is inherently a nanoscale species: Its diameter is about 2 nm, and its helical repeat is about 3.5 nm, consisting of 10–10.5 nucleotide pairs separated by 3.4 Å.

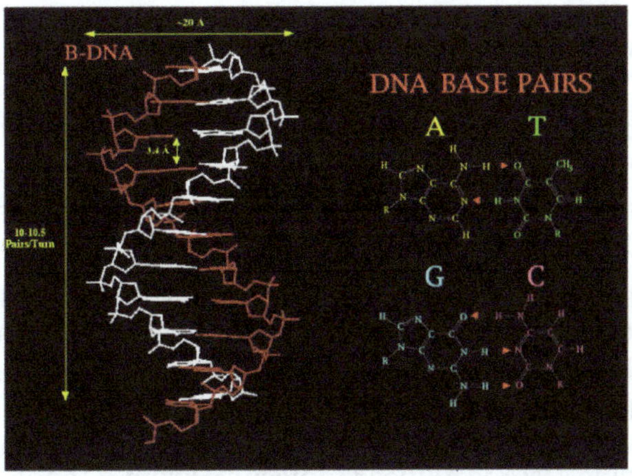

B-DNA. The molecular structure and dimensions of key features of the molecule is shown on the left. The base pairs are shown on the right, where red arrows represent hydrogen bonded interactions.

The early 1970's saw molecular biologists begin to exploit enzymes that modify the DNA molecule to produce new arrangements of DNA sequences. Using restriction enzymes that cleave at specific sites and ligases that fuse nicks in backbones, they were able to cut and splice DNA molecules, in much the same way that a film editor uses his tools to re-order sequences of film, creating a new overall effect. The upshot of this activity was genetic engineering, which has led to the worldwide biotechnology industry. The key feature of their efforts was the use of single stranded overhangs (called 'sticky ends') to specify the order in which molecules were to be assembled. Sticky-ended cohesion between DNA molecules is illustrated in Figure. An important feature of sticky-ended cohesion is that when two sticky ends cohere, they form classical B-DNA, the structure established by Watson and Crick and refined by their successors. This principle is illustrated in Figure Thus, if one knows the positions of the atoms of one component of a cohesive pair one knows the positions of the other component. Hence, sticky-ends provide the most readily programmable and predictable intermolecular interactions known, from the perspectives of both affinity and structure.

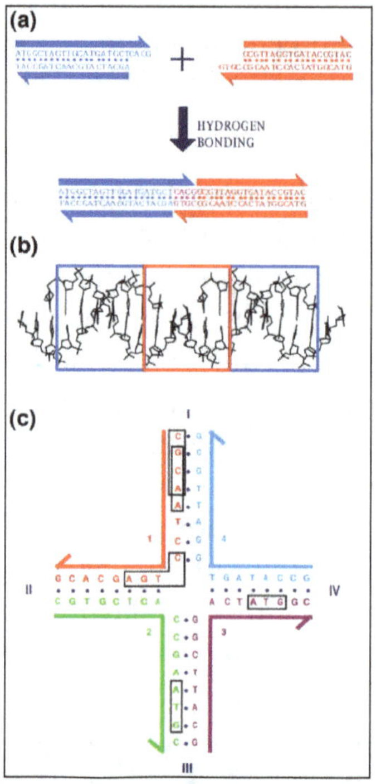

Sticky Ended Cohesion and Branched DNA.

(a) Affinity in Sticky Ended Cohesion. Two double helical strands with complementary overhangs are shown. Under appropriate conditions, they will cohere in a sequence-specific fashion, and can be ligated, if desired. (b) Structure in Sticky-Ends. A portion of the crystal structure of an infinite DNA double helix formed by sticky-ended cohesion is shown. The part cohering by sticky ends is in the red box, whereas the blue boxes surround continuous DNA segments. The DNA in all three sections is B-DNA. (c) A Stable Branched Junction. There is no dyad symmetry flanking the branch point; tetramers, such as the boxed sequences CGCA and GCAA are unique, and there is no TCAG to complement the CTGA flanking the corner.

Of course, naturally-occurring DNA is a linear molecule. From a topological standpoint, the DNA double helix is just a line. The line may be curved, and closed ones can be knotted or catenated; nevertheless, concatenating a bunch of lines together end-to-end just results in longer lines. Specific arrangements of DNA are of significant value to genetic engineers, because this establishes the presence and regulation of gene products, but linear DNA helical structures are not great structural components for nanoscale structural engineering. The notion that rescues DNA from its linearity is to make synthetic branched molecules.

DNA is found to be branched at the level of secondary structure in biological systems; this means that conventional DNA strands associate to produce junction points. Thus, if we think of the strands of DNA as the lanes of a two-lane highway, a junction point would correspond to the lanes going through an intersection. Traffic could flow through four different lanes in a traditional intersection; in the US, northbound traffic would turn east, westbound traffic would turn north, southbound traffic would turn west and eastbound traffic would turn south. A branched DNA molecule would be the same.

Typically, branched DNA is an ephemeral intermediate in DNA metabolism; for example, the four-arm Holliday junction is an intermediate in genetic recombination. Although branched DNA of biological origin usually displays symmetry that allows its branch point to migrate, it is a simple matter to design and assemble synthetic DNA sequences that are stable because they lack this symmetry. An example of a stable asymmetric Holliday junction analog lacking this symmetry is shown in Figure. Branched junctions containing three, five or six arms have also been assembled and characterized; in recent work laboratory has made junctions with eight and twelve arms. This is important, because the connectivity of the graph produced by cohesion or ligation of junctions is limited by the number of arms that flank its junctions. Space-filling lattices built from the classical Platonic or Archimedean polyhedra have connectivities up to twelve. The ability to generate stable branched DNA molecules opens new opportunities for the bottom-up construction of nanoscale objects.

The basic notion behind structural DNA nanotechnology is to combine the concepts of stable branched DNA molecules with sticky ended cohesion. This idea is illustrated in Figure, which shows a branched junction with sticky ends assembling as a group of four units to produce a square-like object. The outside of this object contains further sticky ends, so this arrangement could be extended to produce a lattice in two dimensions. As in Figure, DNA is often drawn as a ladder-like molecule, with parallel backbones, rather than as a double helix. While often convenient, this representation obscures a key property of branched DNA constructs: The relative orientation of two junctions joined by an edge of double helical DNA is a function of their separation; consequently, the system described here is not a two-dimensional system, but rather a three-dimensional system. Once one recognizes that branched DNA molecules can be fused in three dimensions, molecular graphs and connected networks whose edges are DNA helix axes can be constructed in an almost limitless variety.

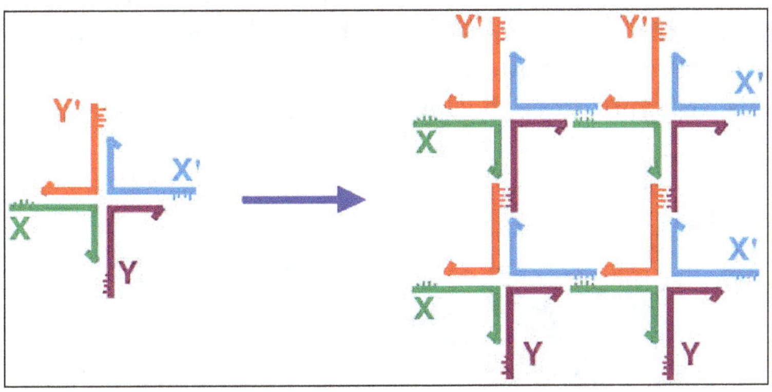

Sticky-Ended Assembly of Branched Molecules.

A branched molecule is shown on the left with four sticky ends, X, complementary to X', and Y, complementary to Y'. Four of them are shown to assemble to form a quadrilateral, with further sticky ends on the outside, so that an infinite lattice could be formed by the addition of further components.

It should be clear that DNA nanotechnology is a bottom-up approach to nanotechnology. The components are a few nm in size, so the exquisite placement of individual atoms or molecules that characterizes some of the top-down work done with scanning probe microscopes is beyond the scope of this methodology. However, it is important to recognize that bottom-up approaches are able to take advantage of the vast parallelism of chemical reactions. Thus, in the worst of constructions, a billion or a trillion product molecules typically will be produced. The parallelism of DNA nanotechnology is related to the parallelism exploited in DNA-based computation, so that the two fields have convergent features.

The Goals of Structural DNA Nanotechnology

Ideally, the systems we have described above should be able to be used like molecular Lego: It should be possible to throw together a variety of branched-DNA components and exploit their well-defined combining properties to get them to self-assemble into a variety of shapes and arrangements.

The key motivating goal in this system has been the desire to conquer the macromolecular crystallization experiment. Although quite a lore now exists, every macromolecule presents different crystallization problems. The problem of crystallizing a macromolecule of unknown structure is that one does not know how to promote the organization of the molecules into a periodic lattice in a rational fashion, because the structure is unknown. Furthermore, in contrast to experiments that are easily debugged, every relevant parameter must be within the correct window for the crystallization to be successful; consequently, the independent factors cannot easily be optimized individually, although incomplete factorial methods have been developed.

The approach we suggest is to use DNA to form periodic arrays that organize the macromolecule of interest. The notion is illustrated in figure. A 'box' of nucleic acid [drawn in magenta] is designed to act as the repeating unit of a 3D periodic lattice. A series of boxes are to be held together by sticky ends. These boxes act as the host lattice for a series of macromolecular guests that are positioned and oriented within the host lattice. If all are ordered adequately, then the x-ray diffraction experiment can be performed on the entire contents of the crystal. In addition to the ability perhaps to organize otherwise molecules that are otherwise intransigent to crystallization, this system offers the advantage that guest molecules that change their structures while binding to ligands would not destroy the lattice, because only DNA would be involved in maintaining lattice contacts.

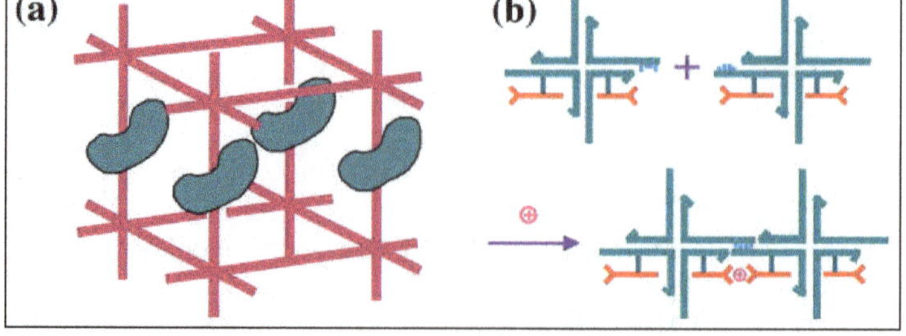

Scaffolding Applications of Structural DNA Nanotechnology.

(a) Scaffolding of Biological Macromolecules for Crystallographic Purposes. A DNA box (magenta) is shown with sticky ends protruding from it. Macromolecules are organized parallel to each other within the box, rendering them amenable to crystallographic structure determination. (b) Scaffolding Nanoelectronics. Blue branched DNA junctions direct the assembly of attached nanoelectronic components (red) to form a molecular synapse stabilized by the presence of an ion.

If one can imagine organizing macromolecules, it is also possible to imagine organizing other species that do not readily self-organize. Prominent amongst them would be nanoelectronic components. The idea of using DNA to organize nanoelectronic components is not very new, but only recently has it started to gain favor. The concept, illustrated for a simple component in Figure, is that the outstanding architectural properties of DNA be used to act as scaffolding for species that are intractable to simple bottom-up organization with nanometer precision.

A further goal of DNA nanotechnology is the use of DNA's structural and informational properties to build nanomechanical devices. The earliest DNA devices were predicated on structural transitions of the molecule. There are two kinds of DNA-based devices, those based on structural transitions triggered by the addition of small molecules to the solution, and those that are sequence dependent, although these can be further subdivided. The latter are more useful, because a variety of them can co-exist in solution at the same time, yet they can be individually activated. This is closer to the notion of acting as a successful basis for nanorobotics, and this is the route that current investigators largely are taking at this time.

It is may be useful to distinguish structural DNA nanotechnology from a less precise form of DNA nanotechnology, termed compositional DNA nanotechnology. Structural DNA nanotechnology uses well-structured components, combined by using both affinity and structure to control geometry, or at least, strand topology; the goal of this approach is structural predictability with a precision (or resolution) of 1 nm or less in the products. By contrast, in compositional DNA nanotechnology, these conditions are not completely met: The components may be flexible or the cohesive interactions by which they are combined are uncharacterized; as a consequence, the composition of the product may be known, but its 3D structure cannot be predicted. Incompletely characterized forms of cohesion such as paranemic cohesion or the osculating interactions of tecto-RNA may eventually be used in structural nucleic acid nanotechnology when they have been as characterized as well as sticky ends have been characterized in a periodic context.

High-resolution structural DNA nanotechnology works on the scale where DNA is both the bricks and the mortar in nano-scale assemblies; by contrast, compositional or low-resolution nanotechnology uses the DNA only as the mortar, as cartooned in Figure. As shown on the bottom of the low-resolution side, DNA mortar can make well-defined structures on a large-enough scale, but the highest levels of control require DNA acting in rigid motifs with rigid connections to other species. It is important to recognize that numerous laboratories have made useful and valuable materials by this approach. For example, this approach has been used in diagnostics, to organize DNA nanoparticles on small and on large scales, in combination with non-DNA organic components, and in the production of DNA-protein aggregates. Although they do not provide the high resolution structural features sought in structural DNA nanotechnology, using smart-glue approaches to organize nanoparticles can lead to organized products, but with lower precision, 10's to 100's of nanometers. In a complementary vein, G-wires are examples of well-structured nucleic acid systems that lack the sequence diversity central to structural DNA nanotechnology.

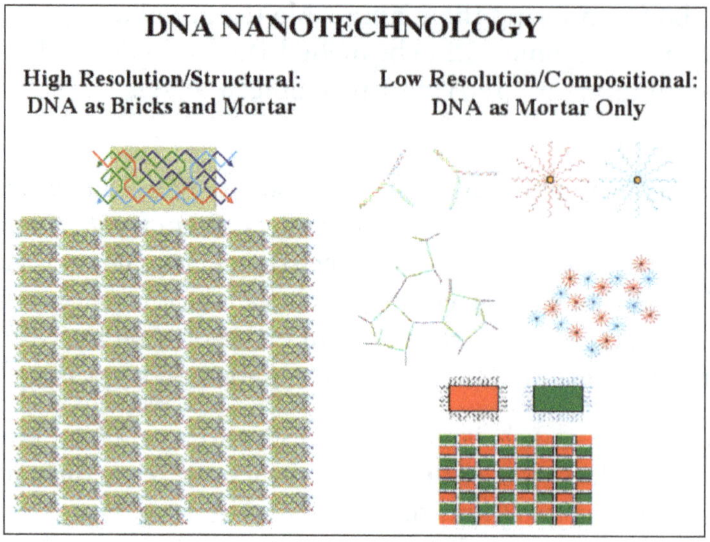

Types of DNA Nanotechnology.

High resolution/structural DNA nanotechnology is shown on the left, where DNA is both the bricks and mortar of an assembly. A TX molecule is shown assembled into a 2D array. Low resolution/ compositional DNA nanotechnology is shown on the right with floppy components or applications using DNA as 'smart glue' (top). This type of DNA nanotechnology can also be used to build structures, as shown at bottom, but the resolution is not as high.

Features of DNA that make if Useful for Nanotechnology

The primary advantage of DNA for these goals lies in its outstanding molecular recognition properties, enabling precise structural alignment of diverse DNA molecules that can scaffold various molecular species. DNA and its congeners appear to be unique among biopolymers in this regard. Antibodies, for example, may lead to specific binding, but the detailed geometry of antibody-antigen interactions must be worked out experimentally in every case. It is only with nucleic acids (and currently only with sticky-ended cohesion) that the detailed 3D geometries known in advance, because sticky ends form B-DNA when they bind to link two molecules together. It is a very good approximation to assume that all sticky ends, regardless of length have the same overall B-DNA structure when they cohere with their complements.

The convenience of chemical synthesis is another advantage of DNA; "vanilla" DNA is available from a number of vendors, and DNA synthesizers are readily capable of generating a number of varied molecules based on commercially available phosphoramidites. It is routine to synthesize DNA molecules whose lengths are 120 nucleotides or shorter. A variety of enzymes are commercially available to manipulate DNA and to trouble-shoot errors: DNA ligases catalyze the covalent joining of complexes held together by sticky-ended cohesion; exonucleases (which require an end to digest DNA) are useful in purifying cyclic target molecules from linear failure products; restriction endonucleases are employed both to trouble-shoot syntheses and to create cohesive ends from topologically-closed species; topoisomerases, which catalyze strand-passage reactions, can be used to correct folding errors.

We noted above several features of DNA that help it to perform its genetic functions. One of these was the persistence length, about 500 Å, leading to a predictable overall structure for the short

(70–100 Å) lengths typically used. In fact, it is possible to design branched species with longer persistence lengths. Thus, the self-assembled structure of DNA constructs and of some of its interactions can be taken for granted, just as one can take for granted the shape of a Lego tile. The external code on DNA can be read, even when the double helix is intact; thus, if DNA is used for scaffolding, absolute positions can be addressed within a pre-designed cavity. The ability to pack nanoelectronics or other scaffolded materials very tightly is likely to be aided by the high density of functional groups (every 3.4 Å or so) on DNA; Consequently, DNA tile motifs with dimensions of 10–20 nm do not place an inherent limit in the close-packing of components that can be scaffolded by DNA.

Most of the work done to date utilizes DNA molecules like the molecules evolved in nature for use as genetic material. However, nanotechnology is not limited to the natural DNA molecule. A vast number of DNA analogs have been produced and analyzed for therapeutic purposes. This means that systems prototyped by conventional DNA may ultimately be converted to other backbones and bases, as required by specific applications. For example, it is unlikely that nanoelectronic components will be scaffolded successfully by conventional poly-anionic DNA molecules. However, there are numerous neutral analogs, most prominently PNA (a molecule with a peptide backbone to which bases are attached), that may be much better suited to act as scaffolds for this purpose.

Motif Generation and Sequence Design

Numerous motifs have been used in DNA nanotechnology. The simplest motifs are those described above, DNA junctions that resemble intersections, from which between three and six [or more] double helical arms emanate. These junctions are easy to design, but, with the exception of the four-arm junction, they do not appear to have well-defined structures. The four-arm junction has a fairly well-defined structure, but it certainly is not the simple cruciform indicated in figure or in figure. The structure is a right-handed cross involving two stacking domains, but even for this molecule, the connection between the domains is somewhat flexible.

Consequently, other motifs have been sought whose structural integrity is greater. This search has proved less difficult than one might imagine, because one of the mechanisms that biology uses to produce the Holliday junction provides a general mechanism to derive stiff motifs. This mechanism is known as reciprocal exchange, and is illustrated in Figure. It can be performed multiple times between strands of the same polarity or of opposite polarity, leading to different products. Although the process of reciprocal exchange is often complicated in living systems, nanotechnological systems do not actually have to undergo reciprocal exchange to produce new motifs; once the structural design process is complete, new motifs are generated directly by using a sequence assignment procedure to design strands that will self-assemble into the motif. figure illustrates the DX molecule (opposite polarity), the DX+J molecule, the TX molecule (opposite polarity) and the PX and JX_2 molecules (same polarity). There are five different isomers of the DX molecule, depending on the polarities of the exchanging strands and on their separations; the complete set of DX molecules is illustrated in Figure. The DX molecule is known to be about twice as stiff as conventional linear duplex DNA; the DX, DX+J and TX motifs have been used to produce patterned 2D DNA arrays and the DX, PX and JX_2 motifs have been used as components of DNA nanomechanical devices. The rigidity of these motifs is key to their utility in these applications.

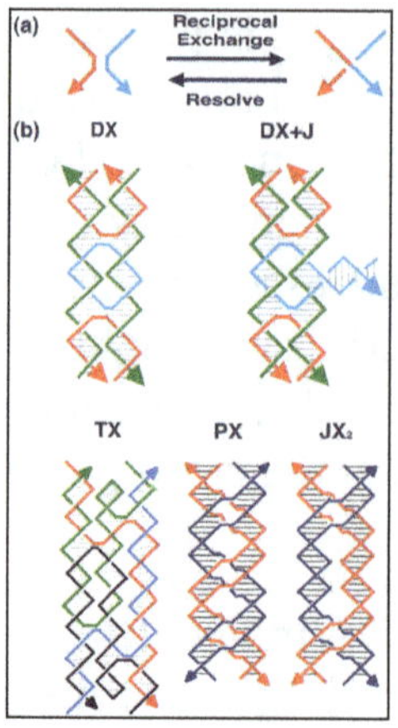

Motif Generation.

(a) The Process of Reciprocal Exchange. A red strand and a blue strand exchange to form a red-blue strand and a blue-red strand. (b) Motifs used in Structural DNA Nanotechnology. Two reciprocal exchanges between strands of opposite polarity yield the DX molecule shown. The 'D' means that there are two double helices. The DX+J (a DX combined with a junction) motif, usually made with the extra helix roughly perpendicular to the plane of the other two, is made by combining a DNA hairpin and a DX molecule. The TX (indicating three fused double helices) motif results from combining the DX molecule with another double helix. The PX (standing for paranemic crossover DNA) motif is derived by performing reciprocal exchange between two helices at all possible positions where strands of the same polarity come together. The JX2 (indicating a PX with two juxtaposed backbone groups) motif is similar to the PX motif, except that reciprocal exchange is omitted at two adjacent juxtapositions.

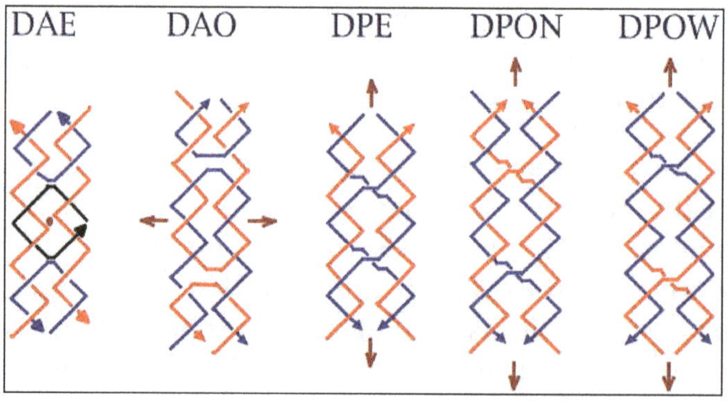

DX Conformational Isomers.

Brown arrows indicate dyad axes, and arrowheads on strands indicate 3' ends. The second character of the name indicates antiparallel (DAE and DAO), with dyad axes perpendicular to the plane of the helix axes, or parallel (DPE, DPON, DPOW), with the dyad axis coplanar with the helix axes. The third character indicates whether there is an even (DAE or DPE) number of half-turns between crossovers, or an odd number (DAO, DPON, DPOW). Odd parallel molecules are further differentiated by a fourth character, N or W, indicating whether the extra half turn is a minor (N -- narrow) groove separation or a major (W -- wide) groove separation.

Stable branched junctions are derived from synthetic molecules, so there must be a method to assign sequences to them. The method of sequence symmetry minimization has proved to be up to the task of designing sequences. The basic idea is that DNA strands will maximize the double helical structures that they form. Although an early approach involved calculating a likelihood of formation from nearest-neighbor equilibrium thermodynamic parameters, no constructs built to date appear to require a calculation this extensive. The method used can be understood readily by reference to figured. This molecule contains four 16-mers, labeled 1, 2, 3 and 4. We break up each of these single strands into a series of thirteen overlapping tetramers, such as the CGCA or GCAA that have been boxed; we insist that each of these be unique. In addition, we insist that each tetramer that spans a branch point, such as the boxed CTGA, not have its linear complement (TCAG) present; this restriction results in these tetramers being unable to form linear double helices. Consequently, competition with the four octamer double helical targets can occur only from trimers, such as the boxed ATG sequences. The reason to go through the exercise of sequence-symmetry minimization is that the molecules that are designed correspond to conformations of DNA that are of higher free energy than the 'ground state', i.e., the simple Watson-Crick double helix. The aim of sequence design is to ensure that the 'excited state' that they system chooses is also the target state sought by the investigator.

The assumption used in all of the sequence design is that molecular geometry and structure formation are sequence-independent. In those cases where it is known that sequences result in structural features, e.g., the bends associated with tracts of several A;s, the issue is handled in an ad hoc fashion by forbidding those sequence within a molecule. Likewise, runs of G's that often form their own secondary structure, are avoided. This approach has largely worked to date. A second issue, of course, is folding. It is difficult to model the folding of and pairing of a large number of DNA strands. Nevertheless, it is often a good idea to examine the associations that one wishes to promulgate in the construction of a given motif. If, for example, it seems that the ends of a strand are likely to pair earlier than the middle of the strand, perhaps because of higher G-C content, then re-design may be indicated, because the middle may not be able to wrap around its complementary strand if its ends are already tied down.

Motif Characterization

Just because a given motif appears as though it will self-assemble successfully does not mean that it will. It is key that one have a battery of techniques available to characterize it. Of all of the motifs discussed here, to date only the structure of the Holliday junction has been established crystallographically. Consequently, it has been necessary to use somewhat less direct forms of structure analysis to establish the proper self-assembly of branched DNA motifs. The first question one must ask is whether each strand participates in the designed complex with the expected stoichiometry. This may be readily determined through the used of non-denaturing polyacrylamide gel electrophoresis. The material is loaded onto one end of a slab of polyacrylamide and driven by an electric field acting on the negatively charged backbone to the other end. Mobility on such a gel is a function of molecular weight and surface properties. The target complex, the individual strands and various incomplete complexes can be loaded onto the same gel in the presence of various markers. A single band near [but rarely exactly at] the predicted molecular weight is a good indication of successful formation. Bands with lower mobility corresponding to dimers, trimers or higher complexes are sometimes seen, indicating a motif whose structure has internal stresses likely resulting from electrostatic repulsion. Likewise, products migrating with the mobilities of various partial products indicate that the complex is dissociating from its target composition for some reason.

Sometimes incomplete stoichiometry is the cause of a gel that does not conform to expectations. Cutting out the target band and re-hybridizing it will yield a product with correct stoichiometry, and will indicate if the single band has re-equilibrated to a multiple product.

The qualitative shape of a DNA motif can be compared usefully and conveniently with standards by means of a Ferguson plot. This is a plot of log(mobility) vs. acrylamide concentration, and its slope is proportional to the friction constant of the molecule. Ferguson analysis has been used to characterize the stacked character of 4-arm junctions, in comparison to 3-, 5-, and 6-arm junctions and to demonstrate that the addition of helices to linear duplex and to DX molecules results in similar changes in friction constant. Likewise, the qualitative structure of PX-DNA has been compared usefully to that of double crossover molecules to establish their similar shapes. Both denaturing and non-denaturing Ferguson analysis have been applied to a DNA cube-like molecule and its sub-catenanes.

The highest resolution technique that is used conveniently to characterize the structures of unusual DNA motifs is hydroxyl radical autofootprinting. Hydroxyl radical autofootprinting is a variation on the technique of hydroxyl radical footprinting, used to establish the binding sites of proteins on DNA. Branched junctions, DX, TX and PX molecules have all been analyzed by this method. The analysis is performed by labeling a component strand of the complex and exposing it to hydroxyl radicals. The products are driven through a denaturing gel (sensitive only to size for a given topology), and the intensities of the bands corresponding to each individual nucleotide are quantitated. The key feature noted at crossover sites in these analyses is decreased susceptibility to attack when comparing the pattern of the strand as part of the complex, relative to the pattern of the strand derived from linear duplex DNA. Decreased susceptibility is interpreted to suggest that access of the hydroxyl radical may be limited by steric factors at the sites where it is detected. Likewise, similarity to the duplex pattern at points of potential flexure is assumed to indicate that the strand has adopted a conventional helical structure in the complex, whether or not it is required by the secondary structure. The technique is particularly powerful as a method to establish whether crossovers occur where they are expected to occur.

Cohesive Interactions

Sticky-ended cohesion can be thought of as analogous to the familiar corner-sharing motifs of inorganic chemistry. However, that discipline has provided us with other models for the cohesion of its polyhedra. One of these is edge-sharing, and another is face-sharing. Face-sharing has yet to be developed in DNA nanotechnology, but it is possible to join triangular motifs in an edge-sharing mode that is mediated by DX molecules. A schematic diagram of one-dimensional arrangements of edge-sharing triangles is shown in Figure. Hairpin loops with complementary sequences have been known to molecular biologists since the 1970's. Recently, researchers have used non-Watson-Crick interactions to employ a motif that they have dubbed 'Tecto-RNA'. A generalization of that type of pairing is found in paranemic cohesion, illustrated in Figure; neither type of interaction requires the opening of closed DNA molecules. Two double helices appear to wind around each other to produce a cohesive interaction. In contrast to sticky-ends generated by restriction enzymes, topologically-closed molecules can be made to cohere with interactions of any strength desired. The importance of this is that only topologically-closed molecules are readily purified by the best method currently available, denaturing gels electrophoresis.

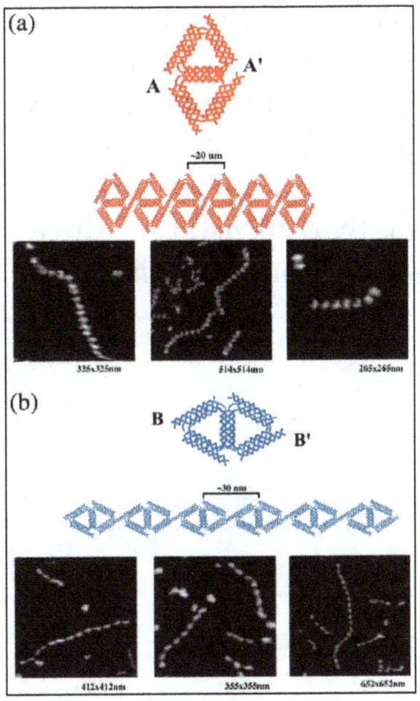

Edge-Sharing Cohesion.

(a) A 1D array of Edge-Sharing Diamonds Joined in their Short Direction. A schematic of the molecule is shown at top in red, as is a schematic of the 1D array. Three AFM views of arrays are shown beneath the schematics. (b) A 1D array of Edge-Sharing Diamonds Joined in their Long Direction. The same conventions apply as in (a), except that the schematics are in blue. Note the difference in the 1D arrays.

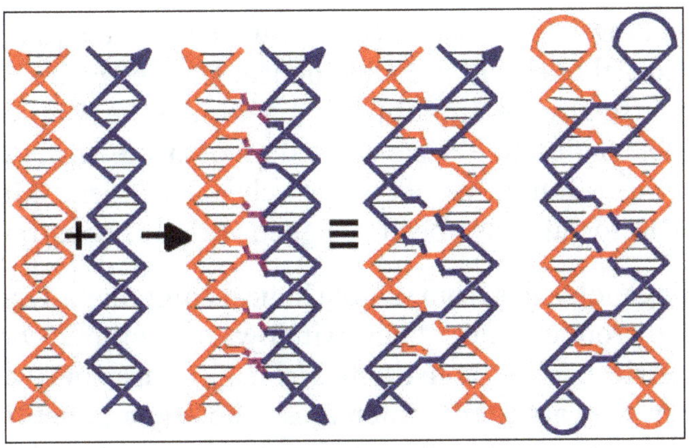

Paranemic Cohesion.

The origin of the motif is shown at the left, where a red and a blue double helix exchange strands at every point, that they come in contact, to produce purple crossovers. The covalent strand structure is shown to the right of the identity sign, where a red and a blue double helix are seen to inter-wrap. The paranemic relationship of the inter-wrapped helices is emphasized on the right where the blue and red helices are capped in hairpin loops.

Topological Aspects of DNA Nanotechnology

We often represent the double helical DNA backbone as a pair of antiparallel lines. This is often a useful approximation, but the true structural flavor of the DNA molecule is lost if the wrapping of the double helical strands around each other is ignored. For example, Figure illustrates an extended version of the lattice shown in Figure, but now with topology taken into account. The left panel shows a junction ligated to form a quadrilateral containing two turns per edge, and the right panel shows the formation of a quadrilateral with 1.5 turns per edge. The lattice with two turns per edge contains a cyclic molecule and four others paired with it directly; an infinite lattice of such squares would be a network of linked circles. By contrast, the lattice with 1.5 turns per edge contains no cycles, but rather is a woven meshwork of long strands.

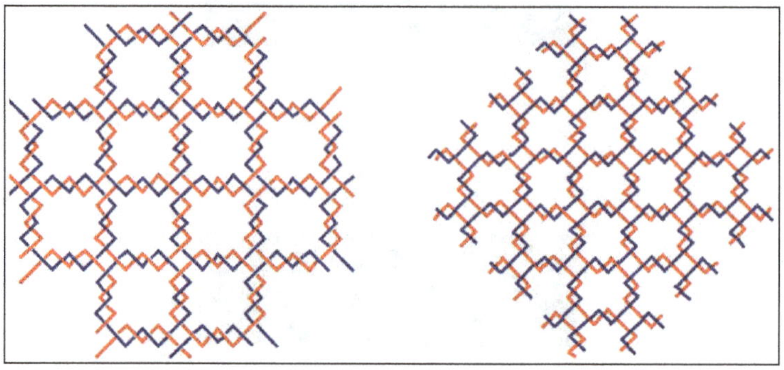

Impact of Topology in Branched Networks.

Both arrays are extended versions of the array illustrated in figure, but the twisting of the helices is indicated. The array on the left contains an even number of half-turns between vertices, leading to an array that consists of molecular chain main, red circles linked to blue circles. The array on the right contains an odd number of half-turns between vertices, leading to an interwoven structure, blue chains going lower right to upper left, and red chains going lower left to upper right.

The double helical nature of DNA also places constraints on the lengths of edges in DNA figures. Two branched junctions at either end of a double helical DNA edge are related to each other like wing nuts on a screw: Their relative orientation is a function of the length of the intervening DNA. Although it appears possible to design objects with edges of any length, the simplest architectures derive from the use of an integral number of double helical half-turns. Most (but not all) geometrical objects built from DNA to date have been designed with this constraint in mind.

The difficulties of designing molecules from double helical components are more than compensated by the properties of the products and by new windows of synthetic control that are opened by this feature. Closed geometrical objects constructed from DNA are not weakly associated hydrogen bonded complexes. Rather, they are robust catenanes, whose components are topologically bonded to each other. If each edge of a DNA polyhedron contains an integral number of double helical turns, each face corresponds to a cyclic single strand of DNA, linked to each of its neighboring faces once for each turn in the edge that they share. The catenation between strands permits rigorous topological analysis of products under denaturing conditions. As a logical extension, it is possible to design periodic arrays from DNA whose topology is that of molecular chain mail.

Polyhedral Catenanes and other Graphs Produced from DNA

The earliest interesting DNA constructions involved polyhedral stick figures whose edges consist of double helical DNA, and whose vertices correspond to the branch points of branched junctions. The formation of the edges was implicit in these constructions, just the result of joining junctions together. Figure illustrates a molecule whose edges have the connectivity of a cube or rhombohedron. Although the DNA in the arms of DNA branched junctions is stuff, the angles between the arms are apparently floppy. Consequently, DNA polyhedra are characterized topologically, rather than structurally. The cube in Figure is a hexacatenane, with two turns of DNA per edge. There are two turns per edge, so each strand corresponds to one of the faces of the cube, and each strand is linked twice to the four strands that flank it. The molecule was constructed in solution, by methods that had little control over the products of the synthesis. Proof of synthesis was by breaking down the final product to sub-catenanes that could be constructed and characterized independently.

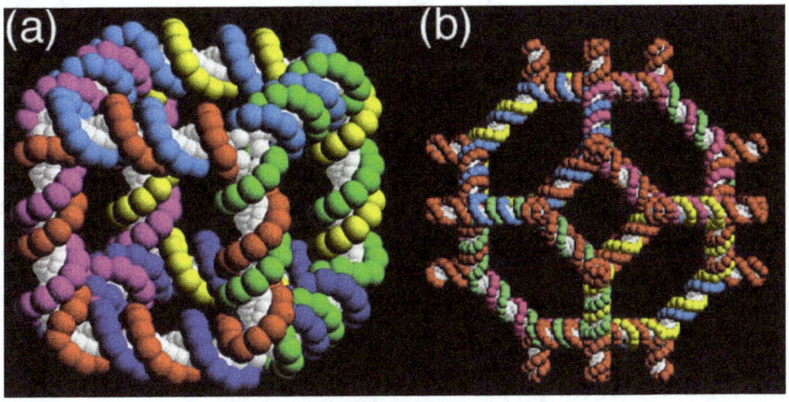

Polyhedral Catenanes.

(a) A Stick Cube and (b) a Stick Truncated Octahedron. The drawings show that each edge of the two figures contains two turns of double helical DNA. There are two turns of DNA between the vertices of each polyhedron, making them, respectively, a hexacatenane and a 14-catenane.

Many of the problems of the synthesis of the cube were eliminated by the development of a solid-support-based methodology, which allowed the use of the full logical control implicit in sticky-ended construction. As a test case, this approach was used to build a truncated octahedron, again with two turns per edge. Thus, the DNA representation of this Archimedean polyhedron is a 14-catenane. Six of its faces correspond to squares and eight of its faces correspond to hexagons. The faces corresponding to squares contain an extra arm at each junction, so that this 3-connected polyhedron was built with 4-arm junctions. Consequently, it might have been able to make a 4-connected lattice, such as the zeolite A lattice, but not enough material was produced to use it as starting material for a lattice construction.

Both of these polyhedra correspond to regular graphs. In an exercise related to DNA-based computation, we have also constructed an irregular graph. The construction prototyped a monochromatic version of a solution to a graph 3-colorability problem. Whereas the two previous polyhedral constructions entailed step-wise syntheses, the nature of DNA-based computation entails doing as many steps as possible at once. Similarly, the requirements of the calculation entail mixing edges of different properties, so that molecules corresponding to edges were added explicitly to the solution containing the junctions. This single-pot reaction resulted in the construction of the irregular graph illustrated in Figure. Unlike the cube and the truncated octahedron, this graph is a single-stranded knot, rather than a catenane. Proof of synthesis in this case requires the breakdown of the final strand to predictable restriction fragments.

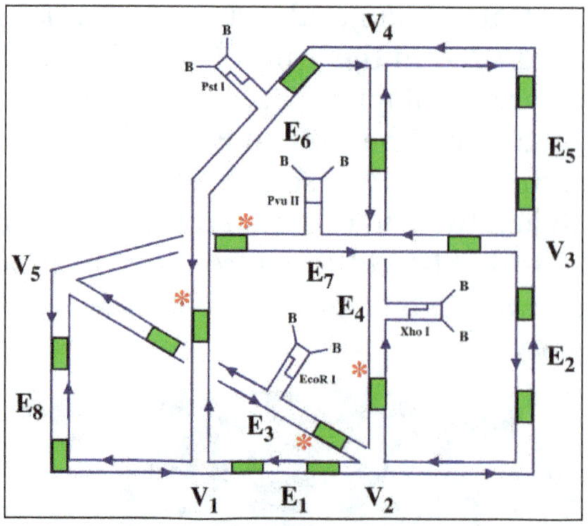

An Irregular Graph.

The graph shown has explicit edges (E1–E8) as well as junctions (V1-V5). The sites of the sticky ends are indicated by patches of green. Restriction sites are indicated by the names of the restriction enzymes, biotin groups are labeled as ‘B’ and radioactive labeling sites are shown by asterisks. Note that this is a knot, not a catenane.

Topological DNA Constructs

Figure illustrates a simple relationship between catenanes and knots. A 5-noded knot is shown in the upper left of the drawing, with arrowheads indicating an arbitrary polarity to the strand. If one imagines a node to be made up of four strands, one before and one after the node on the top and the bottom, one can destroy the node by disconnecting the intact strands, and reconnecting them while maintaining local polarity. This converts the knot to a catenane. The catenane can then converted to a knot using the same operation. Thus, it appears that a system that is successful at making catenanes can also be successful at making knots and other topological targets. Figure shows the relationship between DNA and a node in a knot. The trefoil knot, is shown as a red strand. A box is drawn around each of its three nodes, and the strands of the knot act as diagonals of the box, dividing it into four zones, two between parallel strands and two between antiparallel strands. Whereas the DNA backbone is antiparallel, one can make the transition from topology to nucleic acid structure by placing about a half-turn (six base pairs) of double helix between the antiparallel backbones.

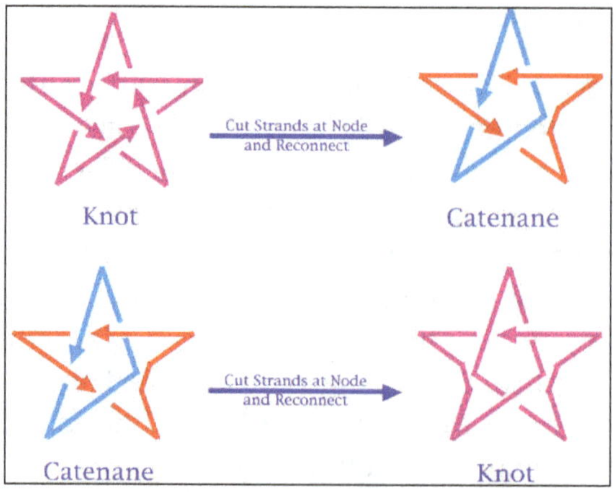

The Relationship Between Knots and Catenanes.

At the upper left is a 51 knot with an arbitrary polarity. In the transition to the upper right, a node is destroyed by cutting both strands and rejoining them, while maintaining local polarity; a catenane results. The catenane is repeated on the lower left. Performing the same operation on a different node converts the catenane to a knot.

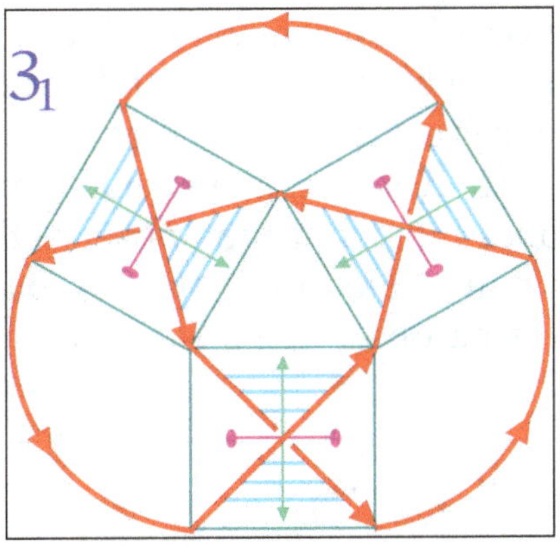

The Relationship of Nodes to DNA Half-Turns.

The simplest knot, 3_1 is drawn with red strands and arbitrary polarity. Blue boxes are drawn about its nodes, and the strands that make the nodes serve as the diagonals of the boxes; they divide the boxes into four regions, two between parallel strands and two between antiparallel strands. Drawing a half-turn of DNA (about 6 nucleotide pairs) between the antiparallel strands effects the transition from strand topology to nucleic acid structure.

The nodes shown in Figure are all of the same sign, corresponding to the negative nodes produced by right-handed B-DNA. However, knots and other topological constructs are not limited to negative nodes. Fortunately, a left-handed form of DNA exists, known as Z-DNA, which can supply positive nodes; the differences between B-DNA and Z-DNA are cartooned in Figure. This molecule requires particular sequences (typically CG_n) to form, and it also requires conditions that promote the formation of Z-DNA, usually high salt or effector molecules, such as $Co(NH_3)_6^{3=}$. By using appropriate sequences and conditions, it is possible to produce a variety of topological targets containing positive nodes, negative nodes, or both.

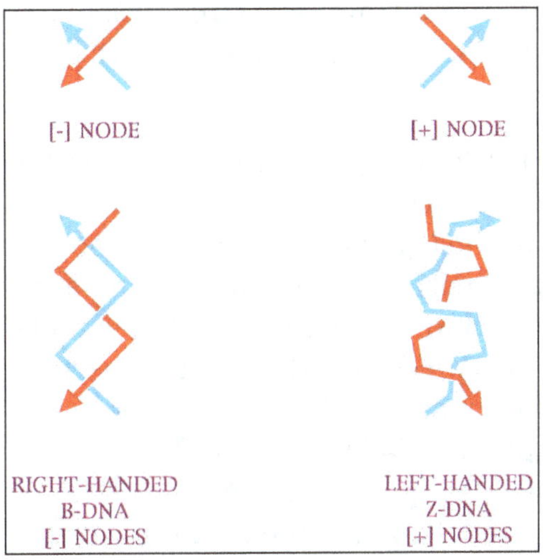

B-DNA and Z-DNA.

The two forms of DNA are cartooned here. B-DNA has a simple helical structure and is right-handed, leading to negative nodes. Z-DNA is a helix with a zigzag backbone resulting from a dimer being the helical repeat unit. It is left-handed, leading to positive nodes.

An example of a collection of topological targets is illustrated in figure. By changing conditions, it is possible to use sequences of different Z-forming propensity to make a circle, a trefoil knot with negative nodes, a figure knot and a trefoil knot with positive nodes, all from the same strand of DNA. Topoisomerases are molecules that transform DNA molecular topologies by catalyzing strand-passage events. It is possible to make one of the knots shown in figure and then change the solution conditions, and then observe the knot identity change if a topoisomerase is present to catalyze the transition. This approach has been used to discover an RNA topoisomerase activity by using observing strand-passage transitions occur in knotted RNA molecules.

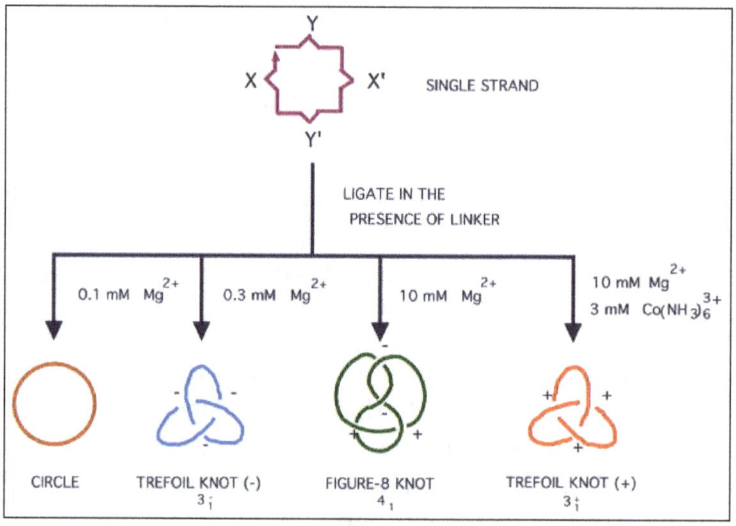

Synthesis of Four Topologies from a Single Strand of DNA.

X and X' represent a turn of DNA and its complement, as do Y and Y'. Both have a Z-forming propensity, but that of Y-Y' is higher. As the Z-forming nature of the solution increase, the main ligation product changes from a circle (very low ionic strength), to a trefoil knot with negative nodes (both turns B-DNA), to a figure-8 knot (the Y turn is Z-DNA), to a trefoil knot with positive nodes (both turns are Z-DNA).

Borromean rings are a complex topological target that has been obtained relatively simply from DNA. figure shows that they consist of three interlocking rings, no two of which are themselves linked. The signs of the three inner nodes are opposite from the signs of the three outer nodes. If one of the rings is cut, the other two will fall apart. To produce this property in DNA, it was easiest to change the individual nodes to triple nodes, i.e., to replace the half-turn of DNA with three half-turns of DNA. If one thinks of the structure in figure as a polar projection, so that the central point corresponds to the north pole and every point on the outside is the south pole, then one can put a B-DNA junction flanking the north pole. If every point on the outside corresponds to the south pole, one can put a Z-DNA junction there. Hairpin loops that do not change the topology are added to the equator for synthetic and analytical purposes. Their assembly is illustrated in figure.

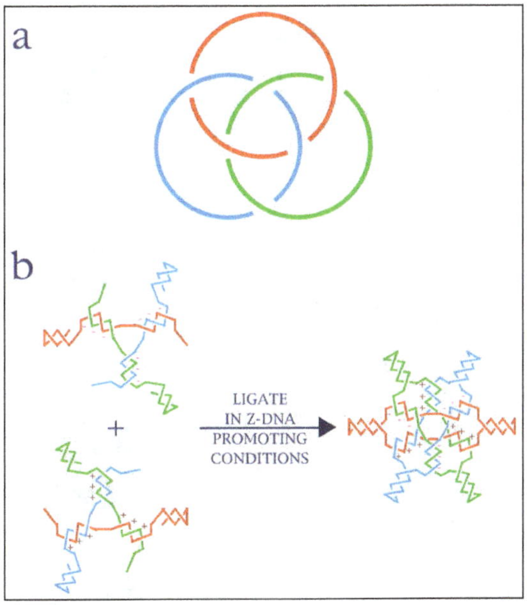

Borromean Rings.

(a) The Topology of Borromean Rings. Note that cleaving any ring leaves the other two rings unlinked; the key is that the nodes on the inside are of a different sign from those on the outside. (b) Forming Borromean Rings from DNA. A 3-arm DNA branched junction made of B-DNA (top) is linked to a 3-arm junction made of Z-DNA (bottom) through hairpins at the equator of the product molecule.

2D Periodic Arrays

Individual objects like polyhedra, knots and Borromean rings may demonstrate the versatility and convenience of DNA branched junctions as a system for the construction of difficult targets. Nevertheless, the goals of structural DNA nanotechnology go beyond the construction of individual objects to periodic arrays, and even to aperiodic arrays. The ability to make specifically-patterned 2D and 3D structures is central to the success of this enterprise. The floppiness of individual branched junctions usually makes them unsuitable as components for such constructions. However, as noted above, DX molecules are branched species that are roughly twice as stiff as linear DNA, and therefore substantially stiffer than simple branched junctions. They, and TX molecules are therefore well-suited to serve as components for periodic arrangements.

The first examples of two dimensional periodic arrays were DX arrays that contained the capability to produce patterns. Figure illustrates an array produced from a DX molecule and a DX+J molecule. The dimensions of these 2 nm-thick tiling components are about 4 nm x 16 nm. The DX+J molecule has its extra domains rotated out of the plane of the array, so that they can act as topographic markers for the atomic force microscope. Thus, a series of striped features should appear on the pattern, separated by about 32 nm, as seen on the right of the drawing. To demonstrate the level of control over the pattern, a second DX array is shown in Figure. Here, three DX tiles are combined with a DX+J tile, to produce a pattern where the stripes are separated by 64 nm.

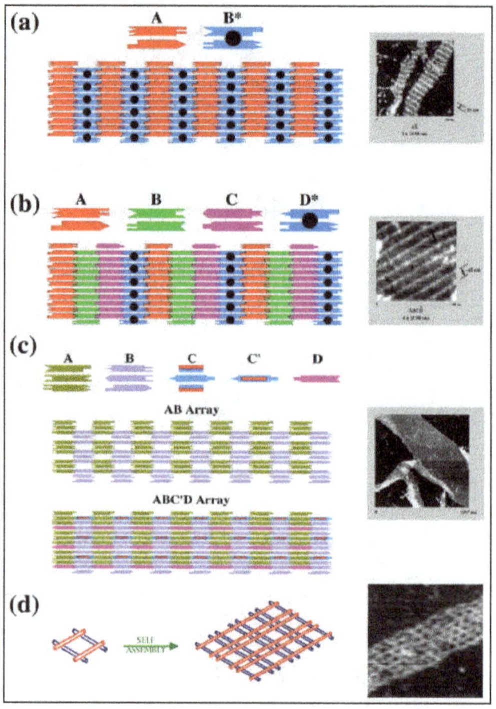

DNA Arrays.

(a) Two DX Molecules Tile the Plane. A conventional DX molecule, A, and a DX+J molecule, B*, are seen to tile the plane. The extra domain on B* leads to stripes. The molecules are 4 x 16 nm in this projection, so the stripes are ~32 nm apart, as seen in the AFM image at the right. (b) Four DX Molecules Tile the Plane. This arrangement is similar to (a), but there is only one DX+J molecule, D*, so the stripes are separated by ~64 nm, as seen on the right. (c) A TX Array. Two TX tiles, A and B are connected by complementarity between their first and third double helical domains, resulting in spaces between the tiles. D is a linear duplex that fits in the yellow rows, and C is a TX rephased by three nucleotide pairs; it fits into the gray rows and extends helices beyond the AB plane in both directions, as shown in the micrograph at the right. (d) A DNA Parallelogram Array. Four Holliday junction analogs form a parallelogram that is extended to produce a periodic array. The sizes of the cavities in the array may be tuned. Those in the array the right are ~13 nm x ~20 nm.

TX molecules also can be used to produce 2D arrays. One can take the same approach to making patterns used with the DX arrays, using a TX+J tile. However, the TX tile offers a convenient way to insert specimens within the array: It is possible to produce a gap in a continuous lattice by connecting individual tiles 1–3, i.e., by designating sticky ends that connect the first domain to the third domain, as illustrated in the AB array of Figure. One way to demonstrate a robust insertion into this array is to re-phase a third TX tile (C in Figure) by three nucleotide pairs (~102), so that it is roughly perpendicular to the AB array (designated C' in Figure). This tile fits into the blue-gray column in the AB array by cohesion of its central domain. A fourth component, a piece of linear duplex DNA (D in Figure) is also shown inserted into the gold row.

An unlikely, but effective system results from combining four 4-arm branched junctions into a parallelogram, as illustrated in Figure. The 4-arm branched junctions assorts its four arms into two double helical domains, which are twisted with respect to each other. The twist can be 40–60 from antiparallel or (with 3',3' and 5',5' linkages in the crossover strands) 70 from parallel, so that a variety of parallelograms can be produced. The parallelograms can be connected through sticky-ended cohesion to produce an array containing cavities, such as the one shown in Figure. It is straightforward to alter the sizes of the cavities, so that the porosity of this system is readily tunable.

In addition, Yan, Reif and their colleagues have developed a motif that they call the 4 x 4. it consists of four DX molecules that flank a central cavity. The molecules fail to stack within the cavity because of loops that proceed from one domain to another. The system is nearly, but not exactly tetragonal. This is a substantially new motif that appears to be robust, and is likely to have a number of applications in DNA nanotechnology.

All of the periodic motifs described so far are basically low-symmetry motifs, corresponding to translations or twofold rotations normal to the plane. Recently, we have developed a trigonal motif that produces a pseudo-hexagonal array. There have been a number of significant failures to make hexagonal structures, the 'stick' version of hexagonal close packing within the plane. We have adapted the unsuccessful bulged junction triangle motif by doubling its height, so that it is made of DX molecules, rather than double helices. This molecule forms an excellent honeycomb structure. Analysis of the system reveals that the key to the formation of the 2D array is the DX cohesion: Blunting one of the sticky ends leads to failure in honeycomb array generation. This approach to DX cohesion has now been applied to a large number of motifs that were intractable or at least recalcitrant to 2D periodic array formation. An approach from Adleman and his colleagues uses DX cohesion in structures confined to a single plane, and is largely successful, but inspection of their results demonstrates that the stiffness of the DX sides to the triangles is apparently partially responsible for the uniquely stiff nature of the DX triangle motif and the other DX cohesive systems being investigated.

DNA-Based Nanomechanical Devices -The Path to Nanorobotics

There are at least three different routes to obtaining nanomechanical devices from DNA-based systems. The first of these is the use of DNA structural transitions. Such transitions are well-known, and they can be used to change the structure of DNA in a reliable fashion. The very first of these devices entailed a DNA cruciform with several nucleotides at the middle that could undergo branch migration. As illustrated in Figure, the addition or removal of ethidium, an intercalating drug (which binds between the bases and in doing so unwinds the helix), changes the position of the branch point. The change in twist in the body of the circle as a consequence of the presence or absence of the drug results in the extrusion or intrusion of the cruciform branch. This device is indeed under the control of an external agent, but it is not a convenient system from which to obtain work.

A second system that is better designed is based on the B-Z transition of DNA. As noted above, Z-DNA has two requirements for formation: An appropriate sequence (called 'proto-Z-DNA') and Z-promoting conditions; lack of either of these keeps the structure in the B-form. The need for a special sequence allows us to control the transition in space (i.e., where, and over how much DNA it will occur), and the necessity of Z-promoting conditions allows us to control the transition in time (i.e., when it happens). A device that takes advantage of the B-Z transition is illustrated in Figure. It consists of two DX molecules connected by a shaft (yellow backbone) that can undergo the B-Z transition. In the absence of Z-promoting conditions, the shaft is in the B-conformation, and both domains unconnected to the shaft are on the same side of it. However, addition of the effector, $Co(NH_3)_6^{3=}$, switches it to the Z-conformation. This in turn rotates an unconnected domain of a DX molecule to the other side of the shaft. The transition is demonstrated by a pair of dyes using fluorescent resonance energy transfer (FRET).

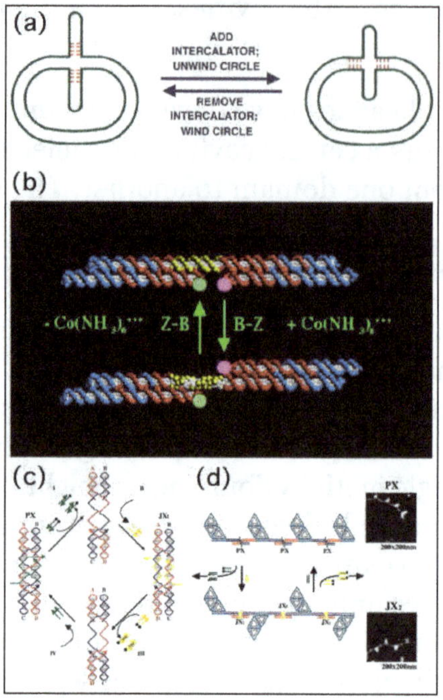

DNA Devices.

(a) A Mobile Control Device. The cruciform structure on the left contains four mobile base pairs at its base. Addition of an intercalator (ethidium) unwinds the circle, moving them into it. Removal of the ethidium reverses the action of the device. (b) A DNA Nanomechanical Device Based on the B-Z Transition. The device consists of two DX molecules connected by a shaft containing 20 nucleotide pairs (yellow) capable of undergoing the B-Z transition. Under B conditions the short domains are on the same side of the shaft, but under Z-conditions (added $Co(NH_3)_6^{3=}$) they are on opposite sides of the shaft. The pink and green FRET pair are used to monitor this change. (c) The Machine Cycle of a PX-JX_2 *Device.* Starting with the PX device on the left, the green strands are removed by their complements (Process I) to leave an unstructured frame. The addition of the yellow strands (process II) converts the frame to the JX_2 structure, in which the top and bottom domains are rotated a half turn relative to their arrangement in the PX conformation. Processes III and IV reverse this process to return to the PX structure. (d) AFM Demonstration of the Operation of the Device. A series of DNA trapezoids are connected by devices. In the PX state, the trapezoids are in a parallel arrangement, but when the system is converted to the JX_2 state, they are in a zigzag arrangement.

The problem with devices based on DNA structural transitions is that they are triggered by a single molecular species, and are not individually addressable. A couple of devices have been developed that utilize a transition between B-DNA and G_4 tetrad structures, or based on pH but these devices suffer from the same problem. The breakthrough in this area was made by Yurke and his colleagues, who developed a sequence-dependent device. This was molecular tweezers that contracted when a particular strand was added to the solution, and then relaxed to an expanded state when the strand was removed. The removal of the strand was achieved by adding a short, unpaired segment (called a toehold) to its end. Addition of the complete complement to this strand resulted in its binding first to the toehold and then ultimately the complement branch migrates its way through the structure until the entire strand is in duplex form; this is a largely irreversible state, because there is much more base pairing with the complement than with the device structure. Numerous devices have been built using this approach, including a bipedal walker and a transcriptionally-controlled device.

A robust two-state device was developed that is based on this principle. That device is illustrated in Figure. It shows a molecule in the PX structure (left), with two green 'set' strands that put it into that state. The addition of the complements to the set strands ('unset' strands) results in their removal, leaving a naked frame. The addition of two yellow set strands puts the system in a different state, called JX_2, because there are two juxtapositions between the helices. The feature of the JX_2 structure that is central to its role here is that its is rotated a half- turn from the bottom of the PX structure. Note that the sequence complementary to the set strands can be varied, so that there are a lot of different frames that can be developed, and which can be addressed independently. Figure shows that this system can do work, by rotating a marker visible in the AFM. In the PX state, the markers should all be on the same side of a 1D array, but in the JX_2 state, they should alternate.

This same notion has been used to build a device that walks on a sidewalk. Figure shows that this device consists of two double helical legs that are connected to each other by a flexible linker. It is connected to the sidewalk by set strands that hydrogen bond it in position. As can be seen in Figure, removal of one set strand by its unset strand releases one foot that can then be bound to a different position by a different set strand. The same position can be used to release the other foot, and then it, too, can be set down on the sidewalk in a new position. It is possible to move this device in both directions. Proof of movement consists in crosslinking the feet in aliquots of the solution at every point along the walk, and then examining them in a denaturing gel, which will reveal bands corresponding to the crosslinked species.

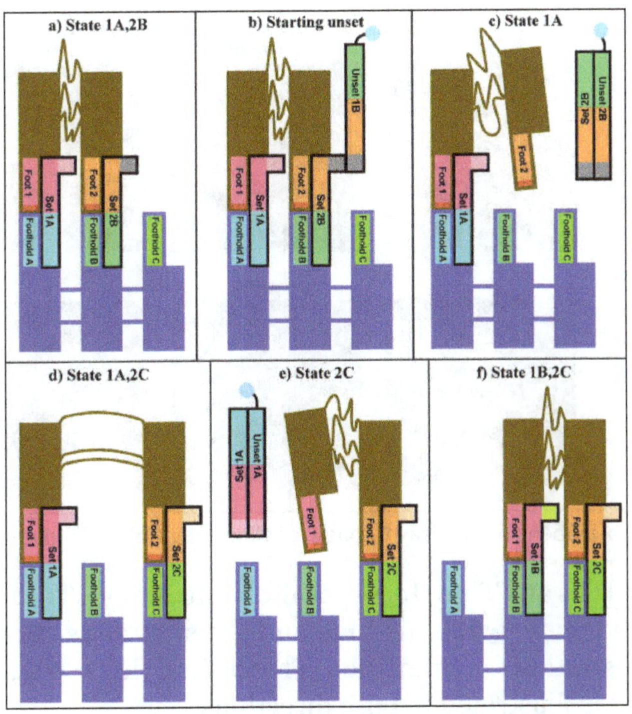

A DNA-Based Walker Walking on a Sidewalk.

The states through which the system passes sequentially are shown in the panels of the drawing. Matching colors indicate complementary sequences between strands. The red section at the end of each foot indicates psoralen, which is crosslinked for analysis. (a) Initial State of the Walker Feet 1 and 2 are attached to the sidewalk by hydrogen-bonded strands. (b) and (c) Removal of the Connection of Foot 2. The Unset complex is removed from solution via a biotin group (depicted as a blue circle) attached to the unset strand. (d) Re-Attachment of Foot 2 at a New Position. (e) Release of Foot 1. (f) Re-Attachment of Foot 1 at a New Position. At this point the device has moved a complete step.

An exciting hybrid of a device and a 2D lattice has been pursued by Yan, Reif and their colleagues. As shown in Figure, a parallelogram-like lattice has been built that contains a hairpin. A strand can be removed from the base of the hairpin, and a complement to the entire hairpin and base can be added. This expands the lattice dimensions of the parallelogram in one direction from four turns to six turns. The dramatic AFM evidence in Figure suggests that ultimately this approach can be used to change the dimensions of crystalline samples.

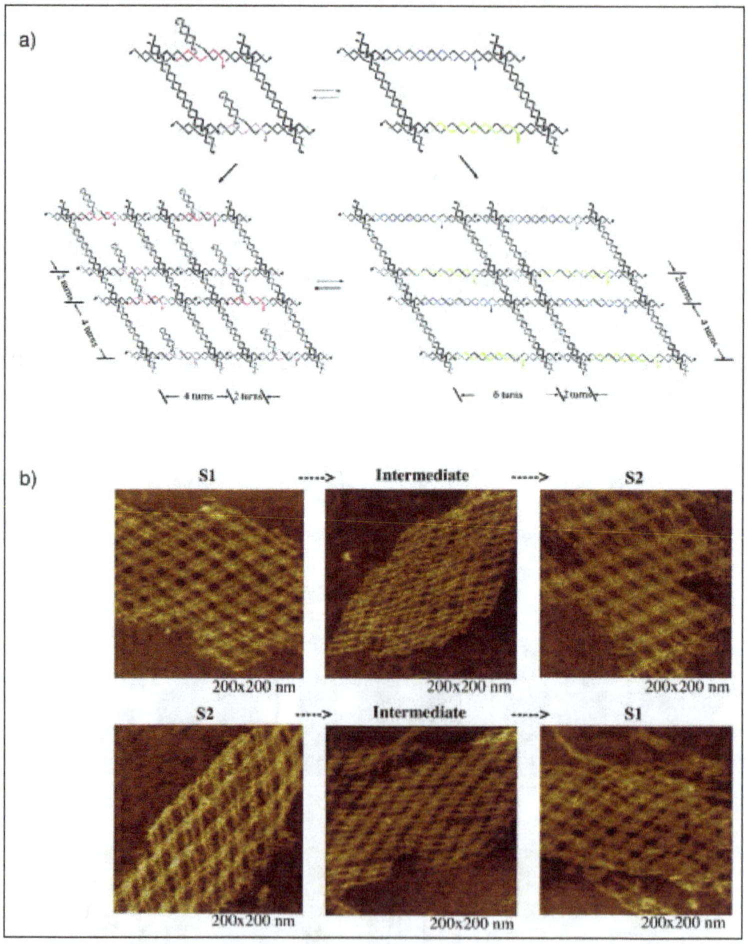

A Two-Dimensional Array Capable of Changing Cavity Dimensions.

(a) Strand Diagram of the Tiles and Their Incorporation into Arrays. The red and purple set strands correspond to a contracted state, while the blue and green strands correspond to the expanded state. (b) AFM Images Illustrating the Transition. The 'before' (left), 'transition' (center) and 'after' (right) states of the array are illustrated in both directions.

A third form of device is the activation of a DNA transition by a protein. This is the closest to a phenomenon that actually occurs within a living cell. This type of device is sequence-specific, but a new protein much be found for each different transition, thereby decreasing the convenience and generality of the approach. This type of device was prototyped recently in a context similar to that of the B-Z device; the proto-Z segment of the device was replaced with the binding site for a DNA-bending protein known as integration host factor (IHF). The binding of IHF to the DNA was monitored by FRET, just as it was in the B-Z device. The interesting feature of this device is that it was used to measure the amount of work that the protein could do upon binding. As schematized in Figure, the protein was required to break a number of hydrogen bonds so as to bind. By titrating the number and type of interactions, it was possible to estimate that the protein is capable of doing 7–8 kcal/mol of work when it binds to DNA.

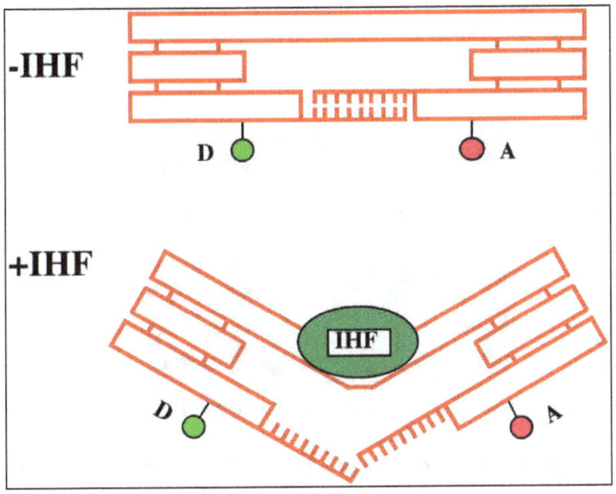

The Action of a Nanomechanical Device Working Against a Load.

Double-helical DNA is shown as rectangular boxes. Triple crossover motifs (TX) are shown as three fused rectangular boxes. The upper domain connects the two TX motifs with the binding site for IHF. IHF is shown as a green ellipse, and in the lower panel its binding distorts the connecting shaft. In the upper panel the lower TX domains are held together by a sticky end. In the lower panel, these are disrupted by the binding. By changing the strength of the sticky end, the amount of work the protein can do upon binding is estimated.

The devices described above are all clocked devices: They respond to a particular stimulus in a way that is designed. However, there is a great deal of interest in autonomously operating devices. One has been developed recently by Mao and his colleagues. He has employed a DNAzyme, a DNA molecule that binds to a given short sequence of RNA, and then catalyzes its cleavage. When the RNA molecule binds, it brings together a tweezer-like molecule. Following cleavage, however, the two RNA molecules no longer hold the tweezers together, so it expands. In addition, the RNA is so short following cleavage that it is no longer bound to the DNA; it dissociates when a new substrate molecule is present, and the cycle repeats. Numerous other autonomous devices are being developed currently.

Self-Replicating Systems

The use of DNA as a building material immediately suggests the notion that a self-replicating system could be developed. One of the key features of DNA in cells, however, is its linearity. The replication of branched DNA by the cellular enzyme, DNA polymerase, will only result in further linear molecules, even if the ends are joined by hairpin loops. A number of schemes have been advanced, but only two have been realized experimentally. The first is a scheme devices by von Researcher, which entails branching at the primary structure level. He has three strands of DNA attached to a small organic molecule at one of their ends. He then adds the complements to the three strands, which bind to the first three strands. They are then bonded to a second organic molecule. Dissociation leads to the ability to continue the replication.

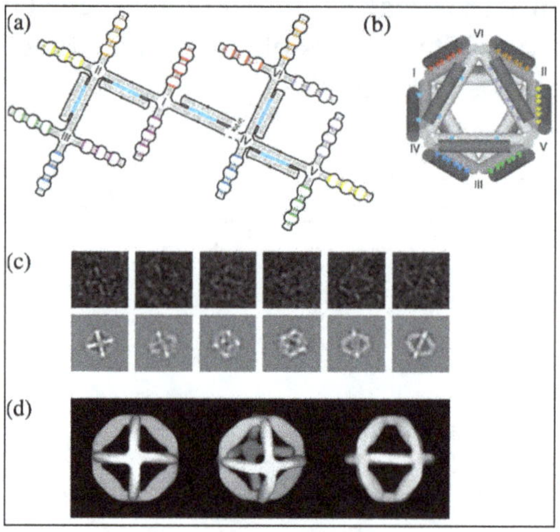

An Octahedron Consisting of One Long and Five Short DNA Strands.

(a) Strand Diagram of the 'Branched Tree' Structure. The structure consists of DX struts (double helical regions with light blue central strands), four-arm junction joints (labeled I-VI) and half-PX domains (complementary domains labeled with the same colors) destined to fold into struts. (b) 3-Dimensional Representation of the Folded Octahedron. The drawing shows the crossovers in the DX molecules as two light blue dots between cylinders and the PX associations as a series of colored dots between cylinders. (c) Electron Microscopic Images. Raw EM data are shown at top and maps onto projected octahedral symmetry of the folded octahedron are shown at the bottom. (d) Views of the 3-Dimensional Electron Density Model Based on EM Data. Reproduced with permission.

Another suggestion by von Kiedrowski was that one could make polyhedra whose were not covalently bonded to all of their vertices, but only to one vertex. This idea was adopted elegantly by Researcher to form a self-replicating octahedron. Seven of the edges of the octahedron are held together by PX cohesion, as drawn schematically in figure. To make all twelve edges structurally equivalent, five helper strands were also added, so that the other five edges could be DAE-DX molecules. The design of the octahedron is illustrated in figure. Electron micrographs of the final molecule are shown in figure, and symmetrized structures are shown in figure. This is a remarkable achievement, which may ultimately remove much of the drudgery from DNA nanotechnology.

Structural DNA Nanotechnology and DNA-Based Computation

DNA-based computation was originated as a laboratory science by Adleman, who solved a Hamiltonian path problem using DNA. The field has grown large, and space does not allow for a description of all branches of the discipline. Its central approach is to use the parallelism of the vast number of molecules in a solution to solve solutions in parallel. Two approaches bring it within the sphere of DNA nanotechnology, and those are described here. We have already noted that it is possible to make irregular graphs from DNA. A goal is to solve 3-colorability problems using branched DNA molecules. The problem is to color a given graph so that no two nodes joined by an edge have the same color. Jonoska has pointed out that if one mixes junctions of all possible colors and joins them specifically with edges that only permit differently colored nodes at then ends, a successful solution will be characterized by a molecule that is cyclic. This is easily detected either as an off-diagonal spot on 2D polyacrylamide gels (gels where the material is driven through two different concentrations of polyacrylamide), or by exonuclease resistance. Although the rigidity of components normally is valued highly in structural DNA nanotechnology, this is an instance in which flexibility must be maximized, because one does not want rigidity to get in the way of structure formation.

The other way in which DNA-based computation is convergent with structural DNA nanotechnology is due to Winfree, who approached DNA-based computation from the perspective of Wang tiles. Figure illustrates a group of Wang tiles in the upper left corner; they have a variety of differently-colored edges. In theory, Wang tiles assemble into a mosaic, such as the one shown at the bottom of figure, according to the local rule that every edge in the mosaic is flanked by the same color. This form of assembly emulates the operation of a Turing machine; indeed, the mosaic shown illustrates a simple calculation, the addition of 5 to 9 to yield 14. Winfree noted that branched junctions with sticky ends are logically equivalent to Wang tiles, and that branched DNA molecules could be a way to reduce Wang tiles to practice in computation. Recently, researcher have demonstrated the successful assembly of a Sierpinski triangle, using DNA self-assembly. This is the first successful 2D algorithmic DNA assembly, although a 1D assembly of a cumulative XOR was demonstrated in 2000.

Wang Tiles and their Relationship to Branched Junctions.

The upper left corner illustrates 16 Wang tiles of various types. The bottom shows a mosaic formed from them using the rule that all edges in the mosaic are flanked by the same color; this mosaic assembly that performs the calculation of adding 5 to 9 to get 14. The upper right is a color coded branched junction that, when compared to the enlarged tile at the center shows the relationship between Wang tiles and branched junctions with particular sticky ends.

Scaffolding other Systems

The goals of structural DNA nanotechnology noted above entail the scaffolding of heterologous species of molecules, for example proteins or nanoelectronic components. Some progress has been made in this direction. Kiehl and his colleagues have been successful in organizing small goal nanoparticles on a 2D DNA periodic lattice similar to those illustrated in Figure. This work lends credence to the notion that DNA can organize species that are not necessarily similar chemically. A different goal has prototyped the use of DNA to direct the topology of industrial polymers. Derivatives of uridine (the variant of thymine found in RNA) have been synthesized that have a branched side-chain on the outside of the helix. One variant contains a di-amino group, and the other contains a di-carboxyl group. These are the constituents of nylon. A knotted DNA chain containing alternating di-amino and di-carboxyl groups would impart this topology to a nylon molecule that was formed from these side-chains. A prototyped system was successful in the condensation of these groups.

Nanostructures from RNA

Since RNA is chemically more labile than DNA, dynamic and modular supramolecular engineering to RNA nanostructures is possible akin to jigsaw puzzle game. RNA techtonics are short modular RNA segments, which can be self-assembled in presence of cation such as Mg^{2+} with well controlled size, shape and pattern to programmable stable 2D architectures known as tectosquares, RNA supramolecules. These tectosquares with sticky ends can further be assembled in well-defined lattice-like structures. Thus, choosing the RNA connectors with complimentary sticky end is very important as it leads to either 1D ladderlike pattern or 2D fishnet like pattern as shown by the AFM images in figure.

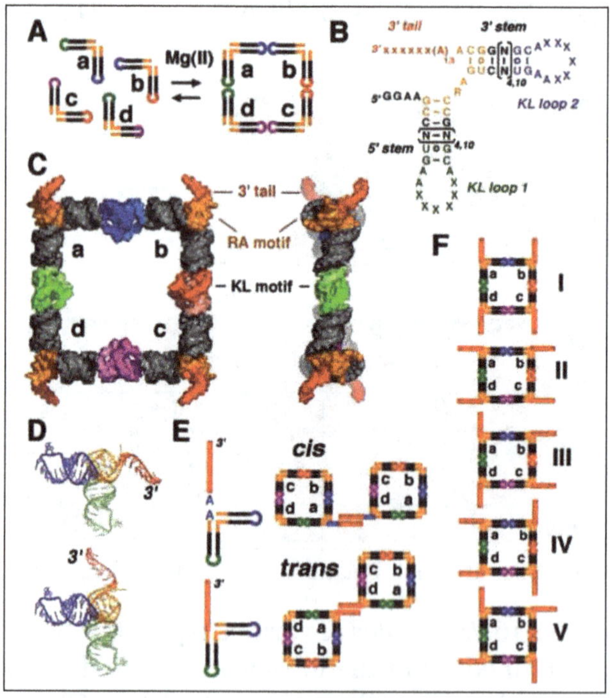

RNA nanostructures: (A) Tectosquares formation from tectoRNA, (B) primary sequence of techtoRNA, (C) front and side view of the tectosquare model, (D) the modifcations in 3' end of the tectoRNA, (E) possible cis and trans configuration of the tectosqure self-assembly and (F) different types of possible tectosquares.

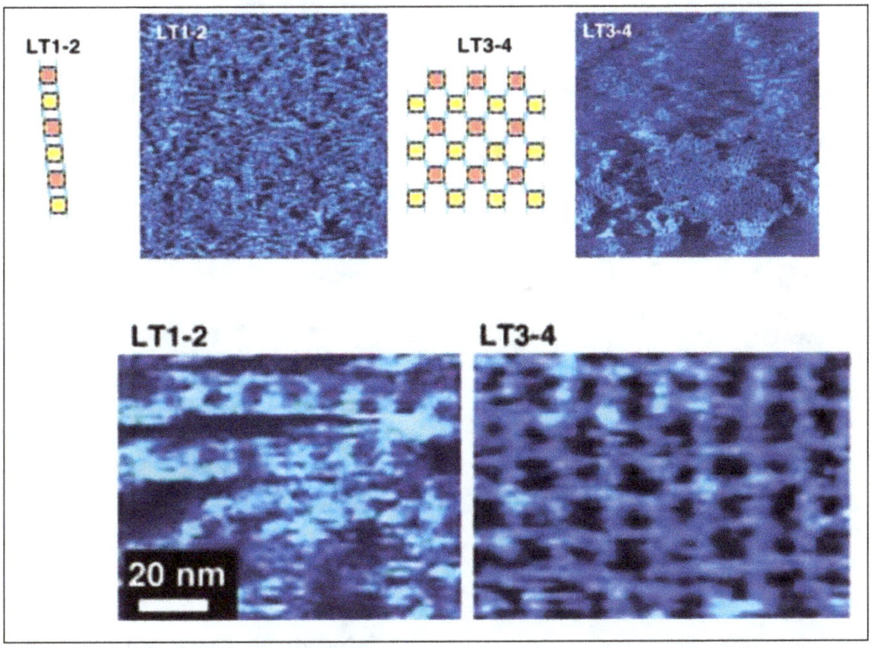

RNA nanostructures from jigsaw puzzle tectosquares: AFM images of tectosquare assembly in ladder (left) and fish net pattern (right).

Hierarchical Structures of Protein

Proteins are very important class of biomolecules as they are involved in almost all physiological functions. Some of the examples are enzymes, hormones antibodies, receptors. Enzymes, hormones and antibodies are involved in catalyzing biochemical reaction, immunogenic reaction, while receptors are important for signal transduction in the cells.

Proteins consist of one or more polypeptides chains arranged in a functional manner, often bound to metal or ligands such as coenzymes and cofactors. Polypeptides are distinguished from proteins on the basis of size and the polypeptide chains are made of amino acids building blocks. Hence, the structural and functional monomer of protein is amino acid and is known as proteinogenic amino acid. There are 21 proteinogenic amino acids, out of which 11 amino acids human body can synthesize, rest of the amino acids needed to be taken up in body by nutrition and diet. These amino acids are called essential amino acids. However, there are more than 100 non-proteinogenic amino acids in nature.

Proteinogenic amino acids have amino group and carboxylic acid group which are connected by an α-carbon atom. At physiological pH of 7.4, the amino acids stay in zwitterion form. The α-carbon atom is chiral in nature (except glycine) as different side chains are connected to it. Based on the side chains amino acids can be classified in different groups as depicted in figure.

There are four different hierarchical structures of protein, which depicts how amino acid building blocks can form the functional units of protein. They are knows as primary, secondary, tertiary and quaternary structure of protein.

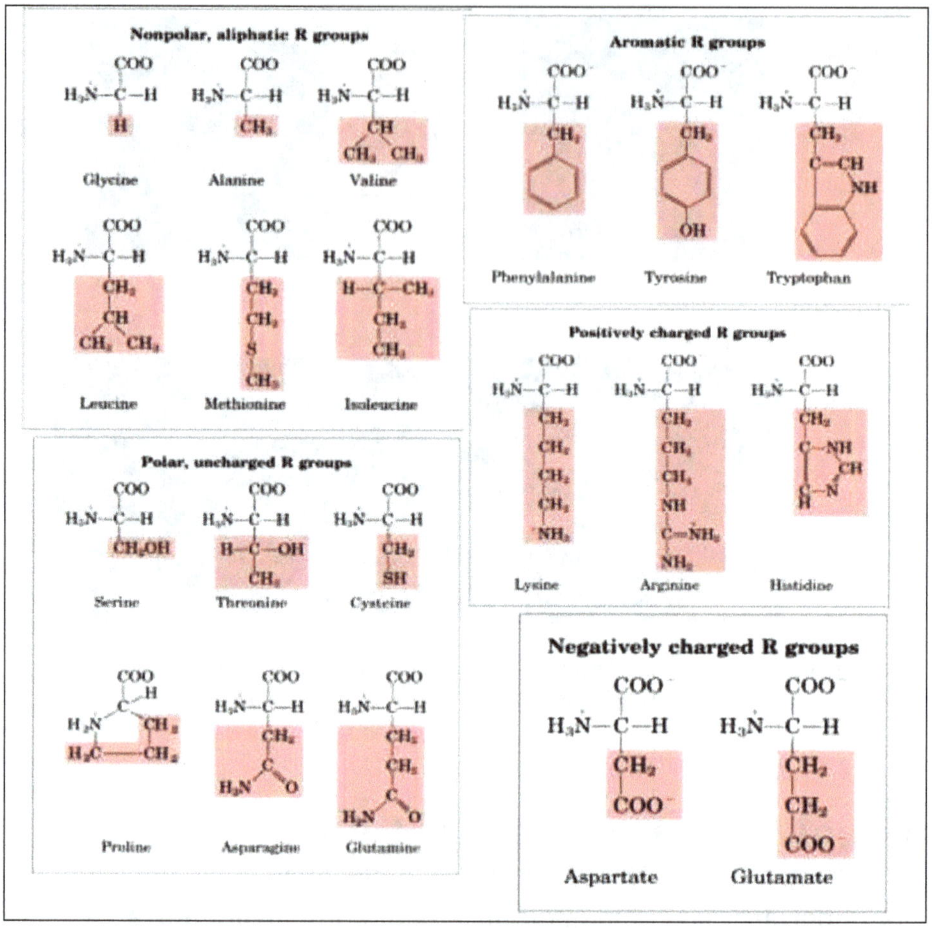

Different types of proteinogenic amino acids classified based on side chains.

Primary structure of protein: This demonstrates the sequence in which the individual amino acids are connected to each other by amide bonds to form peptide chain. The chain has N-terminal with free amine and a C-terminal with free carboxylic acid group. The primary structure can be synthesized by solid phase peptide synthesis.

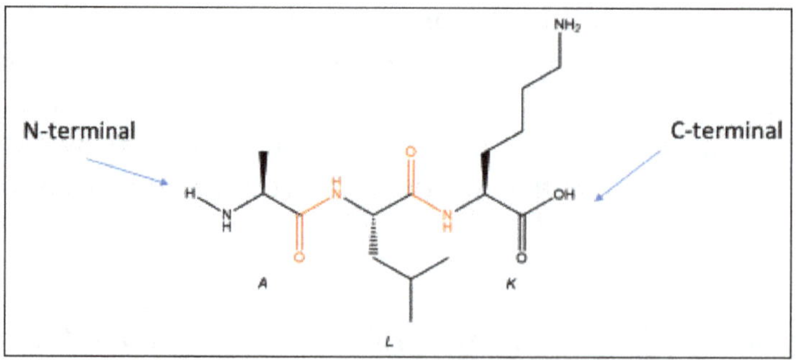

Primary structure of protein by linking amino acids by amide bonds.

Secondary structure of protein: Secondary structures of proteins depict the hydrogen bonding pattern among the amino acids in primary structures in assuming three dimensional form. The most common types of secondary structures are α-helix and β-sheet. Both structures are held in shape by hydrogen bonds, which form between the carbonyl O of one amino acid and the amino H of another. Torsion angles are dihedral angles of proteins, which are defined by 4 points in space. In a polypeptide chain, three torsion angles are defined as φ (phi), ψ (psi) and ω (omega), as shown in the Figure. The resonance and planarity of the peptide bond usually restricts ω to be 180° (stable trans form) or 0° (the rare cis form). The cis isomer is mainly observed in proline rich peptide segments. The torsion angles φ and ψ describe the rotation of the polypeptide chain around between N-Cα (φ) and Cα-C (ψ) as shown in Figure. Most of the combinations of φ and ψ are not allowed due to the steric hindrance (collision of the side chains). However, Indian physicist Prof. G. N. Ramachandran calculated the allowed combinations of φ and ψ and showed that the allowed combinations are clustered in the some definite area of a plot of φ and ψ, famously known as Ramachandran plot.

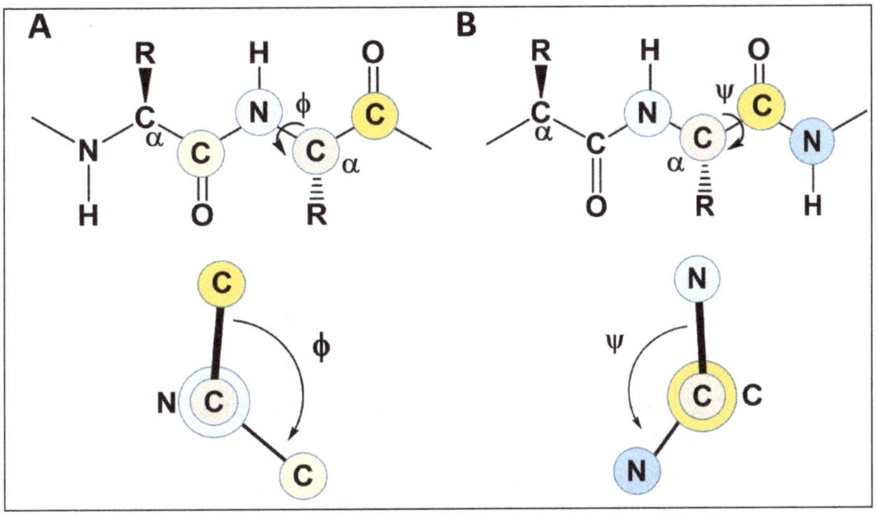

(A) Torsion angles for secondary structures of a dipeptide segment, (B) cis-trans conformations of the peptide bonds.

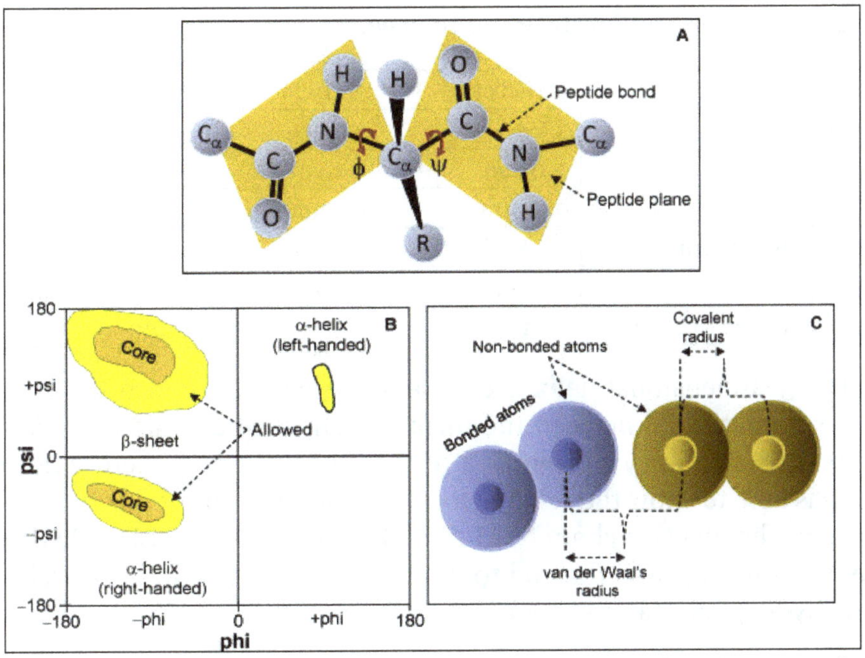

Different torsion angles for a dipeptide segment and allowed combinations of torsion angles (φ, ψ) in Ramachandran plot.

α-helix and other Helices

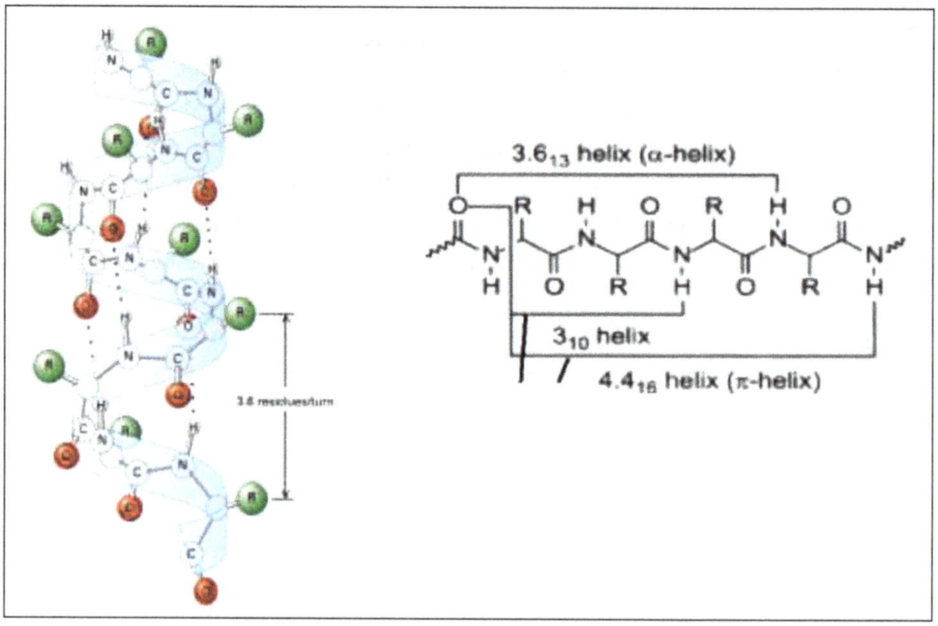

Typical 3.613 helix with the hydrogen bonds and possible other helices.

Figure shows a typical α-helix structure with right handed helical structure and 3.6 amino acids residue per turn. The carbonyl group and N-H bond orient themselves in one direction thereby providing overall dipole moment to the helix. The N-H group of ith residue makes hydrogen bond with carbonyl C = O of (i +4)th residue and involve 13 atoms along the peptide chain from C to N terminus. However, there are other helices e.g. 3.6_{10} and 4.4_{16}, which differ in various parameters shown in table.

Table: Characteristics parameters for different helices.

Parameters	α-Helix (3.6_{13})	3_{10} Helix	π-Helix (4.4_{16})
Per turn	3.6 AA	3 AA	4.4 AA
Translation	0.15 nm	0.2 nm	0.115 nm
Pitch	0.54 nm	0.6 nm	–
H-bonding pattern	i+4 – >i	i+3 ->i	i+5 ->i
(φ, ψ)	(-60,-45)	(−49°, −26°)	–
Sum of the consecutive	-105	-75	-130

However, drawing a 3-dimensional helix is cumbersome process; hence scientists came up with 2-dimensional representation of helix in the form of helical wheel and Wenxiang diagram. The helical wheel represents a helix by a projection of the Cα backbone structure down the helix axis. Wenxiang diagram is able to show the relative locations of the amino acids in α-helix regardless of how long it is. For the oligopeptide chain NIAEAVQQLNHTIVNAAHELHETLGL, the start (N-terminus) of the helix at the edge and the end toward the center. It shows the preferential positions of hydrophilic and hydrophobic amino acids.

Protein crystallography, especially recent atomic resolution X-ray crystallography and cryoTEM are used for characterization of helix secondary structures. The α-helix shows characteristic circular dichroism spectrum with pronounced double minimum at 208 and 222 nm.

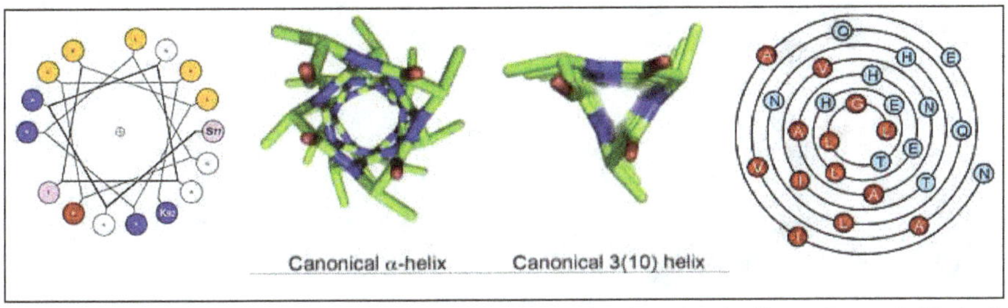

Two dimensional representation of the helix; (A) Helical wheel, (B) α-helix and 310 helix in helical wheel representation, and (C) Wenxiang diagram for the peptide segment NIAEAVQQLNHTIVNAAHELHETLGL.

β-sheet

β-sheets consist of β-strand connected laterally by hydrogen bonds, forming a generally twisted, pleated sheet. A β-strand is a stretch of polypeptide chain consisting typically 5 to 10 amino acids in an extended conformation. The supramolecular association of β-sheets has been implicated in formation of the amyloid aggregates and fibrils observed in many human diseases, notably the amyloidoses such as Alzheimer's, Parkinson's disease.

β-sheets can be of two types: anti-parallel and parallel depending on the direction of β-strands involved in hydrogen bonding. φ and ψ for both anti-parallel and parallel β-sheets lies in the structurally allowed region in the upper left quadrant of Ramachandran plot. Proline and Glycine amino acids are rarely found in β-sheets.

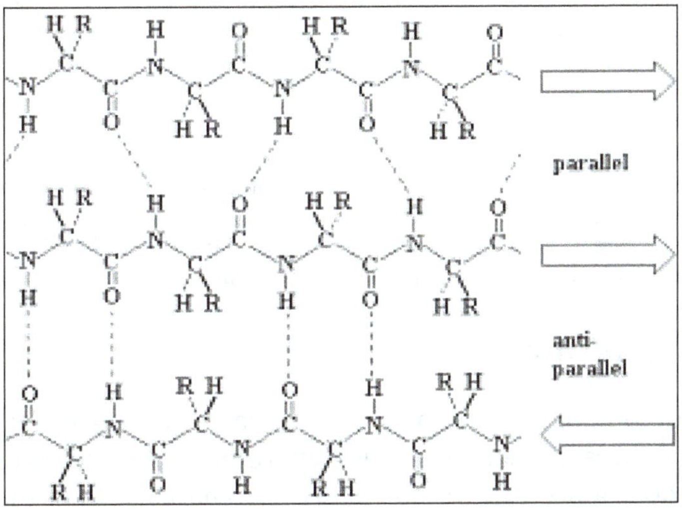

Antiparallel vs. parallel β-sheet with the hydrogen bonding patterns.

Loop and β-turn

In proteins structures, often we find a non-regular loop region which connects different secondary structures. The short loop region between two anti-parallel β-strand takes a sharp turn in a region rich with proline, glycine. This region is known as reverse β-turn. Depending on the conformation of the bonds or the torsion angle, this can be of two types, Type I and II.

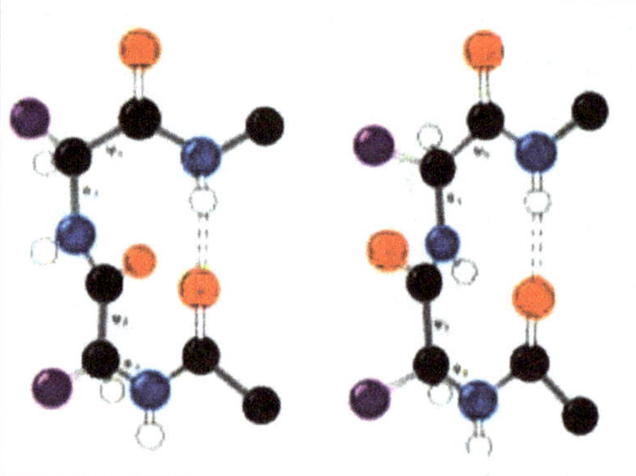

Peptide	Sequence[a,b]
1	Ac-L-Y-V-^{D}P-G-L-W-V-NH_2
2	Ac-^{D}L-Y-V-^{D}P-G-L-W-V-NH_2
3	Ac-L-F-V-^{D}P-G-L-W-V-NH_2
4	Ac-^{D}L-F-V-^{D}P-G-L-W-V-NH_2
5	Ac-L-W-V-^{D}P-G-L-W-V-NH_2
6	Ac-^{D}L-W-V-^{D}P-G-L-W-V-NH_2
7	Ac-^{D}I-W-V-^{D}P-G-L-W-V-NH_2

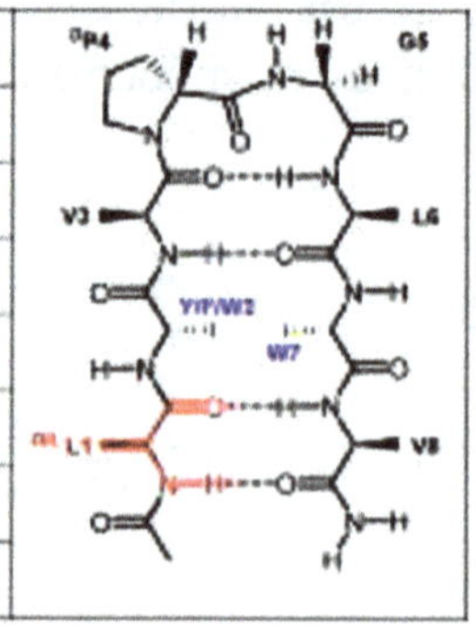

Loop and β-Hairpin secondary structures.

Super Secondary Structures

Super secondary structures (motifs) are combinations of secondary structures such as α- helices and β-sheets connected through loops to form patterns and nucleation sites for protein folding. Figure shows different example of super secondary motifs such as helix-loop-helix motif found in calmodulin (B) β-Hairpin Motifs in Erabutoxin and (C) β-α-β Motifs in triosephosphate isomerase.

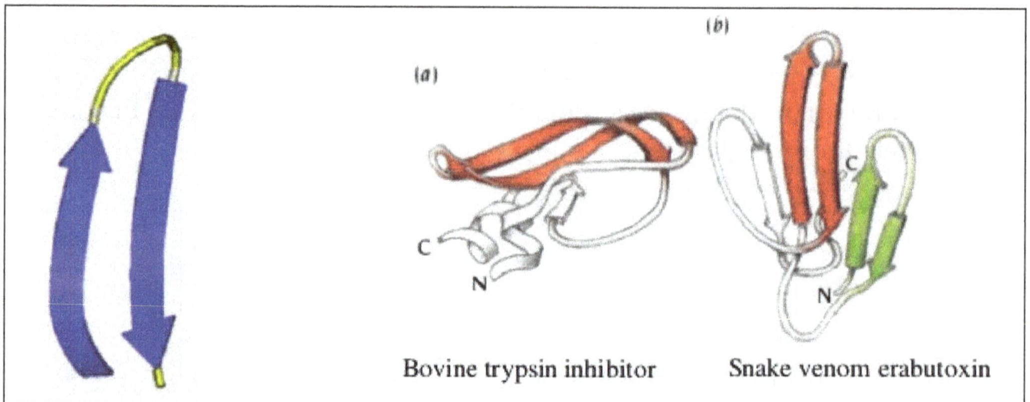

Supersecondary Structures: (A) helix-loop-helix motif found in calmodulin (B) β-Hairpin Motifs in Erabutoxin and (C) β-α- β Motifs in triosephosphate isomerase.

Tertiary Structure of Protein

It refers to special orientation of the amino acids which are far away in sequence from each other and overall describes the three dimensional shape of protein. The tertiary structures are stabilized by a number of interactions among the amino acids side chains such as disulfide bond, hydrogen bond, salt bridge interaction and hydrophobic interactions.

Quaternary Structure of Protein

This describes how folded protein subunits interact with each other and arrange themselves to form a larger aggregate of protein complex. The quaternary structures are stabilized by disulfide bond, hydrophobic interaction, hydrogen bonding and salt bridge.

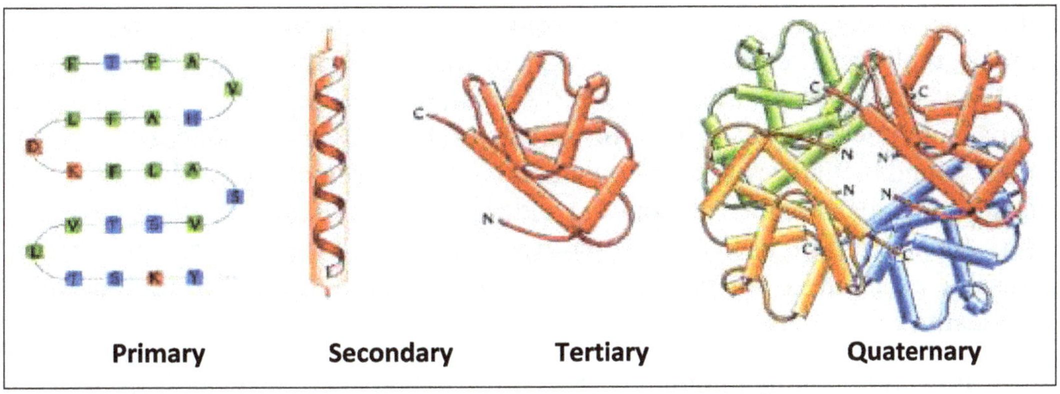

Hierarchical structures of protein: Primary, secondary, tertiary and quaternary structures.

Nanostructures of Protein

Protein nanostructures (PNs) are ubiquitous in Nature and fuel the complex cellular machinery through provision of functions, structural frameworks and molecular recognition. These biomacromolecules are in the first instance prepared in a sequence defined manner through transcription and translation processes, which lead to secondary and tertiary structures that confer functions such as catalytic activity. More complex arrangements such as oligomers, polymers and networks can also be created through protein–protein interactions. In this manner, Nature has evolved its own optimized toolbox through millennia of evolution that allows a plethora of PNs to be constructed. Essentially, such unique functional PNs are formed through precise molecular interactions of the monomeric protein units. For example, viral capsids consist of multiple copies of a monomeric protein unit through non-covalent interactions, resulting in the formation of stable polyhedral structures that are essential for protecting, storing, and transporting genetic information. Another example is the highly potent bacterial exotoxins, such as botulinum toxins, which consist of enzymatic, translocation, and cell binding domains. These discrete structural domains serve individual functions that, in combination, give rise to one of Nature's most potent weaponry. But additional functional and structural diversities can also be conferred to protein building blocks (PBs) through post-translational modifications to expand the repertoire of functional and dynamic nanostructures.

Inspired by Nature's machinery, there has been an emergence of research activities to evolve synthetic strategies that allow the rational design to construct functional PNs. These toolkits have shown great prospects in terms of preparation of PNs for different fields such as catalysis, biotechnology, and biomedicine. In particular, such well-defined nanostructures, which possess bioactivity and attractive materials properties, will be highly relevant for biomedical applications given the stringent demands for stability, biocompatibility, and biosafety. However, the field is fraught with challenges largely due to the complexity of protein surfaces. The exact spatial arrangement of proteins requires strict control of directionality, but the orientation of molecular recognition entities on protein surfaces can often not be predicted from the onset. With the progress in bioinformatics, biotechnology, chemical biology, and analytical tools, there has been an increased understanding in protein structures, folding, and protein–protein interactions.

In terms of applications, mild conditions are required to preserve the activity of the protein components in the complexes. Therefore, non-covalent or dynamic covalent strategies have emerged as valuable synthesis tools to impart molecular recognition units since they are reversible and should have less impact on the tertiary structure and activity of the protein components. In fact, most protein assemblies found in Nature are formed by non-covalent interactions, which allow for the rapid protein assembly and disassembly, responding and reporting to changes in their local physiological environments, such as variations in pH or ionic gradient, ligand concentrations or light. In contrast, chemical crosslinking is limited by the stability of proteins to the reaction conditions, and dynamic features are mostly lost during such processes.

To date, the utilization of biotechnological tools for engineering entirely new protein nanostructures with desired functional features has met with some notable success. For instance, the assembly of protein nanostructures through genetic engineering was reported, whereby natural oligomerizing protein domains were fused together through a rigid, peptide linker to form a defined cage-like structures such as the 12-mer tetrahedral cage. Coiled–coiled peptides have also been used to induce protein dimerization as in split luciferase reporters, and de novo design has been adopted on small modular domains to form distinct 3D structures as in tetrahedral nanocages. However, in some of the designs, greater predictability of the resultant PNs is required, and the introduction of entirely new functions is still challenging. In addition, genetic engineering has limitations if synthetic entities such as dyes need to be introduced, e.g. to further expand Nature's functional portfolio. Likewise, protein aggregation and laborious protein purification are further drawbacks of genetic engineering. Chemical modifications of proteins were considered less attractive for the preparation of protein building blocks due to lack of controlling of the reaction sites. Nevertheless, the rapid progress in terms of site-selective chemical modifications in the last decade has stimulated important advances in the field. Moreover, since Nature uses a combination of genetic and chemical tools to achieve infinite possibilities, the merger of different contemporary strategies in the expanding engineering toolbox has been capitalized to prepare more complex macromolecular structures such as hetero-functional proteins like chemical fusion proteins or higher-order protein conjugates that have been applied for biological applications.

Design Principles of Supramolecular Protein Nanostructures (PNs)

Proteins are attractive building blocks for the design of functional nanomaterials due to their inherent bioactivity and multiple functionalities that provide a rich platform for inter- and intramolecular interactions. In order to form defined PNs, stringent control over directionality and spatial placement of the PBs is essential. Therefore, the surfaces of PBs should ideally be encoded with molecular information, i.e. contain the supramolecular recognition motifs, as anchor points that can interact with the respective SL in a "lock and key"-like mechanism for spontaneous generation of distinct, higher order protein assemblies. For subsequent applications, it is a prerequisite that the PBs retain their structure and bioactivity during the modification and assembly processes. In the following, the design principles are discussed first highlighting (1) the selection of the respective native and modified PBs that contain the recognition motifs as hot spots. Subsequently, (2) the SLs interconnecting the PBs by complementary units that recognize the hotspots on the PBs in an orthogonal fashion are summarized to ultimately achieve (3) their controlled assembly into defined and functional PNs.

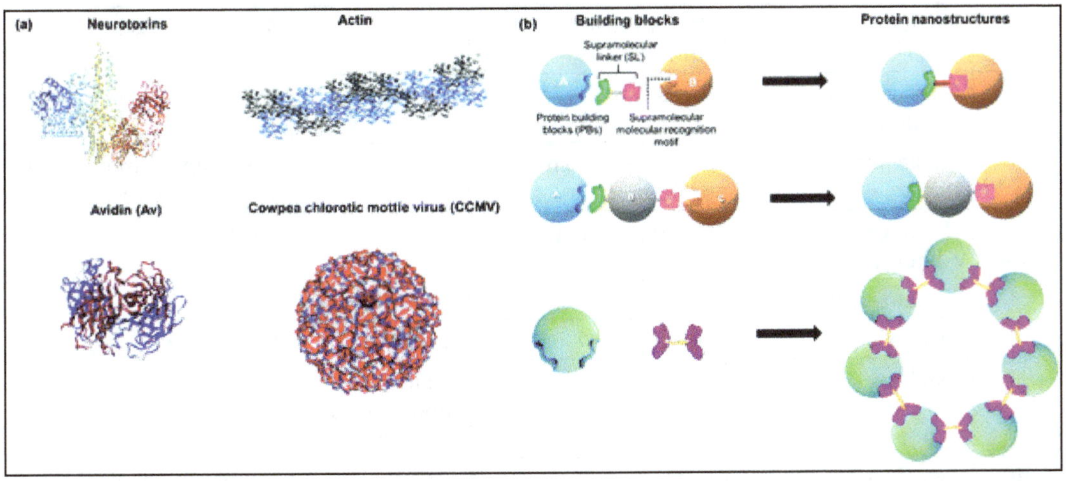

(a) Examples of protein nanostructures (PNs) in Nature: tripartite neurotoxin (PDB: 3BTA), tetrameric avidin (PDB: 1AVE), actin polymer and 3D virus capsids (PDB: 1CWP). (b) Schematic overview and design of distinct PNs formed by precise interactions of protein building blocks (PBs), supramolecular linkers (SLs) and supramolecular recognition motifs yielding e.g. dimeric, trimeric or ring-like nanostructures as examples. Protein images were created using the NGL viewer.

Selection of Monomeric Protein Building Blocks (PBs)

The geometry, functional groups, surface interactions and ligand recognition of the PBs play an important role in the formation of defined PNs. Native PBs that already possess recognition motifs can be applied directly and represent the simplest of these essential components to form PB. In the case where native PBs are not available, chemical or genetic transformations will be required. In order to prepare protein dimers or homoprotein polymers, a single site of the protein is usually modified. In contrast, the formation of higher ordered structures such as rings or nanotubes requires the introduction of two or more recognition motif at the protein surface. The most straightforward strategy is the chemical modification of native proteins at already available single amino acid residues such as cysteines, disulfides, or amines. Otherwise, genetically engineered PBs have to be expressed if site-directed modification of the native protein is not possible. The modification sites for the introduction of recognition motifs are discussed. Table summarizes some of the PBs for nanostructure formation and the introduced modifications.

Table: Summary PBs, type of modification, modification site and the recognition motifs.

	Protein precursors	Function	Modification	Recognition motif
Native proteins	Lysozyme	Enzyme	None	Arg128
	Cytochrome c	Enzyme		Lys4 and Lys100
	Protamine	Nuclear protein		Positively charged surface amino acids
	(Strept)avidin	Tetrameric biotin-binding protein		Binding pockets of protein
	Concanavalin A, lectin A, soybean agglutinin	Tetrameric carbohydrate-binding protein		Binding pockets of protein

	Ferritin	Protein cage for iron storage		Surface charges
	Cowpea chlorotic mottle virus (CCMV)	Virus capsid		Surface charges
	Stable protein one (SP1)	Stress responsive protein		Positively charged surface amino acids and central cavity in protein
Chemically modified proteins	Human serum albumin	Blood plasma protein	Cys34	Biotin
	Somatostatin	Hormone	Cys3–Cys14	Biotin
	Insulin	Hormone	LysB29	Bipyridine
	Catalase	Hemeprotein	Lysine modification (statistical)	ssDNA (multiple)
Recombinant engineered proteins	Cytochrome b_{562}	Hemeprotein	63Cys	Heme binding pocket
	Myoglobin	Hemeprotein	125Cys	Heme binding pocket
	C3 from Clostridium botulinum	Toxin enzyme	N-Terminal Cys	Biotin
	Cytochrome cb_{562}	Hemeprotein	59Cys mutation at i and i + 4 positions: His59/63; His 73/77	1,10-Phenantroline histidine
	Alkaline phosphatase	Enzyme	N- or C-termini mutation	Biotin
	Split luciferase fragments	Enzyme fragments	N- or C-termini mutation	Phe-Gly-Gly
	Cyan or yellow fluorescent proteins (CFP, YFP)	Fluorescent protein	N- or C-termini mutation	Phe-Gly-Gly
	LiDPS protein cage	Ferritin protein cage	Cysteine mutation	Biotin
	Glutathione transferase dimer	Enzyme	N-Terminal polyhistidine	Polyhistidine
			N-Terminal	Phe-Gly-Gly
			137Cys	Phe-Gly-Gly
			137Cys, 138Cys	Histidine
	Chaperonin GroEL	Tetradecameric molecular chaperones	313Cys, 314Cys	Spiropyran

Native PBs

Native PB assemblies form via specific interactions of protein surface patches or protein binding pockets. The interfacial area between interacting proteins plays an important role in protein–protein complexes, which are crucial for self-assembly processes in Nature. For instance, protein cages such as ferritin consist of multiple copies of a subunit assembled into highly defined 3D architectures and topologies. One common feature in these natural assemblies is the presence of charged amino acid residues that can interact via electrostatic forces on the subunit and assemble with the patchy surface subunits of oppositely charged residues to form defined supramolecular complexes.

Researcher devised an elegant strategy to self-assemble binary crystals from natural proteins comprising two different PBs with opposite charges that are presented as surface patches on the proteins. In this way, oppositely charged cowpea chlorotic mottle virus (CCMV) particles (isoelectric point (pI) of 3.8) and avidin (pI of 10.5) were assembled into binary crystals by using electrostatic interactions. Avidin offers the additional advantage that it can accommodate four biotin molecules facilitating the attachment of additional functionalities such as fluorescent dyes, enzymes or gold nanoparticles to impart a next degree of functions to the nanostructure. However, this method only allows the incorporation of PBs that meet the stringent requirements for electrostatic self-assembly and cannot be adopted to assemble two negatively charged PBs, such as CCMV with the green fluorescent protein with an isoelectric point (pI) of 5.

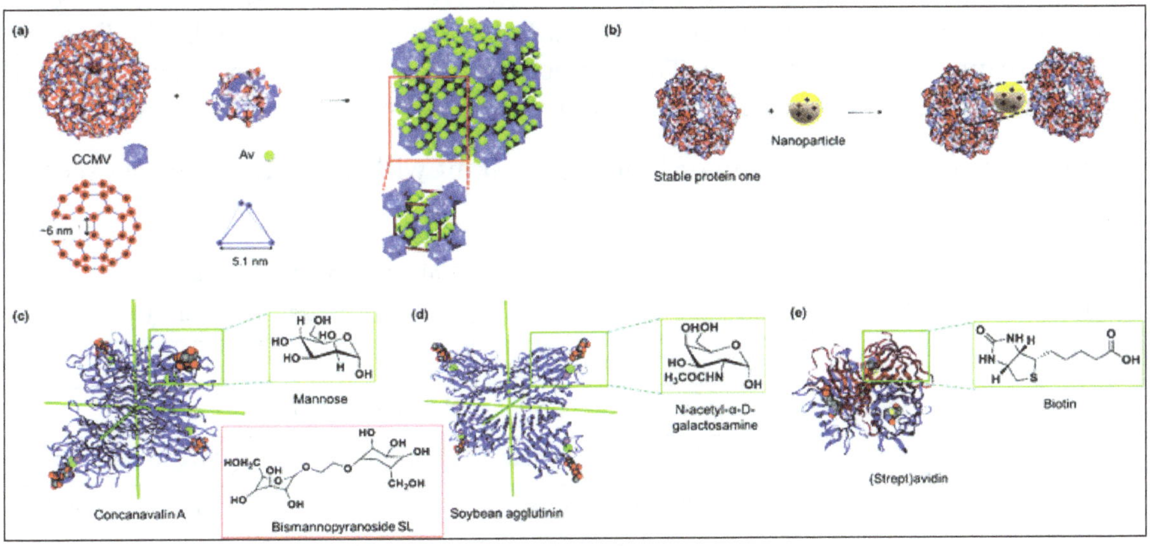

Figure shows (a) The electrostatic surface of native cowpea chlorotic mottle virus and avidin (Av) with the location and geometry of the surface patches, which allow interactions to form binary protein crystals. (b) The electrostatic surface of stable protein one (SP1, PDB: 1TR0) and the central cavity that interacts with polycationic nanoparticles with specific dimensions. Blue denotes positive surface charges, red denotes negative surface charges, and white refer to neutral areas. (c) Tetrahedral concanavalin A (PDB: 1CVN) versus (d) slightly out of plane soybean agglutinin (PDB: 1SBF) that assembles with sugar units. Red inset in (c) shows a bismannopyranoside SL. Both PBs in (c and d) were viewed along the D_2 symmetry axes for direct comparison. (e) (Strept)avidin tetramer (PDB: 1MK5) with the binding ligand biotin. Insets in (c–e) shows the chemical structure of the biotin ligand. Protein images were created using the NGL viewer.

Stable protein one (SP1) forms a dodecameric ring-like protein with a diameter of approximately 11 nm, a central cavity of 2–3 nm, and a width of 4–5 nm. The acidic amino acids are mainly present at the top and bottom surface of the dodecamer SP1, and its unique topology and structural features have been exploited to control the directionality of electrostatic-induced self-assembly.

Specifically, the symmetric concave of SP1 can accommodate globular nanoparticles in the centre of the double-layered nanoring to mitigate unidirectional growth. This method offers the advantage of assembling functional nanoparticles, which could be of interest for numerous applications such as light harvesting antennas.

Overall, the electrostatic assembly based on protein surface topology offers simplicity as native PBs are used without the need for recombinant engineering or chemical modifications, and additional functional features have been introduced into the protein complexes as well. However, the charge distribution at protein surfaces cannot always be predicted easily, which limits access to more complicated 3D protein structures.

Endogenous small molecules are known to bind to various proteins by protein–ligand interactions which has also been exploited to prepare structurally defined PNs. The most prevalent examples for natural ligand-guided PNs are heme proteins, (strept)avidins and lectins that bind heme, biotin, and carbohydrates (mannose, glucose, and galactose), respectively. Most of these interactions are highly specific and exhibit reasonable binding affinities from the micromolar (μM) to the femtomolar (fM) regime as listed in Table. These unique binding features have been exploited in a range of applications and most commonly in affinity purification.

Table: Summary of SLs and stability or binding constant with the corresponding recognition motif.

Interaction motif	SL	Stability/binding constant
Charge-directed assembly	Gly-Val-Gly-Lys-Pro	Complete disassembly [NaCl] > 100 mM
	Newkome-type dendrimer	Photocleavable
	Phthalocyanine	Not determined
	Cationic diblock copolymers	Disassembly at T < 40 °C
	Cationic CdTe quantum dots	Some disassembly at [NaCl] > 250 mM
	G5-PAMAM dendrimers	Disassembly at [NaCl] > 400 mM; pH < 2; pH > 12
	Cationic cross-linked micelles	Not determined
Heme–hemeprotein	Heme	$K_a \sim 10^{12-14}\ M^{-1}$
Lectin–carbohydrate	Galactose	$K_a \sim 10^3\ M^{-1}$
	Tetra-d-galactose	$K_a \sim 10^9\ M^{-1}$
	α-d-Mannose	$K_a \sim 10^3–10^6\ M^{-1}$
	N-Acetyl-α-d-galactosamine	$K_a \sim 10^4\ M^{-1}$
	α-d-Galactopyranoside	$K_a \sim 10^4\ M^{-1}$
(Strept)avidin–biotin	Biotin–hydrazone-linker	Cleavage at pH < 7
	Iminobiotin	pH > 7: $K_a \sim 10^{11}\ M^{-1}$; pH < 7 : $10^3\ M^{-1}$
Boronic acid	Salicylhydroxamic acid	pH > 7: $K_a \sim 10^6\ M^{-1}$; pH < 7 : $10^3\ M^{-1}$
Phe-Gly-Gly	CB[8]	$K_{ter} \sim 10^{11}\ M^{-2}$

Naphthalene–methyl viologen	CB[8]	$K_1 \sim 10^5\ M^{-1}$; $K_2 \sim 10^6\ M^{-1}$
Arg128 (lysozyme)	p-Sulfonato-calix[4]-arene	$K_a \sim 10^6\ M^{-1}$
i + i4 histidine motifs	Zn^{II}	Disassembly pH < 5; in presence of EDTA
Spiropyran	Mg^{II}	Disassembly with mechanical force; in presence of EDTA
5′-GCTACACG-3′ (8-mer)	3′-CGATGTGC-5′	$K_a \sim 0.1 \times 10^6\ M^{-1}$
5′-AGCTACACGATA-3′ (12-mer)	3′-TCGATGTGCTAT-5′	$K_a \sim 9 \times 10^9\ M^{-1}$
5′-AAAAAAAAAAAA-3′ (12-mer)	3-TTTTTTTTTTTT-5′	$K_a \sim 0.1 \times 10^6\ M^{-1}$

Lectins are carbohydrate binding proteins, which are involved in a number of cellular processes including glycoprotein synthesis, modulating inflammatory responses, and cell recognition. Plant lectins such as concanavalin A and lectin A are homotetrameric proteins with four binding sites for mannose, glucose, or galactose. Concanavalin A, a tetrameric protein with D2 symmetry binds to terminal α-d-mannosyl and α-d-glucosyl groups and was first used by Researcher in combination with bismannopyranoside (chemical structure shown in red inset, Figure) as SL to form predesigned, diamond-like protein lattices. Interestingly, the choice of the lectin unit has a strong influence on the resultant nanoarchitecture. For instance, the tetrahedral concanavalin A formed an interpenetrating protein crystalline framework when assembled with a bifunctional linker comprising of a sugar and rhodamine B. However, in the case of the homotetrameric soybean agglutinin possessing a D_2 symmetry with slightly out of plane binding pockets, a microtubule-like structure was obtained. This feature was attributed to the difference in protein geometries of concanavalin A and soybean agglutinin. Besides the plant lectins described above, human Galectin-1, a lectin from animal source, was found to form self-assembled microribbons using this strategy.

Avidin and streptavidin are homotetrameric proteins containing eight β-strands in each subunit, resulting in an antiparallel β-barrel shaped structure. The proteins are proposed to inhibit bacteria growth and bind to vitamin B_7, biotin, in a non-covalent manner with one of the strongest binding affinities known ($K_a \sim 10^{15}\ M^{-1}$). The biotin–(strept)avidin interaction has been exploited for various biological applications including purification and cancer pretargeting as well as a supramolecular "glue" for building up various nanostructures and forming protein networks. Supramolecular assembly to form linear avidin polymers has been achieved using bis-biotinyl linkers and the spacer length affected the stability of the resultant polymer. However, such avidin polymerizations are often uncontrolled, and the resultant materials have no specific functions. More recently, synthetic efforts have allowed the creation of spatially defined supramolecular protein nanostructures with avidin through tailored linker design, lectins and (strept)avidin PBs do not possess any bioactivity and additional functionality has to be incorporated either by surface modifications or by co-assembly with other functional entities such as protein enzymes or synthetic molecules such as dendrons.

Clearly, protein interfaces and pre-existing protein recognition units are valuable to induce non-covalent interactions between PB to obtain PNs with well-defined structures. However, the repertoire of native PBs is limited by the structures Nature offers. In order to further expand the plethora of PBs, alternative approaches are required. In this regard, the incorporation of a recognition unit, usually an endogenous ligand such as biotin, deoxyribonucleic acid, or peptide; or exogenous molecules that undergo host–guest complex formation, is the method of choice to control the assembly process. In this case, site-directed conjugation of a recognition motif to the protein of interest is required to impart the chemical information for assembly.

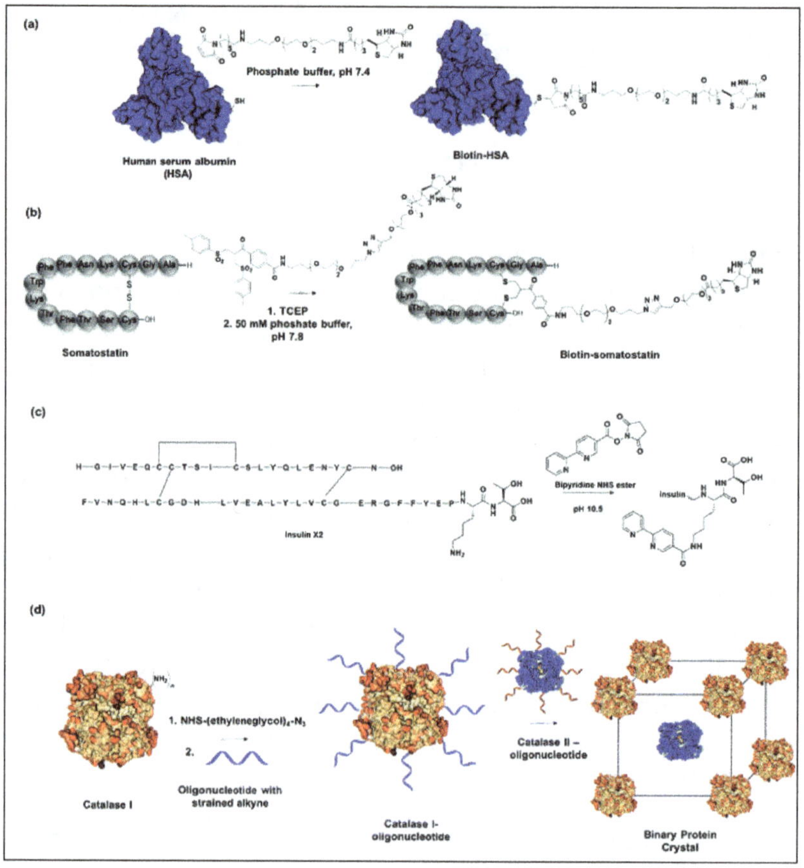

Figure shows (a) Human serum albumin (HSA) modified with recognition motifs at single unpaired cysteine (Cys-34). (b) Biotin recognition motif incorporated through disulfide modification on the hormone peptide, somatostatin. (c) Insulin variant (Insulin X2) modified with bipyridine recognition motif at LysB29Nε. (d) Two different catalase PBs modified with complementary oligonucleotide sequences assembles into a binary crystalline structure.

Chemically Modified PBs

Many proteins provide amino acid side chains that can be employed for chemical functionalization. However, in order to construct defined PNs, chemo- and regio-selectivity of the bioconjugation reactions represent a prerequisite as well as mild reaction conditions. Currently, reengineering protein surface recognition and even imparting novel functions by applying site-selective chemical methodologies under "protein-friendly" conditions represent established methods to form precise bioconjugates. The relatively rare amino acid cysteine is most often exploited due to the selective reactivity of its thiol group towards maleimides under controlled pH. The abundant blood plasma proteins human or bovine serum albumins (HSA, BSA) provide Cys34 modified through maleimide-thiol Michael reaction. For example, a single biotin ligand has been attached to HSA at neutral pH to avoid side reactions with the primary amino groups of lysine residues. Nevertheless, very few native proteins contain a single or even two unpaired and accessible cysteine residues at their surfaces. Most cysteines in proteins are present in the reduced form as disulfides. A broad range of proteins offers accessible disulfide bonds that control their stability and biological activity and that can be modified by disulfide rebridging reagents, such as bissulfone or allyl sulfone reagents. Using this functionalization approach, the cyclic peptide hormones somatostatin and insulin as well as protein enzymes like lysozyme consisting of one, three and four disulfide bonds, respectively, have been equipped with a single recognition motif such as biotin or a boronic acid, which interacts with carbohydrates.

Researcher attached a single bipyridine into insulin through specific modification of the lysine residue in chain B (LysB29). Insulin has one lysine, LysB29, with a pK_a of 11.2 for LysB29N^{ε}, whereas the N^{α} on the A and B chain provide pK_as of 8.4 and 7.1, respectively. By exploiting the difference in pK_a, selective acylation at LysB29 was achieved in basic conditions at pH > 10. Conversely, at physiological pH, the Lys side chain is the least reactive amine but bioconjugation strategies have been reported that also allow N-terminal modification. Tyrosine and tryptophan conjugations have been applied to a broader spectrum of proteins but these residues are normally less accessible as they are often buried in the hydrophobic interior of proteins. Thus, a single site can be easily introduced to place an external recognition motif onto the protein of interest. For instance, bifunctional linkers with click functional groups were used to prepare protein–DNA conjugates through binding onto tyrosine residues. In principle, the site-selective chemical modification strategies mentioned are also applied to introduce SLs to proteins of interest. Dual functionalization strategies have also been developed in a site-directed fashion, for example by capitalizing on reactivity variations of different cysteine residues or in combination with N-terminal protein modification. By the careful selection of the PB and the chemical modification method, it will be possible to attach two different recognition motif in a defined spatial orientation. Consequently, it would allow the generation of nanostructures consisting of two different PBs in a controlled, stepwise manner.

Previously, site-selective incorporation of the recognition motif to induce directionality was considered a prerequisite for nanostructure formation. However, Researcher demonstrated that statistically modified proteins can also be employed as PBs to engineer multi-enzyme crystals due to their distinct pattern of surface-accessible amine groups and the consistency of surface morphology on a rigid protein core. In this manner, they modified two tetrameric heme catalase enzymes with oligonucleotides over two step chemical reactions, first by adding a tetraethylene glycol linker functionalized with N-hydroxysuccinimide (NHS) ester and azide on both ends to connect to the amines on the catalase, followed by a cycloaddition with an oligonucleotide functionalized with dibenzocyclooctyne at the 5′-end. Protein–DNA PBs with functional densities of 30–50 pmol cm^{-2} were obtained in this fashion. By preparing two DNA-modified PBs with complementary oligonucleotide sequences, they demonstrated that multienzyme protein crystals, which retain catalytic function of the individual PBs, were prepared in a straightforward fashion.

Even though a variety of protein bioconjugation techniques exist nowadays, these methods have surprisingly not been used as first choice to reengineer PBs for ultimate assembly of PNs. One plausible explanation could be the widespread application of thiol-maleimide reactions and the ease of engineering cysteine mutants for the desired PB. Nonetheless, chemical toolboxes offer versatile incorporation of non-endogenous ligands into proteins, and with the expanding repertoire of these tools, this strategy will gain even stronger footholds in the future.

Genetically Modified PBs

Genetic engineering is widely applied to express PBs with a single mutation e.g. cysteine. For instance, linear polymers of hemeproteins cannot be formed as this protein only provides one binding pocket. To circumvent this issue, a single point cysteine genetic mutation was introduced onto the hemeprotein and subsequently, a heme group equipped with a site-selective cysteine modification was incorporated into cytochrome b_{562}. This incorporation allowed the formation of the first supramolecular linear hemeprotein polymer. Other hemeprotein PBs, such as myoglobin, have been prepared using a similar approach. Protein homodimers have been formed through dynamic covalent disulfide bonds via recombinant technologies to generate AA type PB, i.e. PB consisting of two similar units, for linear polymerization. In theory, this method could also be applicable to prepare AB-type PB comprising of hetero-heme protein dimers that could be further assembled to generate supramolecular protein polymers with precise alternating arrangement.

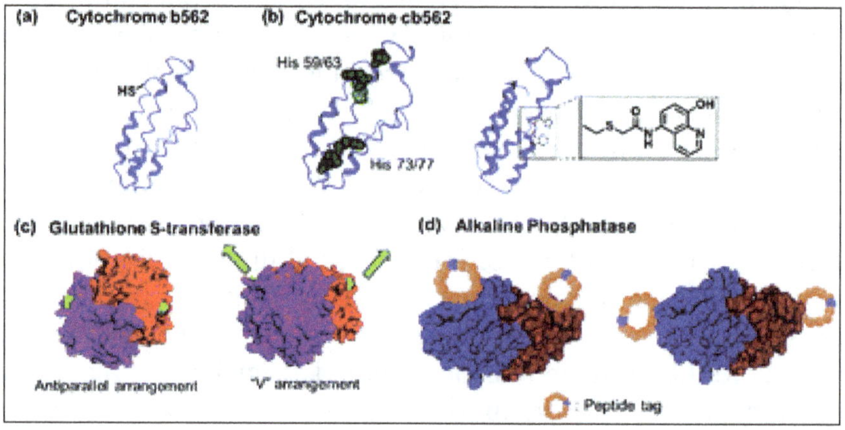

Genetically modified PBs. (a) Cytochrome b_{562} 63Cys mutant with unpaired thiol. (b) Cytochrome cb_{562} variants with i/i + 4 His (left) or quinolate (right) mutations. (c) Glutathione-S-transferase with mutations in antiparallel or "V" orientation. (d) Alkaline phosphatase with mutations incorporating reactive peptide tags along the short or long axis of the protein.

Peptide sequences that interact with external molecular triggers such as host–guest interactions or through metal coordination were introduced via genetic engineering. The short peptide tag, Phe-Gly-Gly, is known to form an inclusion complex with cucurbit[8]uril (CB[8]) in a 2 : 1 ratio. This tag has been introduced into fluorescent proteins such as cyan fluorescent protein (CFP) or yellow fluorescent protein (YFP) via the N- or C-termini. Similarly, insertion of sets of histidine residues via genetic engineering on the proteins of interest, such as within cytochrome C's α-helixes, has proven to be particularly suitable for assemblies formed through metal coordination. Tezcan and coworkers genetically modified a cytochrome cb_{562} variant containing histidine residues capable of coordinating onto a metal ion such as Zn^{2+}. Here, two sets of His-dimers were inserted in positions 59/63 and 73/77 of the helix 3 of cytochrome cb_{562}.

Hybrid coordination motifs have also been achieved using a combination of a natural histidine and non-natural quinolate amino acid as chelating ligands. Mutations of amino acids in the protein sequence were also applied to remove unwanted interactions and guide the self-assembly process. For example, amino acid groups located at C_3 symmetry positions on a ferritin cage were substituted by histidine (Thr122His). To prevent other types of coordination rather than by the histidine residues, the mutations Cys90Gln, Cys102Ala, Cys130Ala, and Lys86Gln were also performed to subsequently generate a highly ordered crystalline metal–organic protein framework. Nevertheless, such strategy requires a profound knowledge of the 3D structures of the proteins and of the active site in order to devise a reasonable design. Additionally, the process from design to executing the mutations could be tedious.

Besides the introduction of a single recognition site, genetic engineering also allowed for the placement of multiple copies of the recognition motifs to direct the spatial orientation of the protein nanoassembly. Tobacco mosaic virus coat protein mutants with two cysteine or four histidine residues at the lateral surfaces were prepared and by controlling the thermodynamics and kinetics of the respective PBs, the crystal structures of the PN were tailored. The cysteine mutant formed triclinic crystals at 4 °C over a month, while the histidine mutant rapidly assembled in the presence of Zn^{2+} to form hexagonal close-packed crystals. Specific cysteine mutations were performed on chaperonin GroEL, a protein that mediates protein folding in cells, to generate GroELCys with 14 Cys residues spatially distributed on the top and bottom of the cylindrical shape protein. Subsequent functionalization with maleimide recognition motif allowed the directionality of protein polymerization to be controlled. Glutathione-S-transferase (GST) from Schistosoma japonicum forms homodimers, and genetic fusion of either the hexahistidine or Phe-Gly-Gly tag on the N-termini yielded an antiparallel arrangement of the recognition motif or the fusion tag was arranged in a "V" shape which affected the final morphology of the protein nanostructures.

Researcher dictates the directionality of the self-assembly through control of the placement of the modification sites on a functional alkaline phosphatase by recombinant engineering. Microbial transglutaminase-catalysed acyl-transfer reaction is known to occur between the side chains of glutamine and lysine amino groups. Thus, the predefined positioning of a transglutaminase-reactive peptide tag at the N- or/and C-terminal allowed biotinylation in a specific orientation on the alkaline phosphatase. However, the chain growth was terminated presumably due to steric crowding since both the N- and C-termini were facing in the same directions. Further optimization of the placement of a microbial transglutaminase tag along the longer axis of the alkaline phosphatase circumvented this issue and allowed intermolecular polymerization. To confer additional self-assembling handles on streptavidin, a twigged streptavidin polymer was engineered by Tanaka and co-workers as a scaffold for hetero-protein assembly. A sortase A recognition site and a horseradish peroxidase recognition site were genetically incorporated into the N- and C-termini of streptavidin, respectively, that allowed the immobilization of two different proteins via biotin–streptavidin interaction and sortase A-mediated ligation.

Genetic engineering provides many opportunities concerning the introduction of functionalities at distinct sites of PBs that chemical modification alone cannot achieve and vice versa. However, recombinant technologies also have some drawbacks in terms of laborious processes, loss of protein activity due to structure changes based on the introduced mutation and lack of possibility to include PBs that have been chemically post-modified to expand and customize the functional profile of the protein nanostructures. Thus, in the long term, the combination of both chemical and biotechnological toolboxes will provide entirely new supramolecular protein nanostructures with customized geometries and features.

Design of the Supramolecular Linkers (SLs)

The supramolecular linker (SL) often functions as a "glue", which interacts specifically with the recognition motif on the PBs to direct the formation of the desired PN. Thus, the assembly and the resultant architectures and stabilities could be strongly influenced by the design of the SLs as well as by the choice of the suitable recognition motif. These two factors are often co-related and are featured together here. In some cases, the SLs attached directly to the PBs to induce PN formation, but in other instances, conjugation to PBs was required and some examples will be given.

SLs based on Electrostatic Interactions

Binary crystals were obtained from oppositely charged PBs but cannot be applied to proteins with similar charges. To expand further on this strategy, engineered PBs, synthetic macromolecules or nanoparticles have been used as SL to form heteroassemblies. In order to induce interfacial electrostatic interactions, these non-natural SLs should possess compatible sizes and shapes (generally globular) to the PBs. This strategy offers flexibility to control assembly and disassembly due to the sensitivity of charged interactions towards pH value and ionic strength. In this approach, the SLs could be applied without the need for covalent conjugation to the PBs. Additionally, due to the nature of the charge interactions between SL–PB, assembly and disassembly of PNs could often be controlled through variation of the ionic strength or pH of the buffer used.

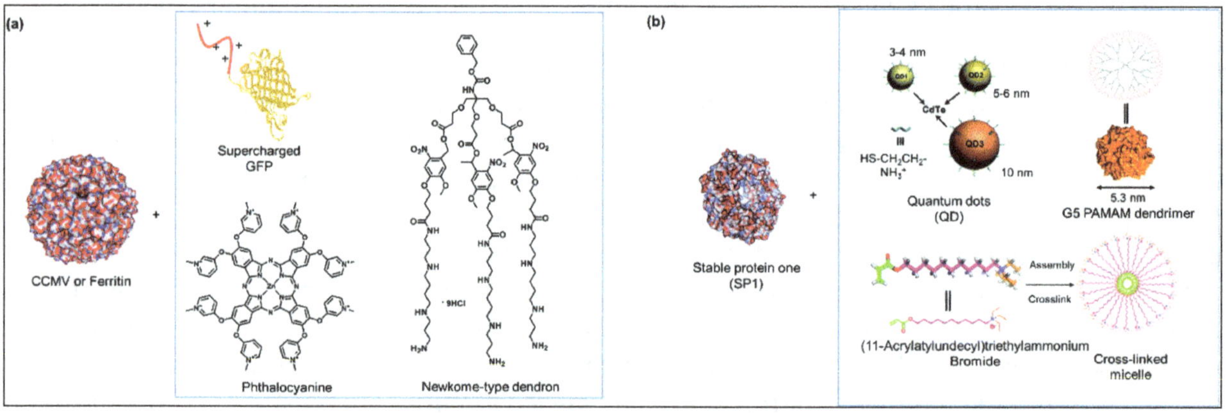

(a) SLs that are applied with negatively charged CCMV or ferritin.
(b) SLs used for forming PNs with stable protein one (SP1).

For instance, new polycationic surface areas were introduced by expressing positively charged peptide sequences into negatively charged proteins or by using positively charged synthetic macromolecules. For instance, the sequence Gly-Val-Gly-Lys-Pro was fused to the sequence coding for a green fluorescent protein (GFP) to form a supercharged cationic polypeptide that formed a new polycationic surface patch at distinct location at the GFP surface. Alternatively, the assembly of PNs with highly branched polycationic Newkome-type dendrons was reported. Dendrons are branches of dendrimers, which are monodisperse globular synthetic macromolecules that resemble proteins in certain structural characteristics such as their sizes, globular architecture and the availability of many polar groups at their surface and unipolar groups within the interior. Positively charged dendrons with photocleavable o-nitrobenzyl have been introduced as SLs that bind to the negatively charged surface patches of CCMV. The cleavage of the SL is an irreversible chemical process and reassembly is not possible. Large ordered PN architectures consisting of (apo)ferritin cages or CCMV were formed that revealed unique characteristics such as light-induced disassembly of PNs. Interestingly, the responsive behavior of the resultant nanostructure was fine-tuned by structural alterations of the cationic inducer. For instance, replacing the dendron with a cationic diblock copolymer introduced a thermo-switch so that the protein nanostructure became responsive to temperature changes. Researcher prepared a tetracationic complex, formed from an octacationic zinc phthalocyanine and a tetraanionic pyrene, as a SL. The SL also produced additional function as a photosensitizers for the formation of singlet oxygen, a crucial process for photodynamic therapy.

PNs consisting of "heteroassemblies" of PBs and macromolecules such as dendrimers or nanoparticles like quantum dots possessing similar sizes and globular shapes as the PBs have been achieved. Researcher have successfully employed cationic, hard nanoparticles such as quantum dots and soft particles such as dendrimers and micelles to derive a series of functional self-assembled heteroprotein complexes, showing the immense potential of this strategy for both biomedical and biotechnological applications. In their earliest report, positively charged, globular quantum dots of various sizes were synthesized, and the impact of the quantum dot sizes on PNs was investigated. Further to this development, an electro-positively charged macromolecule, the fifth generation polyamidoamine (G5 PAMAM) dendrimer was used in place of quantum dots to control the self-assembly process of cricoid SP1 and protein nanorods were formed. Dendrimer component offers the advantage that assembly efficiency and morphology of the PNs could be adjusted precisely by the dendrimer generation (size) and scaffolds. Moreover, dendrimers also provided additional level of functionality and in the above example, G5 PAMAM was functionalized with manganese porphyrin to confer superoxide dismutase activity. In this case, the stability of the SL–PB interaction is affected by ionic strength and pH of the buffer. For instance at pH < 2 and pH > 12 or at higher salt concentrations (400 mM NaCl), the interactions are much weaker and dissociation could be observed. Notably, temperature does not have much effect on the stability of the interactions In order to improve on the ease of preparation and customization of the SL, the same group explored the feasibility of using core cross-linked micelles with cationic surfaces as inducers for self-assembly. They demonstrated that the cross-linked micelles were functionalized in a convenient manner with chromophores to introduce additional functionality to the system.

Overall, SLs that bind via electrostatic interactions offer a convenient and versatile strategy as additional functional features can be introduced to the PN by molecular design and association/dissociation is dependent on and thus, tunable by ionic strengths. However, challenges remain in terms of control of the charge distribution on the complicated 3D protein surface, which can be difficult to predict and to manipulate and if not optimal, it compromises the spatial organization of the PN.

SLs based on Protein–ligand Interactions

In earlier examples, a linker consisting of a protein-binding ligand induced protein polymerization yielding well-defined PNs. However, the morphology of the PNs was determined by the PB, and imparting structural variations was challenging. This has stimulated several attempts to design supramolecular linkers providing directionality in their interactions to control the formation of more complex protein architectures.

The high affinity of heme for apoglobin proteins (equilibrium dissociation constant: 10^{-12}–10^{-15} M) has been exploited for formation of linear protein polymers. To further exploit this protein–ligand interaction, Researcher have adopted a C3-phenyl core to design a heme triad linker to introduce a branching point, resulting in the formation of two dimensional networks with cytochrome b_{562}. Transient thermal stimuli consisting of rigid and hydrophobic tethering group such as azobenzene or stilbene was also incorporated into the artificial heme SL so that a switch between different PNs was achieved. On the other hand, SLs comprising of phenyl and octyl moieties as tethering groups dissociate upon heating. It was proposed that the azobenzene/stillebene moieties offer stabilization of the metastable micellar structure even after cooling, through π–π and/or C–H–π interactions with the heme PB.

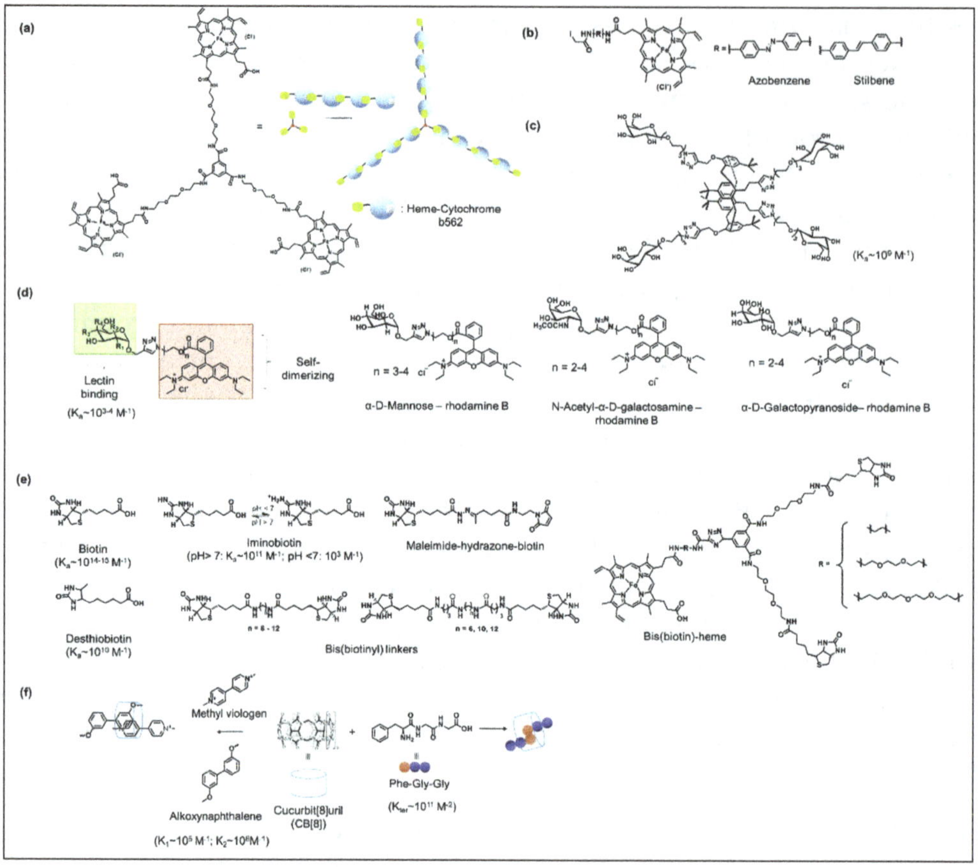

Design of selected SLs described and reference values for binding constants. (a) Tris(heme) SL based on a C3 phenyl core. (b) Thermoresponsive heme SL with rigid stilbene or azobenzene tethers. (c) A tetra-galactosylated glycocluster with increased binding affinity to lectin A. (d) Bifunctional SLs with rhodamine B and different sugar moieties. (e) SL based on biotin and biotin derivatives. (f) CB[8] SLs with Phe-Gly-Gly or synthetic chemical ligands for ternary complex formation.

Multivalent linkers based on a rigid core were also functionalized to control the topology as well as to enhance the binding affinity between the PB and the SL. For instance, the monovalent galactose ligand typically exhibits binding affinity to lectin A in the submillimolar range. Based on multivalent "glycoside cluster effect", a tetra-galactosylated glycocluster was synthesized, in which the galactose ligand reveals significantly enhanced binding (nanomolar) to lectin A. Researcher have further devised a lectin-carbohydrate driven PN formation strategy via a combination of dual supramolecular interactions to achieve variations in PN morphology and to incorporate stimuli-responsiveness. Several SLs have been prepared for the formation of a variety of

PNs. These linkers typically consist of a rhodamine B group, which forms π–π interactions with another rhodamine B molecule, an oligo(ethylene oxide) spacer to control the distance between interacting PBs, and a sugar unit such as α-d-mannose of N-acetyl-α-d-galactosamine or α-d-galactopyranoside for interactions with the corresponding lectins such as concanavalin A, lectin A, or soybean agglutinin.

The inclusion of an oligo(ethylene oxide) spacer also allows varying the SL length, which was found to strongly influence the assembly process. For example, a SL consisting of a monosaccharide (e.g. α-d-mannose) and a rhodamine B group was used to crosslink concanavalin A via both lectin-carbohydrate and π–π interactions resulting in the formation a protein crystalline framework. The incorporation of rhodamine B in the SL design allows reversible association and dissociation of the PNs through competitive host–guest interactions using β-cyclodextrin capable to interact with rhodamine B. In this case, dissociation is achieved and re-association can be induced by using 1-adamantane hydrochloride, a competitive binding ligand.

Of all the systems exploiting protein–ligand interactions, the (strept)avidin–biotin system offers the greatest flexibility in the library of ligands that are available to confer high binding strengths together with stimuli-responsiveness. Binding affinities of biotin to strept(avidin) are one of the strongest protein–ligand interactions known and could be varied through chemical variations. For instance, desthiobiotin ($K_a \sim 10^{10}\ M^{-1}$) and pH sensitive iminobiotin (pH > 7, $K_a \sim 10^{11}\ M^{-1}$; pH < 7, $K_a \sim 10^{3}\ M^{-1}$) reveal altered binding characteristics for (strept)avidin when compared to biotin. The lower binding affinities of these commercially available biotin analogues offer the possibility to control association and dissociation with PB by variation of reaction conditions. In particular, iminobiotin provides dynamic and reversible binding to (strept)avidin as the imine is protonated at acidic pH, which strongly reduces binding to (strept)avidin. Biotin analogues consisting of the redox sensitive S–S or the pH sensitive hydrazone bond are commercially available and offer straightforward customizations of SL for the formation of stimuli-responsive PNs e.g. to react to the more acidic microenvironments of diseased cells. Besides responsiveness, SL design with dynamic covalent S–S and hydrazone linkage offers at the same time, the stability of covalent bonds. Moreover, synthetic heterobifunctional biotin conjugation reagents, such as pH cleavable maleimide-biotin SLs, have also been. Here, the maleimide-biotin SLs are attached to the protein on interest by conjugation through its unpaired cysteine.

Bis(biotinyl) SLs have been prepared with different chain lengths with up to 25 bonds between the carbonyl groups on the biotin. It was found that the number of bonds affected the stability of the resultant linear polymers. When the number of bonds were higher than 12, the SL could crosslink two avidin PBs and yielded linear PNs. PNs with 12–13 bonds were less stable and disassembled while PNs formed with longer chain reagents were more stable in the presence of competing biotin ligands. However, when a critical length of 23 bonds was reached, polymerization did not occur and the SL bound to two ligand sites on a single avidin PB.

Bifunctional SL consisting bis(biotin) and a heme ligand, for instance, have been designed to control spatial arrangement of the individual PBs The bis(biotin) moiety bound specifically to the two adjacent biotin binding sites of streptavidin and thus preorganized the assembly in a linear fashion. Consequently, it was shown that the bis(biotin)-heme SL could be used to crosslink the proteins streptavidin (B) and apomyoglobin (AA) to form a linear AAB type supramolecular PN with precise alternating protein arrangement.

SLs based on Synthetic Host–Guest Interactions

Besides natural binding ligands, synthetic interactions have also been introduced into PBs to control assembly, as exemplified by host–guest interactions such as that of cucurbit[8]uril (CB[8]), a macrocyclic molecule made of glycoluril monomers linked by methylene, with peptides or aromatic ligands. And β-cyclodextrin with lithocholic acid. One could envisage that such exogenous ligands be potentially applied to self-assemble in complex cellular environments due to their bio-orthogonality and high complex stability, while reversibility could be achieved via interplay of binding strengths of SL, PBs and competitive binding ligands.

Protein homo- and hetero-dimerization has been reported using synthetic host–guest interaction. The "ligands" in these cases were either obtained by genetic engineering, as in the case of the peptide such as Phe-Gly-Gly (CB[8] : Phe-Gly-Gly = 1 : 2, K_{ter} = 1.5 × 10^{11} M^{-2}), usually at the N-terminus or in the case of synthetic entities such as lithocholic acid, introduced site-selectively on a single cysteine mutant of the PB, forming a host–guest complexes with β-cyclodextrin. The addition of ligands such as methyl viologen that also interacts strongly with CB[8] (K_a ~ 10^6 M^{-1}), results in competitive binding and PN dissociation. Ternary complexation can also be programmed by the introduction of recognition motifs that interact with CB[8] in a 1 : 1 : 1 manner, for example, with PBs consisting of an electron deficient supramolecular guest molecules such as methyl viologen and a complementary electron rich guest such as alkoxynaphthalene where the CB[8]:ligand binding (methyl violgen and alkoxynaphthalene, respectively) are in the range of K_1 ~ 10^5 M^{-1} and K_2 ~ 10^6 M^{-1}. A combination of chemical and recombinant techniques adopting the cyclodextrin or CB[8] host–guest interaction offers a powerful platform to tailor a variety of multiprotein assemblies with different structural features.

Calix[n]arenes are symmetrical macrocycles possessing four phenolic arms and are typically cone-shaped with defined upper- and lower-rim regions. Their secondary and ternary interactions have been extensively studied and are relevant for various supramolecular assembly and crystal engineering processes. In a biological context, the anionic, water-soluble derivatives such as p-sulfonato-calix[4]-arene and p-phosphonatocalix-[6]arene are emerging as candidates to drive protein assembly by electrostatic interactions due to the possibility for molecular recognition through encapsulation of the cationic side chains of lysine and arginine. For instance, Researcher reported the complexation of the hen's egg white protein, lysozyme, with p-sulfonato-calix[4]-arene to form a linear assembly of protein tetramers. The macrocyclic p-sulfonato-calix[4]-arene serves as SL between the protein units via twofold interactions: (1) encapsulation of the C-terminal Arg128 owing to its steric accessibility and (2) forming a protein-bound complex of p-sulfonato-calix[4]-arene, Mg^{2+}, and a polyethylene glycol, which is present in the reservoir solution for crystallization.

The same group also applied the larger macrocyclic p-phosphonatocalix-[6]arene to form a symmetric C2 protein dimer of cytochrome c. Subsequently, a sulfonate-calix[8]arene SL, which offered greater conformation flexibility, mediated the formation of cytochrome c tetramers in solution. Auto-regulation of assembly and disassembly could be achieved through the control of the SL concentration, without the need to use competitive ligand for inhibition. Researcher used a sulfato-β-cyclodextrin, which has a higher negative charge density, to form spherical nanoparticles (100–160 nm) with the cationic DNA-binding protein protamine. The particle formation is determined to arise from the charge surface rather than any single amino acid residues alone. The sulfato-β-cyclodextrin/protamine nanoparticles could be degraded by trypsin enzyme for controlled release of cargoes.

SLs based on Metal–Ligand Interactions

Metal ions such as Ca^{2+}, Fe^{n+}, Co^{n+}, Ni^{n+}, Cu^{n+}, and Zn^{n+} (n = 1, 2 or 3) play crucial roles in biological systems. They can bind to proteins by metal–ligand coordination with certain amino acids usually bearing a lone pair such as histidine (His), cysteine (Cys), and aspartic acid (Asp) to form coordination bonds. It was proposed that the higher bond strength of metal–ligand coordination could overcome weaker interactions, such as hydrogen bonds, electrostatic, or non-covalent bonds, resulting in more selective and high binding affinities, which could impart greater control of the assembly process and higher stability of the formed PNs. The complex geometry, bond strengths, and functions are affected by the type of ancillary ligands, the oxidation state of the metal ions, and their respective coordination numbers. Thus, the directionality of the coordination bonds, the labilities and the properties of the resultant PNs can be fine-tuned. In the past decade, several approaches have emerged, where metal ions were used as supramolecular "glue" in combination with a suitable coordinating ligand to derive unique PNs.

Four cytochrome cb_{562} variant consisting of two sets of His dimers inserted at the i and i + 4 positions assembled into two interlaced V-shaped dimers in an antiparallel fashion to one another in the presence of Zn^{2+} ions. Dissociation of the nanostructure was induced through a change in pH (≤5) or the addition of strongly chelating ligands such as ethylenediaminetetraacetic acid (EDTA). Variation of the Zn^{2+} concentration in solution yielded monomers and dimers, whereas the formation of tetramers or even higher order polymers occurred at high concentrations. By replacing the Zn^{2+} ions with Cu^{2+} (square geometry) or Ni^{2+} (octahedral geometry), PNs comprising of antiparallel C_2-symmetric dimers of the type $2Cu^{2+}$:2 cytochrome cb_{562} or C_3-symmetric trimers of the type $2Ni^{2+}$:3 cytochrome cb_{562} were formed, respectively, presumably due to the coordination geometry imposed by the metal ions. By variation of the His59 and Cys96 mutation on the cytochrome cb_{562}, the nanotube formation with Zn^{2+} was even induced and the presence of external amino acids that allowed coordination to the metal ions resulted in the formation of 2D nanoarrays.

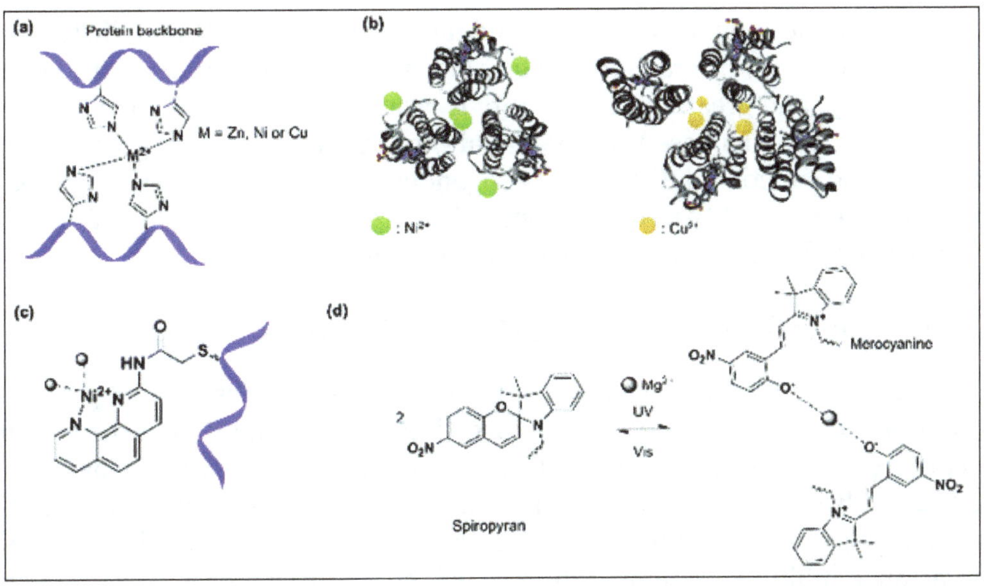

(a) Bis(histidine) complexation to metal ions. (b) Antiparallel C_2-symmetric dimers of the type $2Cu^{2+}$:2 cytochrome cb_{562} or C_3-symmetric trimers of the type $2Ni^{2+}$:3 cytochrome cb_{562} formed due to influence of the respective metal ions on the coordination geometry. (c) Complexation of PB modified with 1,10-phenanthroline to Ni^{2+} ion. (d) Light driven coordination of Mg^{2+} with spiropyran/merocyanine ligands.

Triangular assemblies were directed by the complexation of Ni^{2+} ions to proteins embedded with recognition motifs such as 1,10-phenanthroline attached covalently onto the surface of cytochrome cb_{562} through the Cys59 residue located at the third helix of cytochrome cb_{562}. Ni^{2+} is coordinated by a 1,10-phenantroline group at one protein monomer and a nitrogen of a His77 residue at another monomer. The triangular assemblies were packed into tubular units, and the superposition of different trimer orientations generated an apparent hexagonal hollow geometry. This was in contrast to the dimeric structures obtained from the examples discussed above, indicating that the interplay between different metal SLs and recognition motifs could play a strong role in PNs formation.

However, it was found in some cases that electrostatic interactions may play a more important role on the assembly formation than the metal coordination and the prediction of the protein nanostructure is not always straightforward. By selection of the metal–ligand interaction, light responsive protein PNs were prepared using GroEL. The biological role of GroEL in natural systems is to assist the refolding of denatured proteins by an adenosine-5′-triphosphate (ATP)-induced mechanism, which subsequently releases guest proteins. The barrel-shaped tetradecameric GroEL protein was first modified in the outer part of its cavity with a number of photochromic units, spiropyran/merocyanine, through 14 Cys residues spatially engineered on the top and bottom of the monomeric protein cylinder. The modified GroEL monomers then coordinated to Mg^{2+} in a light dependent fashion to form nanotubes with defined sizes. Other divalent cations such as Ca^{2+}, Mn^{2+}, Co^{2+} and Zn^{2+} were also able to induce assembly on GroEL-spiropyran/merocyanine, whereas monovalent cations (Na^{+}, K^{+} and Cs^{+}) were ineffective. Trivalent cations (Fe^{3+}, In^{3+}, Ce^{3+}, and Eu^{3+}), on the other hand, generated poorly defined aggregates, suggesting that the assembly could be triggered by electrostatic interactions rather the coordination bonds.

The careful selection of the complexing ligands and metal ions as SLs allowed the formation of PNs but the influence of secondary interactions such as electrostatic interactions in the assembly has to be considered carefully, and greater predictability is required, perhaps by involving computation methods. Nonetheless, the SLs based on the metal–ligand strategy offers great opportunities in terms of structural and functional diversity if exogenous metal ions such as Pt^{2+} or Ru^{2+} are considered as well and the possibility to form a dynamic system by the interplay of metal SLs/ligand binding strength, which have mostly been neglected to date.

SLs based on Deoxyribonucleic Acid (DNA) Nanotechnology

In nature, DNA plays an important role in directing protein assemblies. In coat proteins which are components that form virus capsids, the DNA cargo influences formation of the globular capsids. DNA nanotechnology has emerged as the method of choice to construct precise nanoscale architectures, due to the highly specific and predictable binding of complementary base pairs. The success of DNA nanotechnology has consequently led to an upsurge in research activities to exploit DNA as a template to arrange multiple proteins into well-defined nanostructures. The DNA linker can be directly coupled to the protein of interest and then assembled onto the DNA scaffold. Alternatively, the protein of interest is attached directly to the DNA scaffold. Both approaches require either covalent or non-covalent conjugation of the desired PBs. Non-covalent methods include biotin–(strept) avidin, Ni^{2+}–nitrilotriacetic acid, antibody–hapten, aptamer interactions. Here, aptamers are considered as shorter DNA sequences binding other compounds. Covalent strategies to connect proteins and DNA SLs include disulfide and maleimide coupling, protein ligation, biorthogonal click chemistry or enzyme-mediated reactions. DNA binding proteins such as zinc-finger proteins recognize specific DNA sequences, and they were fused to the desired recombinant proteins to guide self-assembly of proteins on DNA nanotemplates without additional chemical entities.

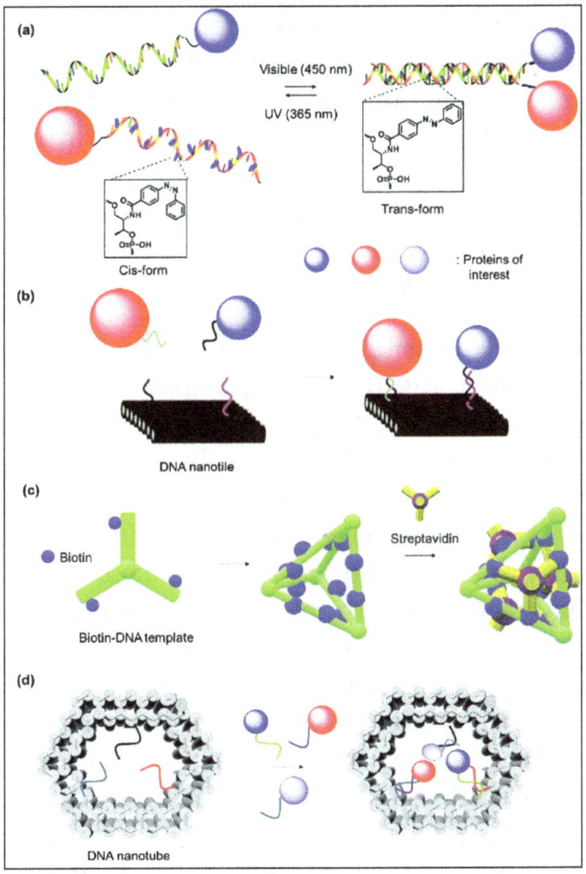

1D, 2D and 3D-DNA SLs. (a) Hybridization of ssDNA with complementary strand. The introduction of azobenzene in the oligonucleotide sequence allowed hybridization to be controlled via photoswitch. (b) DNA nanotile for the precise placement of proteins with defined interspatial distance. (c and d) DNA nanostructures as SLs for organization of proteins in 3D space.

Heterodimeric proteins were prepared by implementing complementary single stranded (ss) DNA sequences in the respective PBs. Interestingly, the spatial arrangement of DNA can be manipulated by applying the DNA origami method to obtain precise 2D or 3D templates, which allowed the self-organization of proteins in a convenient manner. By this method, a long "scaffold" strand DNA is molded into a desired shape "on demand" using hundreds of DNA "staple" strands. To achieve higher order nanostructures, a DNA polyhedral was prepared and in combination with biotinylation of a specific DNA strand to achieve trivalent binding on each of the polyhedral face. In this manner, proteins were organized on each face in 3D space with high spatial precision. Reverse engineering was also implemented by redesigning the binding specificity between DNA strands and coat proteins for the nanofabrication of PNs. Researcher successfully demonstrated that the icosahedral viruses such as CCMV or cucumber mosaics virus can be reorganized to form uniform nanotubes and the length and diameter of the nanotubes controlled through the ratio of coat protein/DNA applied.

Since the DNA staples are usually prepared synthetically, it was also possible to introduce molecular triggers in the design. Azobenzene was incorporated into the DNA linker and the resultant DNA duplex formation was controlled by light irradiation. In this fashion, the activity of the glucose oxidase/horseradish peroxidase cascade system was regulated by a light trigger, which offers unique opportunities in signal transduction controlled by external stimuli. DNA aptamers, which exhibit high affinity in the nanomolar range to a variety of proteins, were engineered into DNA tiles to target and direct the assembly of specific proteins such as thrombin or single-chain variable fragment (scFv) in a regular, periodic arrangement. Additionally, bidentate or tripodal tridentate aptamers were designed for the precise organization of proteins. To improve stability of the DNA origami constructs, DNA linkers were replaced by peptide nucleic acids, where the phosphodiester backbone was replaced by a peptide backbone.

DNA nanotechnology provides the possibility to tune the binding strength by the length or sequence of the DNA SLs For example, a 12-mer DNA consisting of adenosine only displays binding affinity with the complementary sequence that is three orders of magnitude lower than the 12-mer 5′-AGCTACACGATA-3 and comparable binding affinity to the 8-mer 5′-GCTACACG-3′. DNA nanotechnology can overcome some substantial challenges in terms of the precise alignment of individual protein components in multidimensions to form higher order nanostructures. Notably, it allows the exact number of proteins and interprotein distance to be predetermined by design. Significant advancements have been made in this field, and sophisticated nanoarchitectures and functions have been programmed, e.g. logic-gated nanorobots for targeted transport of molecular payloads such as antibodies or 3D multienzyme crystal structures,. Furthermore, DNA sequences are susceptible to cleavage by enzymes (DNAse) and this could confer responsiveness to the system. DNA nanotechnology was also combined with a protein backbone to enable multi-protein labeling by sequence-specific assembly through capitalizing on the unique features of DNA and protein materials. However, scale up and high cost production of DNA sequences as well as their highly negatively charged structures and low stability still limits this technology for several applications.

The design of the SLs for specific interactions with recognition motifs discussed above certainly offer a broad spectrum of tools for nanofabrication of supramolecular protein architectures. It is obvious that the synthetic customization of PNs could be realized through rational selection of SL and recognition motif including electrostatic, protein–ligand and host–guest interactions, as well as metal coordination and DNA nanotechnology. Of all these strategies, DNA nanotechnology offer the highest structural precision, especially with recent breakthrough in the mass production of DNA origami. In addition, there is also parallel development of other synthetic methodologies to contribute to the growth of this emerging field, such as using nanoparticles, peptides, proteins, or polymers as templates to form PNs.

Methods for Quantification of SL–PB and PB–PB Interactions

The SL–PB or PB–PB interactions and their assembly/disassembly control the stabilities of the PNs. Qualitative indirect assessment could be through the "visual" observation of the PN formation and dissociation through imaging techniques such as atomic force microscopy or transmission electron microscopy. To gain greater insights into PN formation, quantitative analysis of the interactions of the recognition motifs on the PBs and the SLs are essential and they are typically determined in the form of enthalpy or binding constants. A number of biophysical methods, which are more sensitive, have emerged recently such as fluorescence spectroscopy, surface plasmon resonance (SPR), microscale thermophoresis (MST) and isothermal titration calorimetry (ITC). Some examples, as well as the advantages and limitations are described herein.

Fluorescence anisotropy/polarization (FP) is one of the common techniques used to unveil the relationship between proteins and their ligands in a quantitative manner due to the easy accessibility to the instrumentation. Fluorescence polarization determines the dissociation constant by measuring the rotational mobility of the ligand/protein before and after binding. It is a sensitive method and offers information on protein–ligand binding down to subnanomolar concentrations. Usually, a fluorescent ligand is titrated against varying concentrations of the PB to obtain a binding curve. Researcher used this approach to study the dimerization of endophilin, a 40 kDa SH3 domain-containing protein, with a dissociation constant of ~5–15 μM in 20 mM HEPES buffer, pH 7.5, and 100 mM NaCl. Although less direct thermodynamic information can be obtained compared to ITC, the method requires lower amount of sample than SPR and ITC. The choice of a fluorescent ligand with a suitable lifetime that does not affect the SL–PB interaction is critical for accuracy. Moreover, the method is only applicable for ligands that are significantly smaller than the protein partner is and can only be applied outside cells. Non-fluorescent native ligands cannot be used directly for measurement and requires modification.

ITC is a sensitive calorimetric technique providing detailed information such as binding affinities and thermodynamic parameters of interacting biomolecules. Out of all the methods discussed here, ITC is a label-free method in which the heat evolved or absorbed during complex formation is measured through gradual titration of a ligand against the biomolecule of interest. It provides the affinity constant, stoichiometry, enthalpy and entropy of reversible biomolecular interactions. Quantification of the affinity range is from nanomolar to submicromolar. However, using competitive technique where the strong ligand displaces weak ligand–protein complex, dissociation constants within the picomolar range can be determined. ITC has been performed to elucidate thermodynamics parameters of the interactions of concanavalin A with SL consisting of α-d-mannopyranoside and rhodamine B. The resultant protein crystalline framework occurs through dual supramolecular interactions.

Typically, protein crystallization is an entropy-driven process with a small change in enthalpy. But in this case, a negative heat of -63 ± 5 kJ mol^{-1} of SL was observed. The result is consistent with the sum of the binding enthalpies of concanavalin A with mannopyranoside and dimerization of rhodamine B (-67 kJ mol^{-1}), thereby confirming the role of the SL to induce crystallization. Besides thermodynamic parameters, ITC analysis was employed to determine the ternary binding constant of CB[8]-(FGG-glutathione-S-transferase)$_2$ as 2.9×10^{12} M^{-2}, while a yellow fluorescent protein fused with FGG peptide displayed a ternary binding constant in the subpicomolar range. The major limitation of ITC is probably the need to dissolve all components in exactly the same solvent as well as low sensitivity towards a change in enthalpy. Therefore, solvents need to be selected carefully and investigations often require that one of the analytes is used in much higher concentrations compared to other techniques. This limitation could be circumvented by nano-ITC, which utilizes lower sample volumes (100 μL) and quantities (nanomole).

SPR measurements correlate the absorption of molecules on a thin, conducting surface ([Au] or [Ag] metal film) through the detection of changes in the refractive index at the surface of the film. SPR measurements give information on binding constants in nM to low mM range. Methyl-α-d-mannopyrannoside was immobilized to a thiol-modified gold surface to determine the dissociation constant to concanavalin A (~μM range). Besides the dissociation constants, the on- and off-rate constants were also evaluated by SPR, for example, between a series of ribonucleic acids to a protein, NS3 protease domain of the hepatitis C virus. In comparison to the other methods, SPR is not conducted in solution and one of the binding partners has to be immobilized on the metal film, which could have an impact on protein structure and ligand binding. Consequently, it could be time consuming to establish a new assay, as site directed labeling of the PB needs to be optimized when different SL–PB or PB–PB interactions have to be investigated. Due to the limitations of mass transport close to the interface, SPR analysis could be complicated and surface immobilization could interfere with the binding event.

MST is a biophysical technique based on the motion of molecules in microscopic gradients and allows the determination of dissociation constant in the micromolar to picomolar range. The thermophoretic movement of the fluorescent molecules along temperature gradients triggered by an infrared laser on a sample in solution placed in a capillary and the mobility of the molecule is detected by fluorescence. The thermophoretic behavior is highly sensitive to variations in conformation, charge and size of the molecules due to a binding event and the binding affinity can be determined using a titration approach. Thermodynamic parameters can also be obtained by assessing the dissociation constants over a temperature range. These measurements can be carried out in minutes, with no limitation on molecular size, and they can be measured in buffer or complex biological media such as cell lysates and has low sample consumption compared to nano-ITC. MST measurement requires that one binding partner is fluorescent but it could also be conducted using the intrinsic protein UV-fluorescence.

MST can record binding constants as low as the picomolar range, without the need of using competitive binding ligand like in ITC. However, the technique is sensitive to the presence of aggregates, provides less thermodynamic parameters compared to ITC and it is not suitable for studying weak interactions, i.e. in the mM range. Ng and Researcher investigated the facile assembly/disassembly of cytochrome c-polyethyleneglycol core–shell architecture formed via boronic acid-salicylhydroxamate interactions and determined the binding affinity to be in the micromolar range at physiological pH and dissociation at pH < 5.0. Similarly, Kuan and Researcher determined the pH dependent association and dissociation of a boronic acid modified lysozyme with a fluorescent dye consisting of salicylhydroxamate group.

Besides the above biophysical techniques which measure the sample "bulk", single-molecule methods, such as single molecule Förster (or fluorescence) resonance energy transfer (smFRET), have emerged that can resolve sample heterogeneity and thus allow probing of real time dynamics. smFRET has been successfully applied to investigate protein–ligand interactions at a single molecule level and even in living cells. It detects the non-radiative energy transfer between fluorescent donor–acceptor pair (<10 nm distance), which gives the intervening distance. In this way, binding rate and dissociation rates can be obtained, which is important for understanding biochemical processes in living cells since they usually occur under non-equilibrium conditions. smFRET can thus reveal more insights into molecular interactions, dynamics and mechanisms compared to other traditional biophysical methods. An additional advantage over other techniques is that smFRET allows investigation of binding with insoluble proteins. For instance, to study the dissociation rates (~μM) of cell-bound T-cell antigen receptors binding to an antigenic peptide-major histocompatibility complex in situ. There are nevertheless some limitations to smFRET. It requires attachment of a minimum of two fluorophores to the analytes as the intrinsic fluorescence of tryptophan in proteins are not bright or photostable for measurement and weakly interacting fluorescent species could be challenging to study.

With these analytical tools, quantitative information of PB–SL and PB–PB over a broad range of interaction strengths and stabilities could be obtained. Consequently, the results give valuable information for the optimization of the chemical design of PB and SLs for PN formation. It should be noted that the examples and the binding constants or thermodynamic parameters given above are based on established literature reports. They should only serve as a general guideline since binding constants and thermodynamic values could vary strongly under different conditions such as buffer used, ionic strength, temperature and pH.

Formation and Characterization of the Protein Nanostructures (PNs)

Here, the formation of PNs based on different permutations of PBs and SLs and the conditions in which they are formed are highlighted. The PN formation is discussed according to the resultant morphology, as well as their subsequent characterization. Typically, the formation of PNs is carried out in aqueous solutions such as phosphate buffer with variations in pH, buffer strength and additives, depending on the type of the PBs, SLs and recognition motifs used. Characterization with gel electrophoresis or size exclusion chromatography (SEC) reveal changes in retention time reflecting variations in molecular size. These are the most prevalent tools but they do not provide detailed information of the morphology of the PNs formed. In addition, it is very challenging to obtain insights on the precise ratio of PB units in heteromeric PNs. With the advancements in microscopy techniques such as atomic force microscopy (AFM), transmission electron microscopy (TEM), fluorescence correlation microscopy (FCS), as well as dynamic light scattering (DLS), more detailed characterization of PNs at the nanometer scale has been achieved. A summary of the PNs is given in Table.

Table: Overview of functional PNs.

PBs	SLs/interaction motifs	Nanostructure (PN)	Application/function
Native PB			
CCMV, avidin	Electrostatic surface patches	Binary crystals	Bioactive protein crystals.
Apoferritin	Phthalocyanine	Crystals	Photosensitizer.
SP1	CdTe quantum dots	Nanowires, nanorods	Light harvesting antenna.
	Porphyrin-G5 PAMAM	Nanorods	Multienzyme cascade.
Concanavalin A	Bismannopyranoside, mannose-rhodamine B	Protein crystalline frameworks	Porous protein material.
Soybean agglutinin	N-Acetyl-α-galactosamine	Microtubule-like	Mimicry of microtubule.
Streptavidin	Bis(biotin)–terpyridine	Polymer	Biomineralization.

Chemically modified PBs			
Insulin	Bipyridine	Trimer	Regulation of glucose metabolism.
Heme catalase	Oligonucleotides	Binary crystals	Cascade protein crystal reactor.
Streptavidin-PAMAM, p53	Biotin	Dimer	Cytotoxic protein transporter.
Avidin-HSA-PAMAM, β-galactosidase	Maleimide-biotin, iminobiotin	Trimer	Transporter of enzymatic protein cargo
Somatostatin, avidin, C3	Biotin–hydrazone-linker	Branched pentamer	Cancer cell-specific transporter, intracellular release.
Antibody, avidin, LiDps cage	Biotin–streptavidin	Trimer	Selective uptake into Staphylococcus aureus.
Her2, anti-CD3 antibodies, avidin	Biotin, Protein A–antibody	Pentamer	T cell-mediated lysis of Her2-positive breast cancer cells.
Horseradish peroxidase, glucose oxidase	Oligonucleotides	Dimer	Enzymatic cascades.
Anti-CD33 antibody; CDw238 FAb	3D DNA hexagonal barrel	Multimer in 3D scaffold	Nanorobot for stimulation of cellular processes.
Genetically modified PBs			
CFP, YFP	CB[8]-Phe-Phe-Gly	Dimer	FRET pair.
Split luciferase N- and C-terminal fragments	CB[8]-Phe-Phe-Gly	Dimer	Signal transduction with on–off switch.
Cytochrome cb_{562}	Bis(histidine)-Zn^{II}	Tetramer; 2D protein array	Antimicrobial protein assembly in cells; template for nanoparticles growth.
Chaperonin GroEL, lactalalbumin	Spiropyran-Mg^{II}	Nanotube	ATP induced release of cargos in HeLa cells.
Glutathione-S-transferase	Histidine-Ni^{II}	Linear or cyclic polymer	Catalytic elimination of cytotoxic compounds in vitro.
	CB[8]-Phe-Phe-Gly	Linear or cyclic polymer	Inhibition of lipid peroxidation; nanospring.
Hemeproteins	Heme	1D, 2D polymer	Oxygen transport.
Cel5A, streptavidin	Biotin	Polymer	Biotemplating of artificial cellulosome.
Streptavidin	Sortase A (G tag); horseradish peroxidase (Y tag)	Twigged polymer	Protein polymer scaffold for immobilization of multprtein.

Dimeric PNs

Protein dimers represent the simplest PNs and most functional examples consist of bi-enzyme cascades or combined fluorescence proteins, with the latter offering ease of characterization using fluorescence spectroscopy. Two identical protein copies interconnected by the SL form a homodimer, whereas the formation of heterodimers of two different proteins is more challenging as orthogonal interactions are required, and there is a more stringent demand on the PB and SL design.

A fluorescence/Förster resonance energy transfer (FRET) pair of Phe-Gly-Gly-CFP (cyan fluorescent protein) and Phe-Gly-Gly-YFP (yellow fluorescent protein) was prepared with the Phe-Gly-Gly peptide tag expressed on the N-termini of both proteins. Heterodimerization was induced in the presence of CB[8] in phosphate buffer at pH 7 and characterized by SEC. It was further substantiated by strong FRET, with an estimation that the protein heterodimerization was accompanied by about 50% homodimerization, which is reasonable since the same Phe-Gly-Gly tags were employed on both proteins. The addition of a small synthetic molecule such as methyl viologen, which is involved in competitive binding, resulted in the dissociation of the dimer. Similarly, two split fragments of the N-terminal (NFluc437) and C-terminal (CFluc398) of firefly luciferase were engineered with Phe-Gly-Gly peptides. In the presence of CB[8], the two non-active fragments were paired and luciferase activity was recovered. The formation of the ternary heterocomplex over the homocomplex was preferred due to higher stability of the former and confirmed by a titration against excess of the weakly binding Phe-Gly-Gly peptide. By using a supramolecular approach, an on–off switching mechanism was implemented by adding a competing ligand, such as the amantadine derivative 3,5-dimethyladamantan-1-amine (memantine), in conjunction with CB[8] for repeated up- and down-regulation of enzymatic activity, which is important for signal transduction.

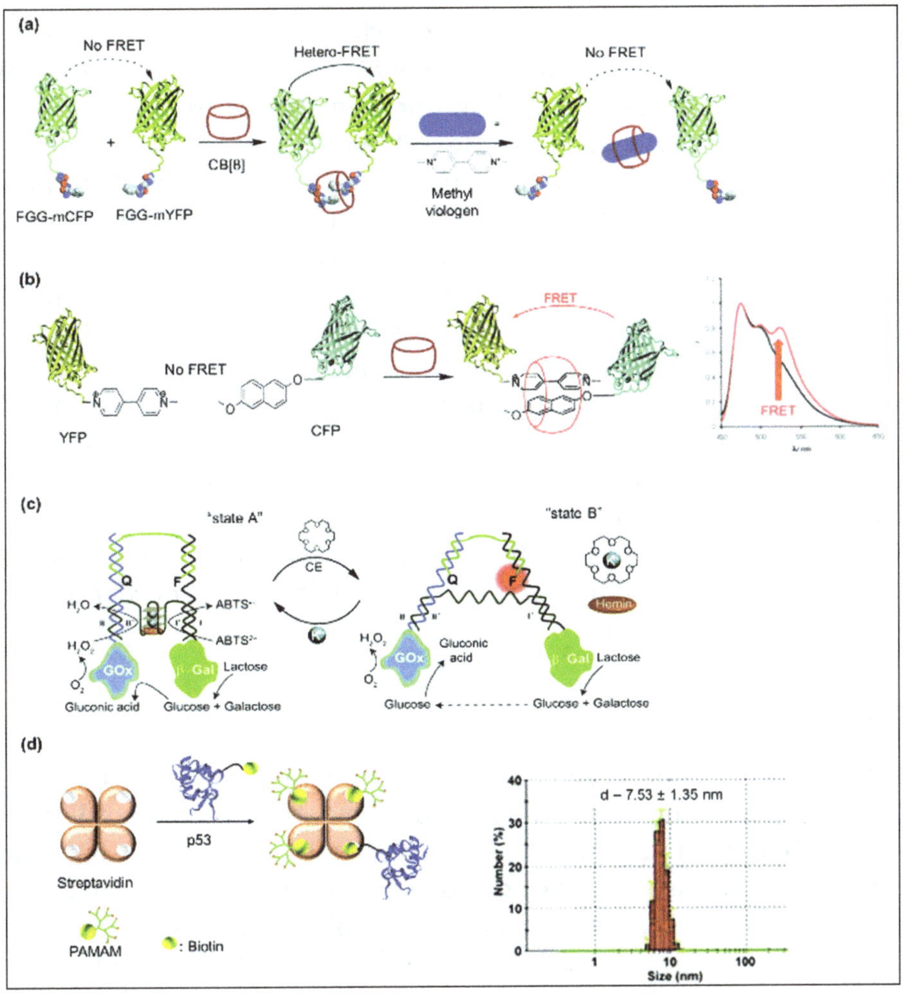

Dimeric PNs and characterization. (a and b) FRET fluorescence pairs formed using CB[8] host–guest interactions. (c) Switchable enzyme cascade using DNA hybridization. (d) streptavidin bioconjugate formed by assembly of three PAMAM dendrimer branches (dendrons) and the respective cargo protein p53. Narrow size distribution was shown in DLS.

To circumvent the formation of homodimers as side products, orthogonal strategies are required. Besides the formation of 1 : 2 complexes with Phe-Gly-Gly peptides, CB[8] can also form stable 1 : 1 : 1 ternary complexes with an electron deficient–electron rich supramolecular guest pairs such as methyl viologen-alkoxynaphthalene to form a charge transfer complex. Researcher successfully assembled alkoxynaphthalene-CFP and methylviologen-YFP FRET pairs using this approach. Through FRET measurements, they determined that the ternary approach using CB[8] eliminated the formation of homodimers. Split protein systems or protein heterodimers were also accomplished using DNA nanotechnology. For instance, switchable enzyme cascades such as glucose oxidase/β-galactosidase pairs with K^+-ion stabilized hemin-G quadruplex horseradish peroxidase mimicking DNAzyme were prepared. First, the proteins were conjugated to two separate ssDNAs and the proteins dimerized on the complementary sequences on the template ssDNA. By using K^+ ions and 18-crown-6, the template DNA switched between tweezers and clamp structures to activate and deactivate the catalytic activity. The occurrence of the reaction cascade to convert glucose to gluconic acid indicated that the bi-enzyme system was successfully prepared.

Researcher first proposed a combinatorial approach where streptavidin was used as a supramolecular "glue" to fuse synthetic entities such as mono-biotinylated polyamidoamine (PAMAM) dendrons, with protein enzymes to mimic binary protein structures of AB-type bacterial toxins comprising of binding and catalytic domains. Here, the optimization of the biotinylated synthetic entity required for saturation of the binding pockets of streptavidin allows more precision in the stoichiometric loading of the cationic polyamidoamine (PAMAM) dendrons and the protein enzyme of interest, with high conjugation efficiency due to the strong streptavidin–biotin interactions. In this manner, the optimal stoichiometric ratio of the PAMAM dendrons and the protein enzymes, cytochrome c or tumor suppressor p53, were mixed with streptavidin at room temperature in phosphate buffer to give the tricomponent heterodimeric protein and dendron-induced intracellular delivery of the protein cargo was successfully demonstrated. Dynamic light scattering (DLS) measurements supported the formation of the AB-type proteins with narrow size distribution. The method offers the advantage of rapid screening and optimization of the biological activity of a broad spectrum of biologically attractive combinations, as well as the possibility to fuse synthetic entities with biomolecules, which could not be accomplished by genetic techniques. However, this does not provide structural precision, especially if higher order nanostructures such as trimers or oligomers have to be achieved.

Trimeric PNs

Fe^{2+} can interact with three units of bipyridine to form a six-coordinated octahedral complex. Capitalizing on this specific metal–ligand interaction, homotrimerization of PBs was achieved. A chemically modified PB, human insulin variant (InsX2) containing a single bipyridine, was added to Fe^{2+}, and the complexation was confirmed due to a distinct colour change of the protein solution from colourless to magenta. A new ^1H NMR signals of the bipyridine ligand appeared in the protein spectrum, and SEC showed a shorter retention time shift compared to the InsX2 monomer. The formation of a InsX2 trimer was corroborated in combination with static light scattering.

The construction of heterodimers requires careful choice of PB and SLs and is challenging since homodimerization should be avoided. Thus, it is intuitive that the formation of trimeric or oligomeric heteroproteins would require much more synthetic efforts. Nevertheless, there has been some recent breakthrough with the development of solid phase preparation of protein nanoarchitectures. Solid phase protein synthesis approaches have been devised to desymmetrize homomeric protein building blocks such as (strept)avidin. Typically, these proteins possess more than one binding site but are symmetrical, and it is difficult to control the placement of different self-assembling protein components on these platforms. Solid phase approach was therefore devised to overcome this challenge. The solid phase exposes only one hemisphere of the protein building block and the other is masked and protected e.g. dynamic covalent S–S linkage or pH-sensitive non-covalent interactions. In this manner, two-faced "Janus-like" PBs were derived to build up non-covalent heteroprotein nanostructures such as a heterotrimer. Notably, chemically post-modified PBs were applied to confer additional functions to the PNs, which could not be achieved by recombinant engineering such as the utilization of HSA incorporated with positively charged dendrons to enhance their cellular uptake.

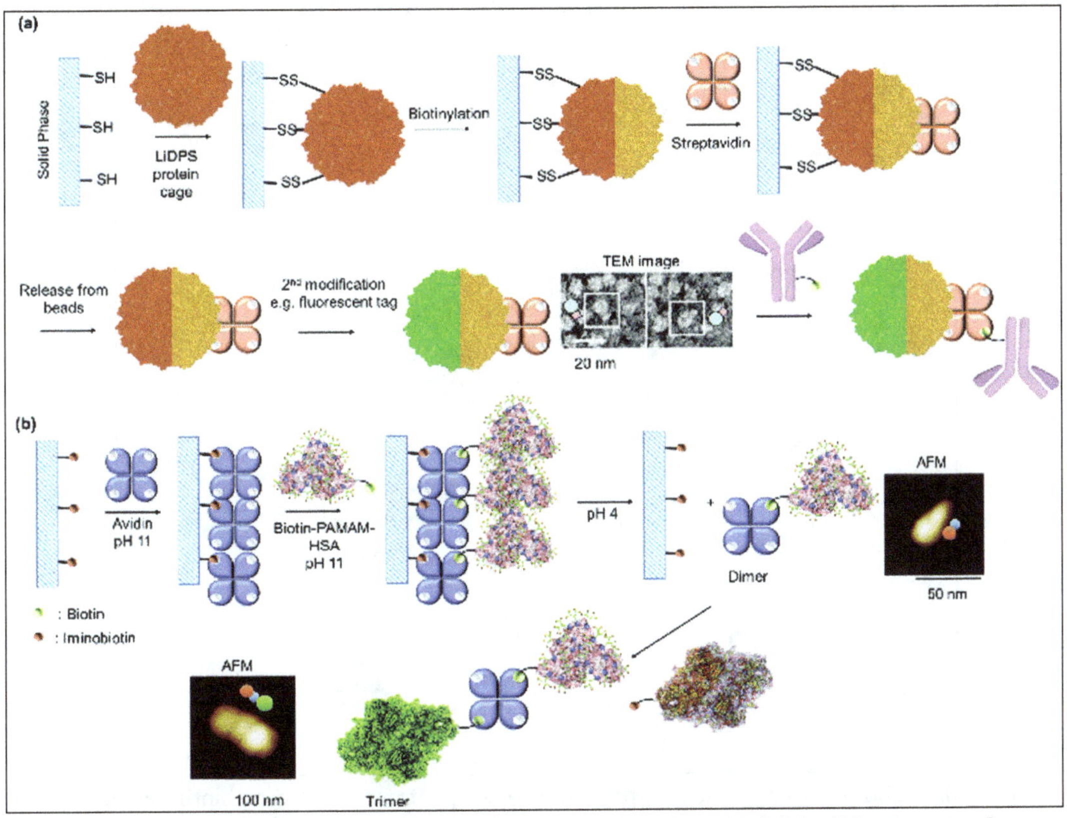

Solid phase approach for preparing heterotrimers based on (a) thiol solid support and (b) iminobiotin agarose and their subsequent characterization.

Douglas and co-workers developed a solid phase using disulfide linkage A cysteine mutant of a protein cage, LiDps (DNA binding protein from Listeria innocua) was immobilized onto the thiol-functionalized solid phase in phosphate buffer at pH 7, followed by toposelective biotinylation of LiDPS, loading of streptavidin and release from the solid phase by reduction of the disulfide linkages with a dithiothreitol solution in sequential order generating a heterodimer platform. The heterodimeric structure was characterized by TEM, DLS, and quartz crystal microbalance. This nanoplatform was further coupled to biotinylated macromolecules such as antibodies. However, one possible drawback of this approach was that proteins, which are sensitive to redox conditions such as dithiothreitol cannot be applied as this could affect their activity.

Researchers proposed a solid phase approach to desymmetrize avidin using the iminobiotin–avidin technology. Avidin was immobilized onto commercially available iminobiotin-agarose at pH 11, which effectively masked and protected one hemisphere of the avidin linker. Thereafter, a chemically post-modified protein conjugate, functionalized with a single biotin group and PAMAM dendrons (DHSA) to enable cellular uptake, was anchored onto the available binding pocket of avidin. Subsequently, the heterodimeric DHSA–avidin conjugate was released from the solid phase by acidification (pH 4) due to protonation of iminobiotin, which abolished binding to avidin. The thus-prepared heteroprotein dimer still consisted of free binding pockets on avidin, which were used for further conjugation to other (imino)biotinylated molecule of interest, e.g. enzymatic proteins such as β-galactosidase and toxin enzymes. Fluorescence polarization is influenced by changes in molecular weight/hydrodynamic radius, thus affecting molecular mobility. The successful assembly of the heterotrimer was therefore confirmed by fluorescence polarization and the formation of a discrete trimeric protein structure was verified using AFM.

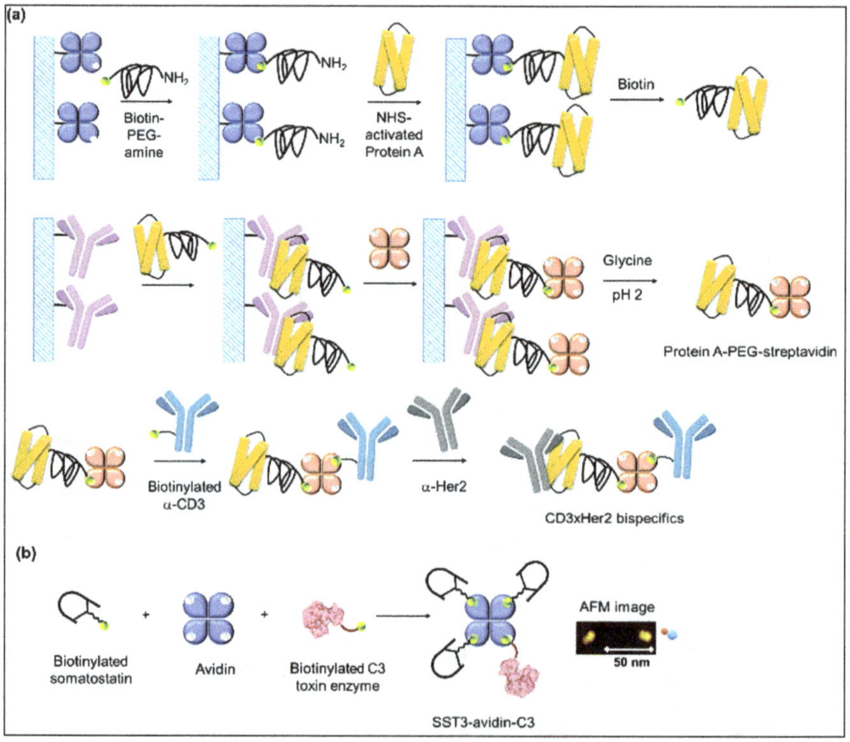

(a) Double solid phases approach for preparing heteropentamers based on biotin–avidin and protein/antibody interactions. (b) Nanoscale assembly of branched oligomeric SST3-avidin-C3 and AFM characterization of the PN.

Oligomeric PNs

Protein homotetramers have been reported using CB[8] as SL with a dimerizing YFP variant and N-terminal Phe-Gly-Gly as the PB. The formation of tetramers in phosphate buffer at pH 7 was characterized by a decrease in fluorescence anisotropy due to homo-FRET and a size increase in hydrodynamic radius from 3 to 4 nm in DLS measurements, as well as SEC.

Homotetramers were also reported using SL based on metal–ligand interactions. For instance, a 4-helix bundle cytochrome cb_{562} PB with two sets of His dimers at the i and i + 4 positions formed tetramers upon addition of Zn^{2+} ions at higher concentrations. At lower concentrations, only monomers and dimers were observed, according to sedimentation velocity measurements, a method used to determine molecular size and shapes based on the rate of molecular movement according to the centrifugal force generated in an ultracentrifuge.

Moving forward from homo- to hetero-oligomers, a solid phase approach was again implemented. Protein A, G, and L are surface proteins that are present in the cell walls of different bacteria and they exhibit high affinities and specificities to Fc region of immunoglobulins found in mammalian species, especially immunoglobin Gs. Using a dual solid phase approach based on Protein A/immunoglobulin G and biotin/(strept)avidin non-covalent interactions, Researcher prepared a Protein A–PEG–streptavidin heterobifunctional adaptor in a defined 1 : 1 : 1 stoichiometry, which was then used to prepare pentameric heteroprotein complexes. First, monomeric avidin resin was loaded with biotin–PEG–amine, followed by addition of NHS-activated Protein A to form monovalent Protein A–PEG–biotin conjugates. After elution of the Protein A–PEG–biotin conjugate from the column, streptavidin was added to Protein A–PEG–biotin immobilized onto a human immunoglobulin G agarose column. The tripartite 1 : 1 : 1 Protein A–PEG–streptavidin adaptor was isolated after elution from the column. Protein A can also be replaced by Protein L or G to expand the library of the available trimeric adaptor. This adaptor was characterized using gel electrophoresis, and it was further functionalized on its two ends with bioactive components.

Polymeric PNs

Linear supramolecular polymeric PNs have been reported with various combination of PBs and SLs including metal–ligand coordination, host–guest interactions and protein–ligand interactions. For instance, the generation of protein nanowires was achieved by using glutathione-S-transferase (GST) dimer from Schistosoma japonicum with a hexahistidine tag inserted at the N-terminus of each monomer. The localization of poly-His tags at opposite directions relative to each other allowed the formation of 1D-nanowires upon their coordination to Ni^{2+} using Tris buffer at pH 7.4, where one metal ion was coordinated by two terminal polyhistidine moieties at each end of two different GST dimers. The presence of Ni^{2+} SL was shown to be pivotal for the formation of nanowires since, in the presence of EDTA, the assemblies were reversed to the homodimers. The nanowires were characterized using AFM and gel electrophoresis. AFM measurements indicated objects with uniform heights of about 4.9 nm, consistent with the height of a single GST PB. Interestingly, a planar network was obtained by increasing the protein concentration through linear assembly, but this network did not display uniform heights. Native gel electrophoresis showed a few protein bands distributed over a range of molecular weights, suggesting that polydisperse PNs with distributions of the polymer lengths were formed.

The protein nanowires presented slightly higher enzymatic activity, being able to catalytically eliminate in vitro cytotoxic compounds such as 1-chloro-2,4-dinitrobenzene when compared to the catalytic capacity of the homodimeric PB. Similarly, GST protein nanowires were prepared using the supramolecular interaction of CB[8] with two tripeptide Phe-Gly-Gly fused to the N-termini of dimeric GST with a C2 symmetry. To achieve a functional assembly, a glutathione peroxidase mimic was prepared with a single site mutation of selenocysteine (Y6C). Remarkably, the nanowires exhibited 20% inhibition in lipid peroxidation compared to the monomer in a mitochondria oxidative stress assay. Other functional features were also introduced by selection of an appropriate genetically modified PB. For instance, a recoverin domain, which is an allosteric protein responsive to Ca^{2+} with an N-terminal Phe-Gly-Gly tag was fused to dimeric gluthathione-S-transferase. The resulting protein nanowire exhibited conformational changes between a contracted and an extended state in the presence of Ca^{2+}, thus functioning like a nanospring.

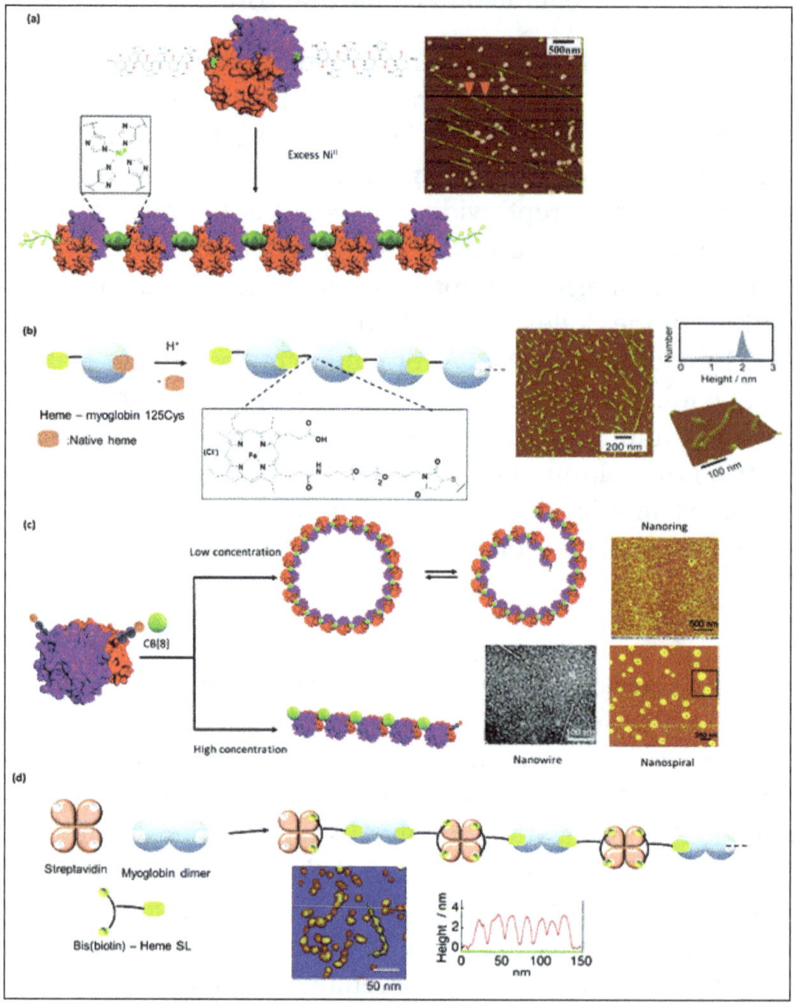

Formation of linear and non-linear supramolecular protein polymers. (a) Linear glutathione-S-transferase polymers through histidine-nickel coordination. (b) Myoglobin polymers from maleimide-modified myoglobin. (c) Nanoring PNs formed using recognition motif aligned in "V" arrangement and control of PB concentration. (d) Formation of A_2B supramolecular protein polymer from streptavidin, myoglobin dimer and a bifunctional SL. AFM images of the PNs were shown.

Hayashi and co-workers made use of the heme-hemeprotein interactions to generate linear supramolecular protein polymers using cytochrome and myoglobin. More recently, sperm whale myoglobin with a single cysteine mutation at the 125 position was prepared, and an external heme ligand was introduced by maleimide-thiol reaction. Following the previously established method, the solution was acidified to generate an apo-myoglobin protein. Linear supramolecular protein polymers were achieved by switching the pH to neutral to induce interprotein heme–heme interactions, and the supramolecular PN was characterized by SEC and AFM. Interestingly, the myoglobin PN retained its bioactivity to transport oxygen. The addition of exogenous ligands such as CO or CN, which assume the axial position of the heme, could regulate the stability of the resultant supramolecular PN. Furthermore, the researchers noted that a 3D protein network was achieved by cross-linking through the tyrosine residues using H_2O_2 and this network was characterized by scanning electron microscopy.

The combination of lectin A with a tetraglyco-SLs with a rigid calix[4]arene core preorganized the PB so that a 1D filament network with defined branching points imposed upon by the defect in the geometry in the glycocluster core was obtained. On the other hand, 1D nanoribbon wires of lectin A were obtained by using SLs containing α-d-galactopyranoside and rhodamine B Interestingly, the resultant nanostructures can be fine-tuned to form 2D nanosheets or even 3D structures by applying different lengths of the tethering ligands.

Aside from linear protein polymers, the formation of other morphologies such as ring structures were also achieved by the selection of a PB with a suitable orientation. By using glutathione-S-transferase with the arrangement of a recognition motif (Phe-Gly-Gly) arranged in a "V" shape the resultant morphology of the PN was controlled to form nanorings. Bis-histidine metal chelating sites were introduced into the GST dimer from Schistosoma japonicum and addition of Ni^{2+} ions induced the formation of nanorings, as determined by AFM. Nanorings with different diameters were obtained by variation of the ionic strength of the buffer used, suggesting that the assembly process was driven by both protein–metal coordination and protein–protein interactions. Conversely, CB[8] induced polymerization was also achieved with a GST engineered with Phe-Gly-Gly tag in a "V" arrangement. Due to the relatively large size of the CB[8] SL, the protein–protein interaction was reduced compared to the Ni^{2+}-histidine system, and PN formation was controlled by the ring-chain mechanism, which was affected by protein concentration. Consequently, a nanoring was prepared at low protein concentrations, and the assembly was transformed into a linear morphology by increasing the applied protein concentrations. Notably, nanospirals were achieved when a high ratio of CB[8] was added, as imaged by AFM.

By astute chemical design, a bifunctional SL comprising bis-biotin and heme can be applied sequentially to connect two different PB monomers, namely, a natural PB (B), streptavidin and a genetically modified PB (A_2), apomyoglobin dimer in alternating arrangement, to obtain the $(A_2B)_n$ linear protein polymer in K^+ phosphate buffer at pH 7.0. The SL was first appended to apomyoglobin dimer formed by a disulfide bridge, followed by subsequent addition of streptavidin and the formation of a 1D copolymer analysed by SEC. The molecular weight distribution of the obtained protein copolymers is affected by the spacer length of the SL and a short spacer prevented the formation of large copolymers. The size distribution was also controlled by varying the apomyoglobin dimer to streptavidin ratio. A smaller copolymer with narrower size distribution was obtained when using higher ratio of apomyoglobin dimer which, presumably terminated the polymer growth. SEC traces showed that a dodecameric $(A_2B)_{12}$ was the largest copolymer that was achieved and the formation of the alternating copolymer was further confirmed by addition of a known disulfide reducing reagent in order to convert the polymer back into the ABA trimer, which was characterized by AFM. Interestingly, the protein copolymer retained the dioxygen binding function of the heme cofactor. This supramolecular polymerization process was thermodynamically controlled and increasing concentrations of the PBs led to the formation of larger one-dimensional heterotropic assemblies. Dual supramolecular interactions were adopted together with dynamic covalent linkage (S–S) in apomyoglobin dimers, and this could serve as versatile platform to tailor a functional heteroprotein PN structures, which could be reorganized using multiple triggers.

2D and 3D PNs

Besides one dimensional PNs the formation of 2D- and 3D-PNs is attractive since they are often employed in Nature for scaffolding, as nanovessels, or cell components. DNA nanotechnology is the most prevalent strategy but suffers from scalability and thus there were also alternative strategies proposed. here, we highlight some examples such as micelle-like structures, protein nanotubes, and 2D protein arrays.

Researcher installed an azobenzene or stilbenzene group in the design of their heme SL which served as a transient thermal stimulus to form thermoresponsive hemeprotein micelles. The heme SL was incorporated into a cytochrome b_{562} protein. SEC showed the formation of large assemblies, as previously observed for similar linear hemeprotein structures reported by the same group. Contrary to earlier examples, in which hemeprotein polymer dissociated into the monomer upon heating, a micelle-type structure was formed instead. The transitions between the morphologies were investigated via DLS and circular dichroism. The researchers proposed that the switch could not have occurred via dissociation into the monomer but rather by the formation of a larger assembly, presumably a micelle, initiated by the denaturation of the protein at higher temperature (>80 °C), to eventually yield the kinetically trapped metastable micelle-type structure observed in TEM and DLS (diameter = 14–16 nm).

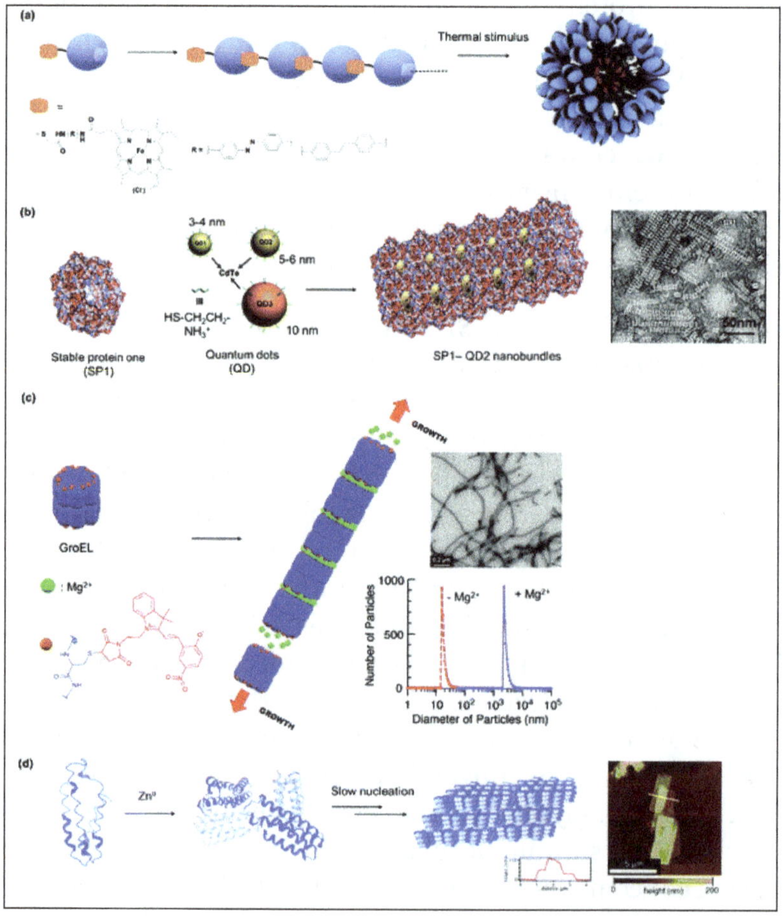

(a) Formation of micelle-type structures from linear hemeprotein polymeric PN driven by a transient thermal stimulus. (b) Nanobundles formed from charge driven assembly between cadmium/telluride quantum dots and stable protein one (SP1) as seen in TEM image. (c) Soft protein nanotubes formed by supramolecular polymerization of chaperonin GroEL. Characterization by TEM and DLS were shown. (d) Formation of 2D crystalline arrays from Zn^{2+} driven assembly of cytochrome cb_{562} as shown in AFM image.

By employing a cricoid PB such as stable protein one (SP1), charged induced assembly was initiated to form nanostructures ranging from nanowires, subsequent bundles, and irregular networks in aqueous solution. For example, nanobundles were obtained by selecting the right dimension of positively charged, globular quantum dots, as shown in AFM Notably, the regular arrangement in the supramolecular structures allowed high efficiency (up to 99%) energy transfer, thus mimicking a light harvesting antenna. Nevertheless, a multienzyme-cooperative antioxidative system was created using this SP1 by incorporating selenocysteine to the PB and manganese porphyrin to the fifth generation polyamidoamine (G5 PAMAM), which was used a positively charged SL for self-assembly. In this manner, the SP1 acts as a glutathione peroxidase mimic while G5 PAMAM serves as superoxide dismutase mimic. The enzymatic cascade exhibits significantly enhanced biological activity over the individual components. SP1/core cross-linked micelle complexes self-assembled to form nanowires of more than 120 nm in length, where the extent of growth was adjusted by the electrolyte strength. At higher concentration, bilayered or trilayered large-scale nanorods were formed through staggering of the protein nanowires. Mimicry of the energy transfer process in natural photosynthetic bacteria in vitro was achieved where donor and acceptor chromophores were attached to stable protein one and spherical micelles, respectively, to obtain an overall energy transfer of 52%.

A protein-based soft nanotube was formed by supramolecular polymerization of the molecular chaperone GroEL through metal–ligand interactions. The barrel-shaped tetradecameric protein assembly was obtained in pH 7.4 Tris–HCl buffer by Mg^{2+}-directed self-assembly in which GroEL was first spatially site-modified on the top and on the bottom of the cylindrical shape protein with a number of photocromic units, spiropyran/merocyanine. Monomers of GroEL-spiropyran/merocyanine coordinated to Mg^{2+} formed cylindrical fibers with a uniform diameter of 15 nm as shown by TEM. Interestingly, GroEL PBs were shown to not polymerize in the presence of Mg^{2+} suggesting that the photochromic units in spiropyran/merocyanine played a significant role in the nanotube assembly. The dependence of Mg^{2+} onto the formation of the assembly was attested upon the addition of EDTA. In the presence of this strong chelator, the cylinders were cut into short chain oligomers as well as into monomeric GroEL-spiropyran/merocyanine PB.

Additionally, the formation of the nanotube was light-dependent and responded to mechanical force generated by adenosine-5′-triphosphate (ATP). Furthermore, GroEL-spiropyran/merocyanine presented similar binding affinity towards denaturated proteins, such as lactalbumin, when compared to native GroEL. To control nanotube growth, an engineered half-cut GroEL variant that firmly bound to the nanotube termini for end-capping can be employed. By variation of the end-capper to PB ratio, protein nanotubes ranging from 40 to 320 nm were obtained in a controlled manner. A DNA cleavable GroEL nanotube was reported by first replacing the metal-coordinating ligand with a ssDNA, e.g. 15-mer. Thereafter, a ssDNA incorporating the complementary sequence, e.g. 20-mer was used as a SL to induce supramolecular polymerization to form protein nanotubes in Tris–HCl buffer, as shown in TEM, DLS and SEC. The protein nanotubes were highly thermodynamically stable due to the multivalent interactions. Dissociation was achieved by adding an external ssDNA which consists of a full length complementary sequence to the 20-mer SL. In this way, the stronger DNA hybridization allowed the displacement of the SL, resulting in dissociation as observed in TEM, DLS and SEC.

Cytochrome cb_{562} variants have also been employed for the formation of 2D-protein arrays. Researcher prepared Zn_8PB_4 units with cytochrome cb_{562} variant PB in which Cys and His residues were incorporated at positions 96 and 59, respectively. It was envisioned that His59 could contribute to Zn^{2+} coordination alongside with proteins' natural weakly coordination domains such as glutamic acid, aspartic acid and alanine. The presence of Cys96 residue allows dimerization of the PB by Cys96–Cys96′ bridges to form PB_2. The efficient formation of Zn_8PB_4 units was accomplished by the reaction between two units of PB_2 and four equivalents of Zn^{2+} in the presence of Tris as a metal-coordinating buffer. The crystal structure of Zn_8PB_4 showed a 2D array in which it is possible to observe the formation of two sets of four internal Zn^{2+} complexes. The nanostructures resulting from Zn^{2+}-directed self-assembly were shown to be dependent on time, Zn/PB ratio, and pH, as determined by TEM. Specifically, nanotubes were formed under high pH, high concentration of Zn/PB, and fast nucleation while the formation of 2D or 3D arrays was reported under low pH, low concentration of Zn/PB, and slow nucleation. The intermolecular interactions can also be tuned by chemical modification with a small molecule such as rhodamine B to induce crystalline arrays formation even under fast nucleation.

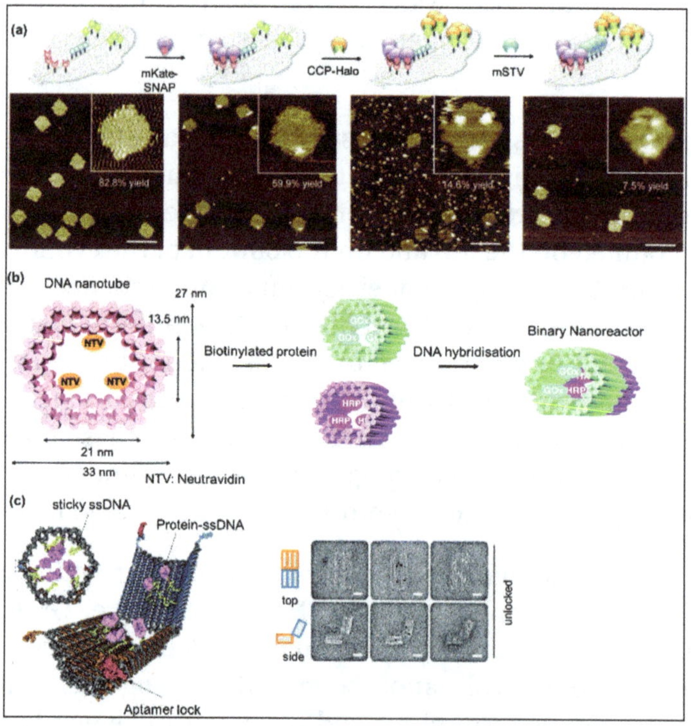

2D and 3D PNs formed using DNA nanotechnology. (a) Orthogonal decoration of proteins on 2D face-like DNA scaffold and AFM characterization of labelling yields at each step. (b and c) DNA nanovessels for precise spatial placement of multiproteins in defined 3D arrangements. TEM image in (c) showed the protein loaded on the 3D hexagonal barrel.

DNA nanotechnology has been exploited for precise organization in 2D- and 3D-protein assembly and the simplest arrangement are represented by bienzyme cascades assembled on DNA tiles, with straightforward characterization by AFM. Notably, the spatial distance in multiprotein systems can be tuned with a great degree of control to modulate spatial interactions between the different protein components. By combining DNA nanotechnology, biotin–streptavidin, Snap tag and Halo-tag chemistry, Researcher functionalized DNA strands for orthogonal complex multiprotein assembly on a biomolecular template. To demonstrate their concept, a 2D DNA face-like scaffold was designed using a software to ensure the correct folding of the M13mp18 ssDNA using 236 staples and the scaffold was characterized using AFM. Twenty three staple strands were biotinylated to create the eyes, nose, and mouth on the 2D face-like template. Monovalent streptavidin was allowed to bind and the obtained facial features were determined by AFM analysis. To introduce multiproteins, three chlorohexane (twice), four biotin, and four benzylguanine containing staples were immobilized on the origami scaffold as the eyes, nose, and mouth, respectively. Sequential addition of the proteins, namely mKate-Snap; CCP-Halo, followed by monovalent streptavidin led to formation of the intended smiley face structure, with an overall 7.5% yield. Each step of the assembly was characterized using gel electrophoresis and AFM analysis. The yields obtained were calculated with respect to the total of defined structures from AFM images. Furthermore, both sides of the quasi-2D plane could be decorated.

DNA nanocontainers were prepared to encapsulate proteins in a modular fashion inside the cavity of DNA nanostructure. A hollow 3D DNA tube was employed as a nanocontainer to anchor NeutrAvidin, a deglycosylated form of native avidin through biotinylated DNA staple sequence protruding in the cavity. In this way, biotinylated enzymes such as glucose oxidase or horseradish peroxidase were attached in the cavity through the NeutrAvidin binding sites. Distinct units consisting of different functional proteins were then stepwise stitched together through design of the DNA base pairing to create, for example, a nanoreactor using glucose oxidase/horseradish peroxidase as a proof-of-concept system. In addition, proteins were also organized into 3D origami nanostructures in a spatially precise manner by DNA hybridization. A hexagonal DNA barrel with dimensions of 35 × 35 × 45 nm was prepared using a 7308-base filamentous phage-derived scaffold strand with 196 oligonucleotide staple strands. Notably, a hinge opening mechanism can be implemented with two domains covalently attached in the rear and noncovalently clasped together in the front by DNA aptamer-based lock mechanism that responds to binding antigen for unlocking. PBs such as antibody fragment, FAb, can be premodified to attach a DNA recognition motif, e.g. on the 5′ end of a 15-base ssDNA and guided to the sites of interest in the inward ring of the barrel nanostructure through hybridization with the DNA SL consisting of staple strands with 3′-extensions in the complementary sequences. TEM analysis showed that three FAbs were loaded and the 3D nanostructure was subsequently used as nanorobot in cell biology.

Crystalline Protein Frameworks

Single CCMV particles were assembled in aqueous solution into larger hierarchical structures based on multivalent electrostatic interactions with Newkome-type dendrons with a nitrobenzyl core. Both DLS and TEM investigations showed that the assembly was affected by the generation and concentration of the dendron as well as the ionic strength of the solution. The supramolecular architecture could be disassembled by the cleavage of the positively charged amine arms from the nitrobenzyl core by photolysis. To demonstrate the broad applicability, the assembly and disassembly of the dendron with a negatively charged protein cage magnetoferritin was also accomplished. Subsequently, Researcher reported the ternary face-centered cubic (fcc) packed cocrystals of apoferritin and a supramolecular complex of octacationic zinc phthalocyanine and a tetraanionic pyrene formed by electrostatic and π–π interactions. The resultant structure preserved the fluorescent and singlet oxygen production properties. In this way, the resultant PN was able to produce singlet oxygen with a high quantum yield (0.72) upon irradiation, which is valuable for photodynamic therapy.

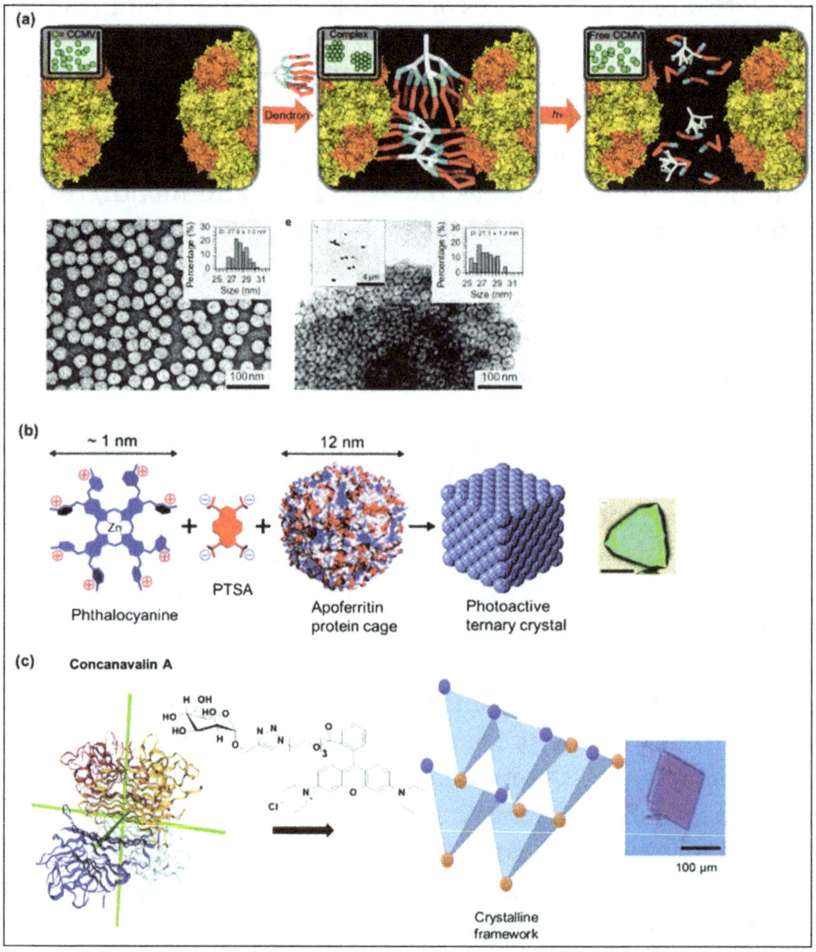

Protein crystalline frameworks. (a) Charge interactions of Newkome-type dendrons on CCMV surface patches and dissociation via photocleavage. TEM image of crystals are shown. (b) Photoactive ternary crystals formed from three-component systems with phthalocyanine. (c) Interpenetrating protein crystalline framework formed from concanavalin A and a dual interaction-SL.

Concanavalin A in combination with bismannopyranoside SL formed a diamond-like protein lattices. The three-dimensional crystalline array was confirmed with X-ray diffraction, as well as TEM which show distances of 6.9 nm between molecules of the PB. A protein crystalline framework was also prepared using concanavalin A and the SL with dual molecular interactions consisting of α-d-mannose and a rhodamine B. Interestingly, the degree of interpenetrability of the crystalline framework was tailored by the linker design. The resultant concanavalin A protein crystalline framework was characterized by X-ray crystallography which showed that the mannose-rhodamine B SL binds to each of the four monomers of concanavalin A in a 1 : 1 binding. AFM images further revealed a square crystal with flat surfaces and sharp edges with a height of 200 nm, which was also substantiated by TEM. In these assemblies, the crystallization process was mainly driven by entropy, i.e. the protein–sugar binding occurred first, followed by rhodamine B dimerization to drive the entire self-assembly process to completion. Calorimetric measurements, circular dichroism and fluorescence lifetime measurements were employed to determine the kinetics and mechanism of protein crystal formation.

Clearly, the combination of suitable PBs with SLs and recognition motifs allowed the synthetic customization of structurally defined, supramolecular PNs. In several instances, it was shown that the innate activities of the PBs are maintained or additional functions can be incorporated through appropriate chemical design of the SL. With the toolbox to engineer a variety of functional PNs in hand, the next step is to explore the applications of these synthetic PNs.

Functional Supramolecular Protein Nanostructures (PNs) and their Applications

Here, the applications of PNs are discussed highlighting the functional activity of the resultant PNs and potential applications, together with future perspectives of these exciting new materials.

PNs for Protein Delivery into Cells

Enzymes are emerging candidates for molecular targeting in diseased cells since they are usually highly specific in their mode of actions. However, applications are often limited by their low cellular uptake and proteolytic stability. The expression of fusion proteins consisting of enzymes and cell targeting entities such as antibodies (Abs), translocation domains of toxins and cell penetrating peptides (CPPs) has enabled cellular delivery of the enzyme cargo. Since the fusion takes place at the genetic level, it is not feasible to include chemically modified proteins or customize linkers equipped with various cleavage groups between the proteins in the design. Chemically engineered fusion proteins represent an attractive strategy to expand the repertoire of natural or recombinant fusion proteins particularly for the incorporation of post-modified proteins. Linker groups with pH- or light-cleavable groups allow the controlled release of the protein cargo in the microenvironment of diseased cells with external stimuli. In this context, dimeric and trimeric PNs have been prepared consisting of a pH cleavable linker connecting a transport protein with a cargo protein to achieve efficient delivery and controlled release of the protein cargo in cancer cells.

Heterodimer proteins consisting of streptavidin containing a polyamidoamine (PAMAM) shell and a cargo protein have been assembled by the biotin–streptavidin technology. As cargos, biotinylated cytochrome c, the tumor suppressor protein p53 or the C3 toxin, a specific Rho-A, B and C inhibitor, have been selected and connected to PAMAM-streptavidin. Both p53 and C3 are relevant for cancer therapy but they are not taken up by cells. The formed fusion proteins were internalized into A549 lung cancer cells by clathrin-mediated endocytosis due to the positively charged PAMAM dendrimer branches. The fusion protein consisting of PAMAM-streptavidin and p53 resulted in cell death via caspase 3/7 activation whereas the heterodimer comprising the C3 toxin enzyme induced changes in cell morphology through inhibition of the Rho A protein.

More sophisticated, trimeric fusion proteins have been achieved consisting of blood plasma protein HSA that was decorated with a PAMAM corona (HSA-PAMAM) to enable efficient uptake into mammalian cells such as A549 lung cancer and HeLa. As cargo, different protein enzymes such as

C3 toxin, β-galactosidase and the enzymatic subunit of the Clostridium botulinum C2 toxin, C2I were assembled with HSA-PAMAM by an avidin (Av) linker. pH-controlled release of the cargo protein was achieved by conjugating the pH-sensitive iminobiotin to the cargo enzymes to enable cleavage of the enzyme cargo from the avidin carrier under acidic conditions. Notably, intracellular release of the cargo proteins under acidic endosomal conditions was demonstrated, and both the internalized C3 toxin and β-galactosidase preserved their enzymatic activity in cells. However, these chemical fusion proteins were not specific to cell-types and, given their potencies; it would be desirable to deliver the enzymes specifically into cancer cells in order to avoid adverse effects.

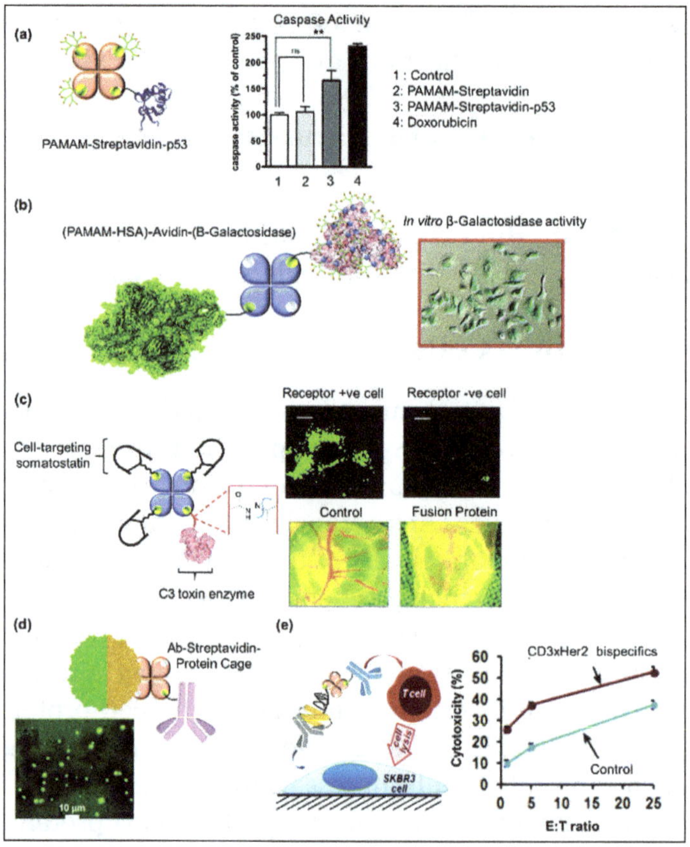

PNs for delivery of active proteins into cells. (a) Delivery of tumor suppressor p53 to induce apoptosis via Caspase 3/7 activation. (b) Intracellular delivery of β-galactosidase into A549 cancer cells. The blue coloration of the cells showed that the enzyme preserves its bioactivity in vitro. (c) Selective delivery of C3 into cell lines overexpressing somatostatin receptors was shown compared to a receptor negative cell line. The application of the fusion protein in chick embryo chorioallantoic membrane model showed an antiangiogenic effect. (d) The internalization of a fluorescently labelled LiDPS protein cage into the microbial pathogen, Staphylococcus aureus, was mediated by assembly with an antibody. (e) CD3xHer2 bispecific antibody target Her2-positive human breast cancer SKBR3 cells increase cytotoxicity due to T-cell mediated lysis.

To impart cell-type selectivity, peptide targeting entities recognizing membrane receptors at the surface of cancer cells would be highly desirable. A structurally refined PN was assembled combining a central avidin decorated with a single C3 toxin cargo somatostatin and on average three copies of the peptide hormone somatostatin that targets specific cancer cells such as lung carcinoma A549 that overexpresses the somatostatin receptors. As another level of structural complexity, a pH cleavable hydrazone group was introduced in the linker interconnecting Av and C3 toxin to achieve cargo release in acidic intracellular compartments such as endosomes. By redirecting cell surface interactions, intracellular uptake and cytosolic release of a therapeutically relevant enzyme by three somatostatin peptides was obtained. This protein assembly showed anti-angiogenic effects and 100-fold improved potency compared to the therapeutic Ab Avastin in an in vivo chick embryo chorioallantoic membrane model of advanced non-small cell lung cancer. The PN was also applied in combination with the marketed chemotherapeutic drug doxorubicin, which enhanced the efficacy of doxorubicin by about three-fold, both in vitro and in vivo. These examples demonstrate the versatility of constructing customized protein assemblies with structures that could not be created with the cellular machinery and unique biological activities.

Protein cages are often employed as platforms for transport of therapeutics or imaging agents by loading the cargoes into their cavities. They are attractive for passive targeting whereby larger nanoparticles show preferential uptake through the leaky vasculature of the tumour microenvironment by the enhanced permeation and retention effect. However, the passive targeting effect is often not substantial and offers only moderate transport to cancer tissue and cells. Consequently, strategies to attach antibodies to address membrane receptors overexpressed at tumor cells have been developed to enhance their therapeutic efficacy. A protein trimer consisting of an anti-Protein A Ab connected via streptavidin to the DNA binding protein from Listeria innocua (LiDps) was prepared by solid phase synthesis. Flow cytometry clearly showed that the Ab fusion protein mediated the uptake of the LiDps cage into the microbial pathogen, Staphylococcus aureus, which was otherwise not possible.

Moreover, the preparation of "bispecifics" that comprise two Abs or antigen binding fragments (FAb) able to simultaneously bind two different antigens have been introduced to develop highly specific and potent biotherapeutics. For example, Ab bispecifics are important for immunotherapy to direct T cells against cancer cells to eliminate tumors. Furthermore, the anti-CD3 (cluster of differentiation 3) Ab binds to CD3 on the surface of T cells and acts as immunosuppressive drugs to direct T cells against cancer cells. Both Abs, anti-Her2 and anti-CD3, were assembled on the Protein A–PEG–streptavidin adaptor. To form the bispecific Ab. Flow cytometry assays confirmed that the Ab bispecific construct was able to target both Her2-positive human breast cancer SKBR3 cells in combination with CD3-positive human peripheral blood mononuclear cells, which served as the effector cells. Therefore, the CD3xHer2 bispecific Ab had an effect on T cell-mediated lysis of Her2-positive breast cancer cells. It increased cytotoxicity by 15–20% compared to the control using a mixture of anti-CD3 and anti-Her2, which could be attractive for tumor-targeted therapy.

PNs as Nanorobots in Biology

In Nature, a variety of biomolecular nanoscale devices comprising of proteins have been formed and are known to play vital roles in important functions e.g. cell division and signal transduction. Inspired by such natural nanomachines that act in a concerted fashion, Nobel Laureate Richard Feynman envisioned smart nanomachines that could respond to molecular cues in a cellular context to regulate biochemical process or effect changes in the cellular biochemistry. Consequently, synthetic PNs have been devised that could serve as smart nanorobots in a cellular context.

Cylindrical, tube-like PN have been developed as "nanorobots" for delivery and programmed to specifically release cargoes in response to a biological signal. Compared to spherical structures, protein tubes provide several attractive features as both ends of the tubes are open, allowing material exchange from the inside to the outside by simple diffusion. By capitalizing on the innate feature of GroEL to assist the refolding of denatured proteins and high ATP concentrations inside cells, which range from 1–10 mM, intracellular delivery and release of a fluorescent dye using a GroEL protein nanotube was shown. Aida and co-workers prepared a boronic acid modified GroEL protein nanotube, which was taken up into human epithelial carcinoma HeLa cells through preferential binding to the glycoproteins and glycolipids at the outer HeLa cell membranes. The nanotube was used to incorporate the fluorescent dye cyanine, bound to denatured lactalalbumin via an esterase-cleavable linkages. It was applied for the delivery and controlled release of imaging agent or even drug molecules, in HeLa cells, in the presence of ATP and esterase. Moreover, the nanotube preferentially accumulated in the tumor tissue when compared to other tissues, except liver tissues, which is very attractive for in vivo applications.

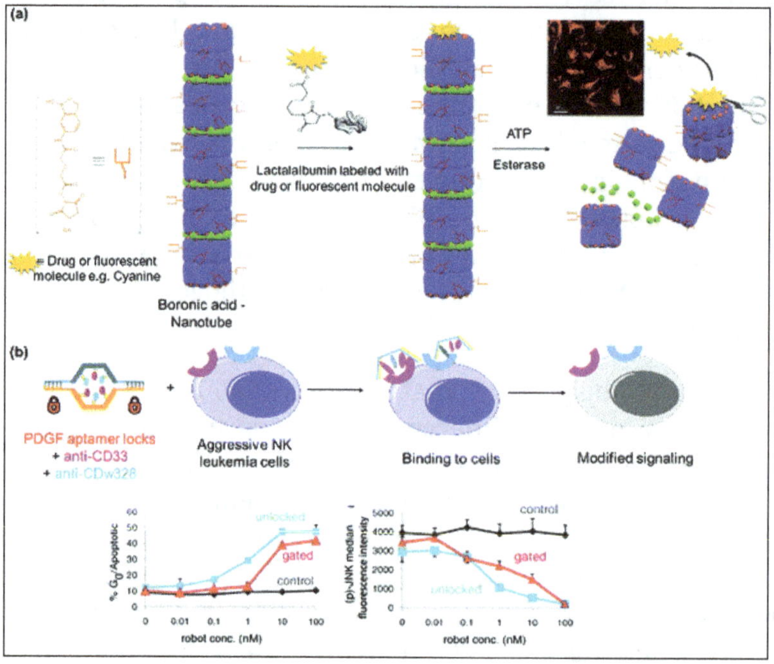

(a) Boronic acid-functionalized nanotubes deliver a dye conjugated to denatured lactalalbumin, which was released in the presence of ATP and esterases. The confocal image showed the uptake and release of the fluorescent cargo. (b) 3D nanorobot loaded with anti-CD33 and anti-CDw328 antibodies modified signaling in aggressive NK leukemia cells. The cell cycle distribution and phosphorylation level of Jun N-terminal kinase as a function of robot concentration after 72 hours was determined.

A 3D nanorobot able to transport molecular protein cargoes, such as a combination of antigen-binding fragments (FAb), a region on an Ab that binds to antigens, and respond to a variety of molecular cues in the cellular context to regulate the offloading of the FAb cargoes has been realized using DNA nanotechnology. A lock was designed in the DNA barrel based on aptamer recognition (Section 2.3.5) where two simultaneous molecular events have to occur to activate the opening, giving rise to a logical AND gate. In this manner, combinations of different molecular cargoes such as FAb and aptamer lock could be implemented to tailor the delivery and release of the molecular payloads in a highly selective biological environment. For example, the molecular cargoes containing Abs against human CD33 and against human CDw238FAb′, which induce growth arrest in leukemic cells, were loaded with precise spatial organization in the hexagonal barrel nanostructure. A pair of aptamer sequence 41t against platelet-derived growth factor was implemented in the design to address large granular lymphocytic leukaemia, aggressive NK type (NKL) specifically. A dose-dependent induction growth arrest of NKL cells through the suppression of Jun N-terminal kinase and protein kinase signalling was observed. On the other hand, incubation of a nanorobot consisting of Ab against human CD3εFAb′ and Ab against flagellin FAb′ led to the recruitment of flagellin at low concentration (100 pg mL^{-1}) and induce augmented T cell activation, which is a convenient tool to bring about changes in cellular behaviour in a controlled way.

PNs as Templates

Amyloid fibers consist of very stable β-sheet structures made up of self-assembling peptides. They are often employed in Nature as a synthetic template or as efficient storage modules. Based on this phenomenon, there have been various groups who used self-assembling peptide fibers as templates for nanoparticle growth, neuronal cell growth, stem cell differentiation, or for polymerization such as formation of poly(dopamine) on the surface of peptide fibers. However, there are not many examples known to date for PNs as biotemplates. One of such few examples includes using a linear streptavidin polymer as a template for biomineralization. The linear polymeric PN formed bundles with micrometer diameter and a millimeter length in $CaCl_2$ solution, due to chelation to the aspartate and glutamate side on the surface of streptavidin. Biomineralization of calcite microcrystals was accomplished using the streptavidin PN as a template in the presence of ammonium carbonate as a carbon dioxide vapor source. This process mimics collagen processing with a hierarchical assembly from nanoscale PB to millimeter supramolecular structures.

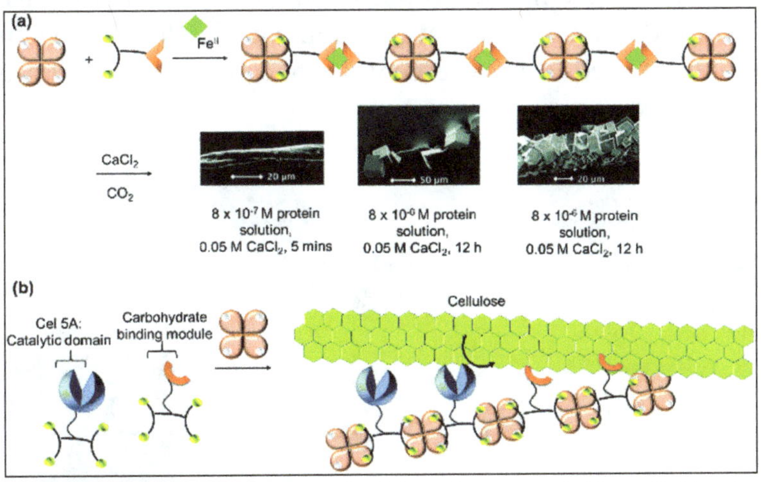

(a) Streptavidin-templated biomineralization of calcite microcrystals (from $CaCl_2$ solution) over time.

An artificial cellulosome was created using streptavidin–biotin polymerization as a template. Glycoside hydrolase family 5 (Cel5A) from Thermobifida fusca was selected as a model enzyme for biomass degradation to hydrolyse β-1,4-glycosidic bonds of internal cellulose chains in the amorphous region. The catalytic domain of the enzyme was fused with a microbial transglutaminase tag (Met-Arg-His-Lys-Gly-Ser; K-tag) and a tetravalent bis(bis(biotin-Gly-Gly-Gly)-Lys)-Lys-Gly-Leu-Gln-Gly ligand was introduced site specifically to interact with the cis dimer of streptavidin.

To mimic the natural cellulosome system where different types of enzymes are recruited through cohesin–dockerin interactions to optimize the hydrolysis activity depending on the biomass type, a heteroprotein complex comprising of Cel5A with a K-tagged cellulose binding module (CBM) from Bacillus halodurans was co-assembled. The resultant supramolecular complexes, $(Cel5A)_n(CBM)_m$ – $(streptavidin)_{n+m}$ demonstrated enhanced activity for saccharification compared to the self-assembled complexes of the individual components alone. Although the actual spatial order of the individual proteins in the 1D structure cannot be precisely controlled, the catalytic activity can be optimized by the loading ratio of the building blocks.

Tezcan and coworkers have developed 2D crystalline arrays based on Zn^{2+}-coordinated cytochrome cb_{562} variant where the Fe heme cofactor is replaced by zinc porphyrin and used it for the growth of platinum nanoparticles. Although this has not been applied in a biological context, the same group has recently reported the possibility to perform the self-assembly of cytochrome cb_{562} in Escherichia coli cells and such system could hold immense promise to induce nanoparticle growth in living cells.

Constructing Bioactive PNs in Cells

Most of the PNs demonstrated are carried outside the cells but in Nature, protein assemblies occur naturally in cells to form nanostructures or architectures that support the essential functions of organisms such as compartmentalized structures like organelles and nanovessels for molecular storage or transport. Hence, there is a burgeoning interest to translate the various strategies to achieve assembly in living cells to come up with entirely new structural and functional features. In recent years, this has met with some success by constructing PNs using genetic tools or synthetic molecules such as N-heteroaromatic quaterthiophene analogues. For example, genetically encoded amphiphilic block-domain proteins were reported to form cellular compartments or organelles in Escherichia coli. By introduction of non-natural amino acids, they can be further functionalized with chemical functionalities such as a fluorescent dye to confer additional properties to the artificial self-assembled organelles.

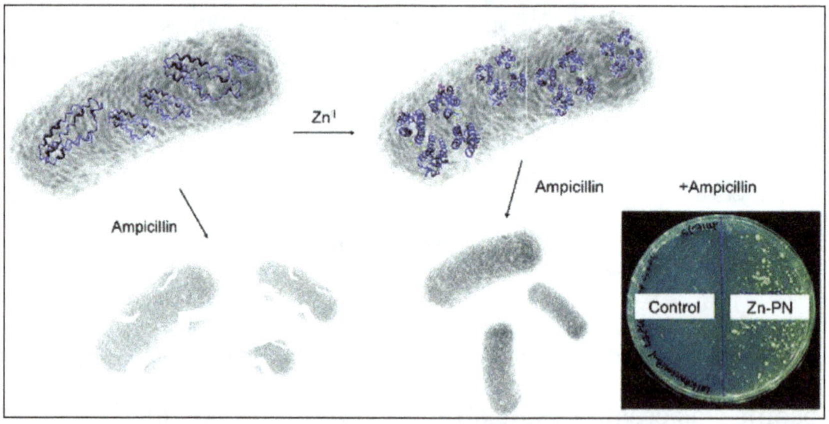

Self-assembly of tetrameric cytochrome cb_{562} variant in E. coli in presence of Zn^{2+} ions forming Zn-PN. The Zn-PN exhibited in vivo metallo-β-lactamase activity, even in the presence of ampicillin, a β-lactam antibacterial chemotherapeutic drug.

However, there are far less examples to construct PNs prepared in cells using synthetic approaches discussed herein. The only example until now involves an artificial metallo-β-lactamase based on the Zn_8PB_4 motifs where a tetrameric cytochrome cb_{562} was assembled by complexation of Zn^{2+} to the PB. Researcher investigated extensively the assembly of such PNs based on metal–ligand SLs and studied the effect of different amino acid mutations on the antimicrobial effect to derive the optimal cytochrome cb_{562} mutant. They next implemented assembly in Escherichia coli cells by including an N-terminal leader sequence in the cytochrome cb_{562} mutant in order to translocate them to and mature in the periplasm of the cells and performed the protein expression in LB media supplemented with 50 μM of $ZnCl_2$. The localization in the periplasm is instrumental for the formation of disulfide bridge between the monomers for subsequent binding to Zn^{2+}. The periplasmic contents were extracted, and SEC showed ~70% tetramer formation. The in vivo β-lactamase activity of the PN was evaluated and it was shown that it is functional in the periplasm and allow the Escherichia coli cells to grow with ampicillin (a β-lactam antibacterial chemotherapeutic drug) concentrations in the range of 0.8 to 1.1 mg L^{-1}, in contrast to a negative control.

As shown in the examples discussed here, there is a rapid development in devising functional PNs for various biological applications, ranging from simple biocatalysis, to biotemplating, drug/protein delivery, and even assembly of bioactive PNs in vivo to control cellular functions, which is an exciting prospect for the further development in this field.

References

- Nanoparticle-synthesis-applications-and-toxicity, applications-of-nanobiotechnology: intechopen.com, Retrieved 19, August 2020
- Functional-protein-nanostructures- a-chemical-toolbox: rsc.org, Retrieved 22, June 2020

Chapter 3

Medical Nanobiotechnology

Nanotechnology has a wide range of applications in the field of medicine. These fall under the umbrella of medical nanobiotechnology. It is used for the development of nanodiagnostic tools as well as drug discovery and development. The topics elaborated in this chapter will help in gaining a better perspective about the different aspects of medical nanobiotechnology.

The use of nanotechnology in medicine offers some exciting possibilities. Some techniques are only imagined, while others are at various stages of testing, or actually being used today. Nanotechnology in medicine involves applications of nanoparticles currently under development, as well as longer range research that involves the use of manufactured nano-robots to make repairs at the cellular level (sometimes referred to as nanomedicine).

The related field of nanomedicine is concerned with the medical applications of nanotechnology, such as the development of clinically useful nanomaterials, nano-sized biosensors, drug delivery using polymer-based nanoparticles and other methods, and molecular approaches to detecting and treating disease. Other problems for nanomedicine involve understanding the toxicity and environmental impact of nanoscale materials, as well as the use of nanotechnology to improve diagnostic imaging and for cancer treatment.

Nanodiagnostic Tools

An early or timely diagnosis of a disease is the key to recovery from that disease. Diagnostic techniques for early detection of diseases, particularly chronic and life threatening diseases, have been eluding the medical practioners for long time. When any pathological condition occurs in the body, the affected tissue undergo some biochemical changes. The clinical diagnostic tests actually see those biochemical changes through blood, or other biological samples. The most reliable and efficient diagnostic method is to directly visualize the affected tissues. Advances in medical, engineering, and more importantly in nanobiotechnology has given an impetus to growth in facilities to accelerate the disease diagnosis, to provide targeted drug therapy, to improve pharmacokinetics of the drugs, to increase specificity of the drug treatment, to minimize the surgical interventions, to identify relevant biomarkers, and to improve tissue imaging techniques for a specific medical treatment of a disease or disorder. Most of these recent advances are attributed to the advent of nanotechnology. So, when nanotechnology is applied in biological system, it is termed as nanobiotechnology. The various nano-sized materials have brought revolutions in diagnosis as well as medical and surgical treatments. Various nanomaterials (NMs) developed for the diagnosis of different diseases include metallic or magnetic nanoparticles (NPs), Quantum Dots (QDs), Silica nanoparticles (SNPs), Carbon nanotubes, silicon nanowires, nanopores, cantilever and graphene. The use of these NMs based nanodiagnostics is cost effective and provide the increased sensitivity.

NMs of different shapes and compositions are fabricated to meet specific purposes. NPs are spherical bodies fabricated from different inorganic and organic chemicals. Accordingly these are classified as Inorganic NPs and Organic NPs. According to the material used for its construction, these are called as Gold, silver, iron, copper, Silica, and in general metallic NPs under the inorganic category, and organic NPs include dendrimers, micelles, liposomes and ferritins. Quantum dots are also a types of NPs. Cantilever is a flat strip of Silicon carbide which is flexible. Nanowires are very thin fibers composed of carbon atoms. Graphene is a single atom thick sheet of carbon atoms. It is very light but very rigid and much stronger than steel. It has very good thermal conductivity. Rolling of a graphene sheet at a certain angle gives a tubular shaped NM called as Nanotube. From these different types of NMs, NPs, quantum dots, and cantilevers are being used extensively in diagnosis.

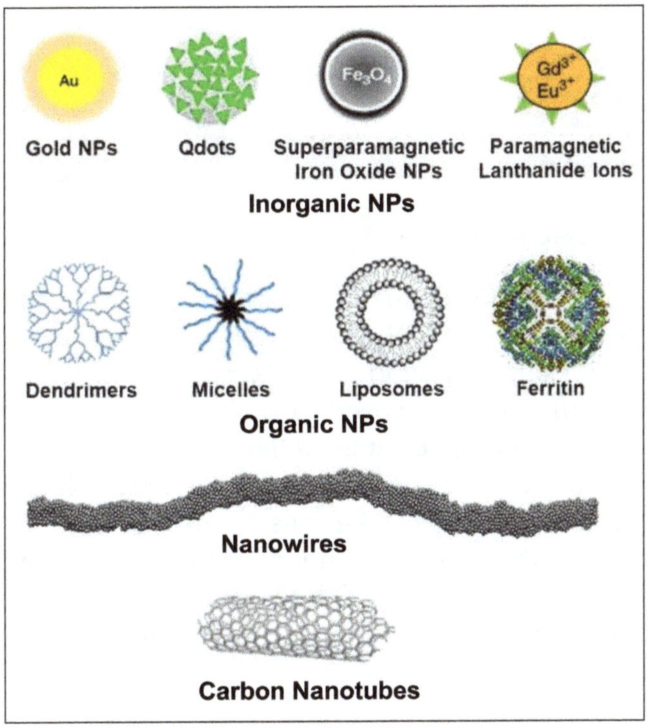

Some Nanomaterials used in diagnostics.

Nanoparticles (NPs)

These are classified into four main categories i.e., metallic nanoparticles (0D), metallic nanowires and rods (1D), metallic sheets and platelets (2D) metallic nanostructures (3D). The most extensively explored metallic nanoparticles (MNPs) are derived from gold and silver, though MNPs are derived from almost all stable metals in the periodic table. These can be synthesized and modified with appropriate functional groups that would allow them to bind with drugs, antibodies, ligands: substances of high interest in biomedical field. Over the last decade the nanoparticles such as magnetic nanoparticles (iron oxide), gold and silver nanoparticles, nanoshells, and nanocages have been continuously used and modified to enable their use as a diagnostic and therapeutic agent. A distinct property of MNPs is surface plasmon resonance (SPR), which is used widely in rapid tests for example, pregnancy test kits. MNPs are also used for surface enhanced Raman spectroscopy (SERS), which is an excellent assay method for high sensitivity and specific detection of low concentration molecules like biomarker detection for various diseases diagnosis. The fabrication and working of a typical MNP (iron NP, for example) is schematically shown in figure.

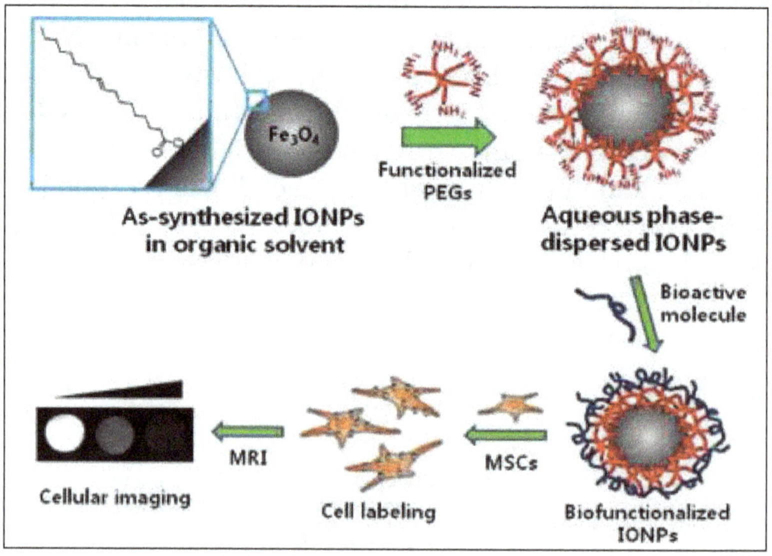

Fabrication and working of a typical iron oxide nanoparticles.

In step 1, the surface of iron NPs (IONPs), is functionalized with Polyethylene Glycols (PEGs) to make them hydrophillic as well as to make them versatile for appendage of suitable bioactive molecules. A variety of PEGs, for example amino-, thio-, carboxyl-, methoxy- and hetero-functionalized, are available to coat the NP surface. A specific PEG is chosen to append a specific biomolecule on the NP. In step 2, the aqueous phase dispersed IONPs are treated with a suitable bioactive molecule like protein, oligonucleotide, Abs, and others to make the IONPs biofunctionalized. In step 3, the biofunctional IONPs are injected into the cells. Inside the cells, they bind with the specific target and get localized. Finally, the cells are scanned through appropriate imaging technique like MRI to visualize the cells through detection of the localized IONPs. Functionalization of IONPs with specific radionucleotide, for example, Copper-64, Indinium-111 and Iodine-124 is used in imaging of different tumor cells.

Gold NPs (AuNPs) and silver NPs (AgNPs) are most widely used because of their interaction with light, radiofrequency and X-rays. On interaction with a specific wavelength these metals exhibit surface plasmon resonance (SPR), the oscillation of the free electrons in a particle's surface, which can be combined with conventional methods like colorimetry or absorption spectroscopy. AuNPs exhibit distinctive extinction bands in the visible region because of the SPR. The fluorescent color exhibited by these AuNPs is dependent on their size. Small sized NPs (about 30 nm) exhibit blue fluorescence and as the size increases the fluorescence wavelength increases and it appears red. AuNPs are fabricated in different sizes, and their application is guided by their size. AuNPs of size 10-100 nm primarily accumulate in all organs of the reticuloendothelial system (RES). AuNPs of size 15-20 nm may cross the blood-brain or blood-retinal barriers, and those of size 1-2 nm are highly toxic (possibility of irreversible binding to some key biopolymers). These NPs are rapidly cleared from the body by hepatobiliary system.

Silver is a historically significant metal used in medical treatments. AgNPs are one of the most heavily used nanomaterials in the world. Numerous shapes of AgNPs are synthesized depending on the applications. The commonly used NPs are spherical silver nanoparticles & metal particle dispersion. The colloidal suspensions of spherical AgNPs are bright yellow in color, typically showing maximum absorption around 420 nm. They are most commonly prepared by reduction of silver salt into zero valent state using reducing agents. AgNPs are found to have a lot of industrial applications. Their extremely large surface area of these both metals permits them to coordinate with a vast number of ligands in order to bind some biomarkers. The major biomedical diagnostic applications of AuNPs include molecular imaging in living cells, as biosensors and intracellular analysis, as biocatalysts and biomarkers, and in fabrications of medical devices such as catheters, prostheses, vascular grafts.

NPs have brought significant advancements, particularly in the diagnosis of cancer, which is otherwise diagnosed at a very late stage. In the first step, a dispersion of magnetic NPs is injected into the body. Secondly, it is distributed evenly in the body. Depending upon the surface proteins specific to the cancer cells, these get localized in the cancerous cells as shown at the third stage. Now here they play multiple roles depending upon their fabrication. If the location of the tumor is to be ascertained, the body is scanned through MRI. The concentrated NP being magnetic in nature will be easily detected and location of the tumor is identified. The other roles are that they release a specific drug in a controlled manner or they can cause hyperthermia upon incident of low intensity IR radiotion to kill the cancer cells without affecting the normal cells.

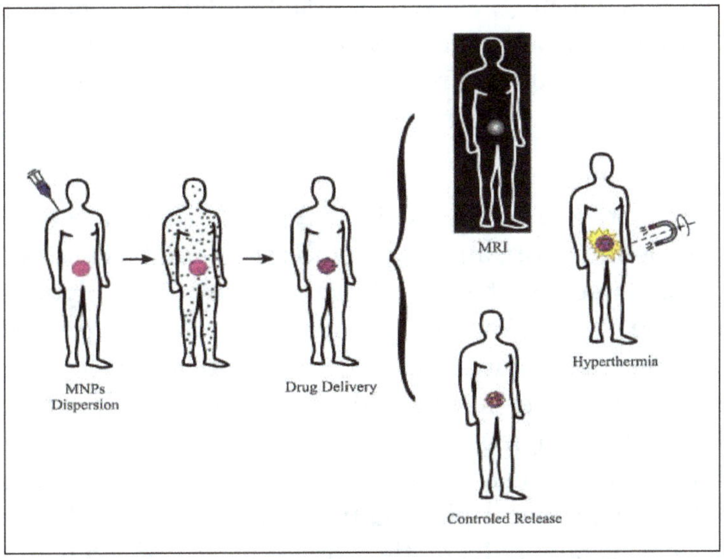

Application of NPs in cancer diagnosis and treatment.

Dendrimers

These are organic NPs, which are a group of highly branched spherical polymers and often referred to as “artificial proteins”. A monomer is polymerised in a controlled fashion around its own core to generate layers of mononers in the form of a branched polymeric network. The number of layers define the generation of dendrimer. A G4 (Four generation) dendrimer has three layers of polymerized monomers around the core monomer. By controlling the degree of polymerization, dendrimers can be constructed in various sizes, molecular weights, and chemical compositions. Due to the high loading capacity of payloads and the capability to control the polymer structure, dendrimers are favorable platforms for the construction of multimodal imaging probes.

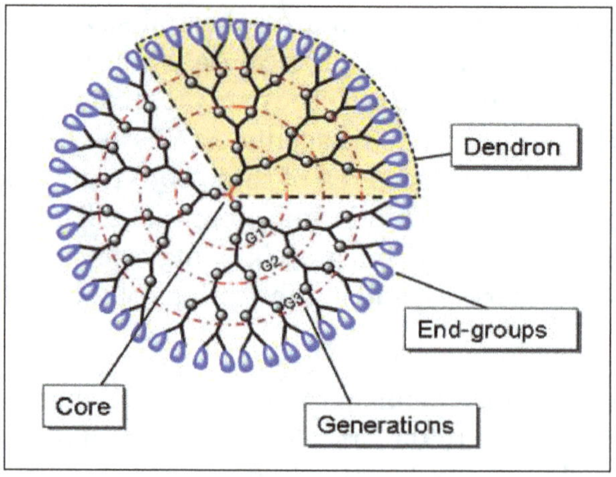

A four-generation (G4) denrimer.

Imaging moieties can be encapsulated in the internal cavity or on the surface of the dendrimers with high loading efficiency. Through the modification of cores, interiors, and surface groups of dendrimer, the properties of dendrimer-based imaging probes can be optimized to reach favorable physical characteristics, biodistribution, receptor-mediated targeting, and controlled release of the payload. Such modifications enable imaging probes to target the areas or organs of interest for facilitating target-specific imaging, as well as assume better excretion pathways that do not interfere with desired applications. Medium sized dendrimers (5 nm) are used as MRI contrast agents, and highly branched dendrimers are used in tissue engineering.

Micelles

These are another type of organic NPs used in diagnosis. These are self-assembling colloidal nanoparticles with a hydrophobic core and hydrophilic shell composed of surfactant (like sodium lauryl sulfate and different Tweens). These are functionalized with polymers and targeting molecules, and may be drugs too. These can passively accumulate in the areas with leaky vasculature such as tumors, inflammation, and infarction. In pharmaceutical development, micelles have been successfully used as carriers for water-insoluble drugs. Recently, polymer micelles are gaining an increasing attention for tumor imaging due to their high stability and good biocompatibility. A special group of polymeric micelles can be synthesized by the conjugation of soluble copolymers with lipids (such as polyethylene glycol-phosphatidyl ethanolamine, PEG-PE). The modified micelles can carry various specific targeted moieties on the surface, which may be suitable as imaging probes for multimodal tumor diagnosis.

Quantum Dots (QDs)

These are nanocrystals of size 2-10 nm, which contains a semiconductor core (usually selenides or sulphides of metals like Cd or Zn), enclosed within a shell of another semiconductor that glow or fluoresce when excited by a light source such as a laser. QDs have fairly broad excitation spectra from Ultraviolet to visible depending upon their size and composition. Smaller QDs (~2 nm diameter) gives blue fluorescence (380-440 nm), while larger particles (~5 nm diameter) give red fluorescence (605-630 nm). The QDs with the sizes < 5 nm diameter can be rapidly cleared by kidneys. Due to (i) broad wavelength range for excitation, and (ii) narrow and symmetric emission spectrum properties, QDs are well suited to optical multiplexing in which multiple colors and intensities are combined to encode genes, proteins, and small molecule libraries for imaging tumors.

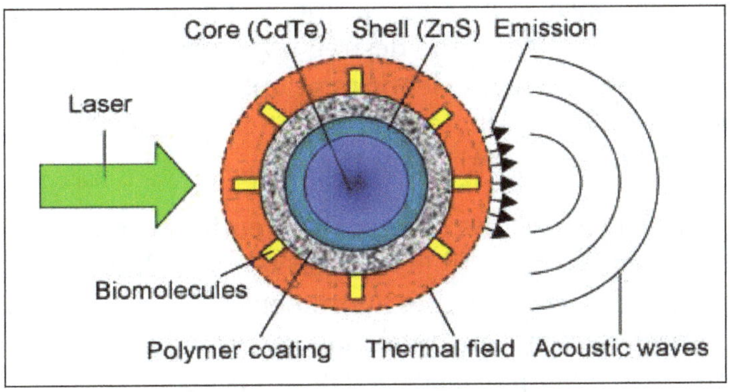

Basic construction of Quantum Dots.

The QDs offer the following advantages:

- Simple excitation, no need of laser.
- Simple instrumentation.
- Availability of red/infrared colors enables whole blood assays.
- High Sensitivity.

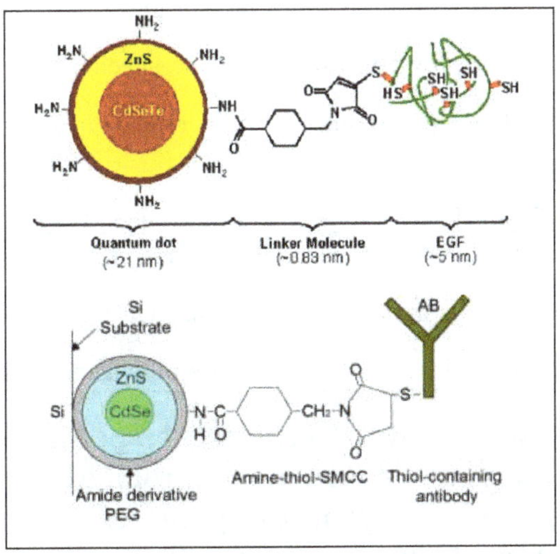

Specifically biofuctionalized Quantum Dots.

Only Hydrophilic QDs can be used in a biological environment. So, to make QDs biocompatible, different methods of preparation are used including silanization and surface exchange with bi-functional molecules, the hydrophobic side (e.g., a thiol group) binds to the ZnS layer of the QD and a hydrophilic side (e.g., a carboxyl group –COOH). Two such biofunctionalized QDs are shown in Fig. 6. The encapsulation of QDs within phospholipids micelles, polymer beads or shells, amphiphilic polysaccharides or block–copolymer micelles, which are composed of synthetic polymers containing hydrophilic and hydrophobic parts also enhance their applications to the biological system. The use of silica in coating of CdSe QDs results in sustained water-solubility of the QDs due to the multivalency of extensively polymerized polysilane. The interaction with the surrounding environment can be controlled by modifying the silica surface on the QD with different groups. Mercaptohydrocarbonic acids such as mercaptoacetic acid can be used for coating QDs to make them biocompatible. QDs are adapted to the desired application by conjugation to a recognition moiety, e.g., antibodies, peptides, oligonucleotides or aptamers, or by coating with streptavidin. Several functionalizations have been adapted to the ZnS layer coating the CdSe core, making the CdSe/ZnS QDs most adaptable for biological applications.

QDs emit the light for a lifetime of 5-40 ns, whereas conventional organic dyes emit in the range of 0.5 -2 ns, which results in a strong and stable fluorescence signal. A number of methods are used for the tracing and detecting the QDs including fluorometric detection, different types of microscopy like fluorescence, confocal, total internal reflection, wide field epifluorescence, atomic force and multiphoton microscopy. The technique to be used depends upon the type of application for which the QDs are being used. The different applications of QDs are summarized in Table.

Table: Applications of QDs.

Application	Description
Diagnostics	• Detection of Her2 (hairy-related 2) on SK-BR-3 breast cancer cells by humanized anti-Her2 antibody, a biotinylated goat antihuman IgG, and streptavidin-coated QDs. • Immunofluorescence labeling of mortalin using QDs show different staining patterns between normal and cancer cells. • Detection of ovarian cancer marker CA125 using streptavidin-conjugated QDs. • Fluoroimmunoassay for detection of prostate-specific antigen using streptavidin-coated QDs. • QD-based FISH labeling to detect specific repeats in Y-chromosome in fixed human sperm cells. • Antibody-conjugated QDs to detect prostate cancer cell marker PSA.
Imaging	• Imaging skin and adipose tissues in mice by injection of water-soluble QDs. • Mapping sentinel lymph nodes at 1 cm tissue depth using oligomeric phosphine-coated QDs. • Tracking diffusion dynamics of glycine receptors using QDs.

Growing concerns of toxicity associated with heavy metals, specifically Cd, Ni, Pb, Hg, have guided various legislations across the globe to recommend for prohibition of use of these metals in fabrication of anything related to household appliances, sports, leisure and other similar purposes. On the basis of same grounds, QDs are being fabricated without Cd, which are called Cadmium Free QDs. The latest trends are now in fabrication of Silicone based QDs.

Cantilevers

Cantilever are very thin flat flexible strips of Silicon carbide coated with probe molecules for a specific target molecule. When a biological sample containing target as well as other interfering non-target molecules are passed through this cantilever, the target molecules conjugate with the probe molecules. As a result the weight on the cantilever is increased and it bends. The extent of bending is quantitatively related to the concentration of the target molecules in the sample. A schematic construction and working of a cantilever is shown in figure.

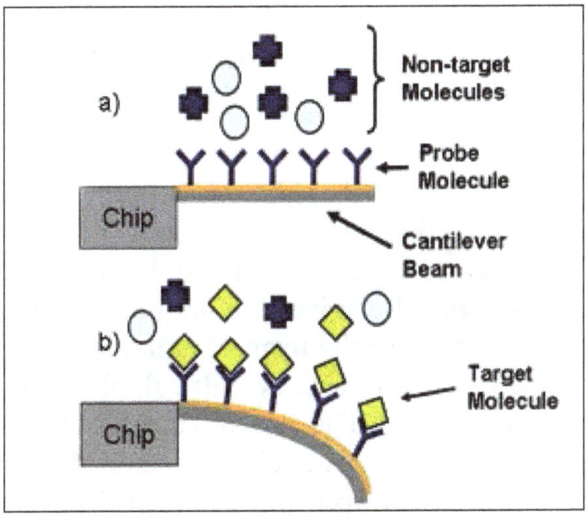

A cantilever beam with target-specific probe molecules in a sample lacking target molecules. (b) The same cantilever in a sample containing target molecules. The latter binds with the probe molecules causing a bend in the cantilever.

Cantilevers developed by Concentris GmbH (Switzerland) are beams of small size similar to those which are used in Atomic Force Microscopy (AFM) that utilizes nonmechanical deflection to screen biological samples for the presence of a particular genetic sequence. The various molecular events like DNA hybridization are monitored using the micromachined silicon cantilevers. The cantilever carrying a particular DNA sequence will bind to the specific target DNA sequence after scanning the whole sample and hybridization occurs with the single stranded DNA sequence present on the cantilever. The deflection of the beam takes place due to the surface stress by approximately 10 nm as an indication of hybridization. The binding event upsets the balance with the resulting nanodeflection proportional to the amount of DNA hybridized which results in an optical signal. Cantilever arrays can detect molecular targets without labeling target.

Cantilevers have been used for the detection of prostate-specific antigen through a single-reaction test in the presence other proteins. These may also be used for the detection of microorganisms such as Salmonella enterica, in-situ by using silicon nitride cantilever. These may also be used for the detection of cardiac troponins, an important class of biomarkers of myocardial damage. In addition, this assay technique could be used for breath analysis to detect acetone and dimethylamine.

Role of Nanobiotechnology in Drug Discovery, Development and Molecular Diagnostic

The application of nanobiotechnology has an impact on diagnostics and drug delivery. Currently, researchers are starting to use nanotechnology in the field of drug discovery and development. Nanocrystals (QDs) and other nanoparticles (gold colloids, magnetic nanoparticles, nanobarcodes, nanobodies, dendrimers, fullerenes, and nanoshells) have unique advantages for the development of drug and its discovery i.e. QDs and magnetic nanoparticles are being used for barcoding of specific analytes which are key components of the bio-barcode assay, can be future alternative to the protein chain reaction.

The term "Nanomolecular diagnostics" is the use of nanobiotechnology at molecular level to diagnose and treat chronic diseases", extending the limits of molecular diagnostics to the nanoscale. The nanomolecular diagnostic field can meet the rigorous demands of the clinical laboratory and can be cost-effectiveness.

Nanoparticles have emerged as promising nanoplatforms for efficient diagnostics and therapeutics possessing characteristic properties at the nanometric scale. The feasible immobilization of specific ligands on the surface of biological site have become ideal candidates for molecularly sensitive detection, molecular imaging, and novel carriers for targeted drug and gene delivery, targeted photothermal therapy. The nanoparticle in the form of nano-vesicle surrounded by a membrane or a layer holds convenient surface i.e. spherical, cylindrical, plate-like for molecular assembly of inorganic or polymeric materials and drug. The size and size distribution of nanoparticle becoming extremely critical and plays very significant role for penetration through a pore structure of a cellular membrane. The Biomarkers for occupying their distinguished color requires efficient fluorescent probes which depends upon the narrow distribution of average particle sizes. These narrow sizes particles emit a very wide range of wavelengths. The multifunctional nano particles core with several luminescent layers like magnetic nanoparticles can detect and manipulate the particles. Thus the Nanoparticles provides combine platform of biotechnology, nanotechnology and information technology facilitating molecular, biochemical and biological processes, e.g. genetics and pharmacogenomics.

The surface morphology of nanoparticles can be redesigned to reach at their target like blood brain barrier and dermal tight junctions more efficiently, overcoming the efficacy issues of drug on the physiological barriers. In case of malignant tumors Nanoparticles with leaky vasculature structure may penetrate the lesion. Nanoparticles may contain natural and synthetic polymer for coating an inorganic core of superparamagnetic materials. Some examples of nanoparticles as contrast agents in magnetic resonance for molecular diagnosis are given here.

Gold Nanoparticles

Gold nanoparticles are the most typically used nanomaterial in diagnostics; they have many different uses as properly. They may be extensively utilized as a connecting factor and to construct biosensors to hit upon ailment DNA. Rather than a fluorescent molecule, a gold nanoparticle can be attached to an antibody and other molecules, consisting of DNA, can be brought to the nanoparticle to supply bar codes. The gold particles are very chemically reactive and make excellent catalysts at nanometer scale. Recently, gold nanostructures have found a very important role in a wide variety of applications, including bio-imaging, drug delivery, toxic gas detection, biosensors and to demonstrate multiphoton absorption-precipitated luminescence (mail).

The Quantum Dots for Drug Discovery

Quantum dots (QDs) are regarded as a unique class of fluorescent labels, with unique optical properties such as high brightness and long-term colloidal and optical stability; these are suitable for optical imaging, drug delivery and optical tracking, fluorescence immunoassay and other medicinal applications. The unique optical property of QDs allows one to investigate the real-time dynamic events in living cells and such events include interaction between intracellular proteins, the mechanisms of intracellular signal transmission and cell growth. Some benefits and downsides have been investigated more suitable optical residences compared with natural dyes like incredible imaging outcomes using organic dyes, absorption of numerous drug molecules for an extended time frame.

Conjugates of Nanoparticles and DNA Protein

Conjugation Nanoparticle-DNA protein hold great promise in biomedical applications. Many Diverse strategies have been developed to conjugate nanoparticles and DNA proteins to assemble and purify nanoparticle-protein link. First, stable and biocompatible nanoparticles are synthesized. Conjugation of the nanoparticle to the DNA protein is then achieved via two different approaches that do not require heavy chemical modifications or cloning, cysteine-gold covalent bonding, or electrostatic attachment of the nanoparticle to charged groups of the protein. Co-functionalization of the nanoparticle with PEG thiols is recommended to help protein folding. Finally, structural characterization is performed with circular dichroism, as this spectroscopy technique has proven to be effective at examining protein secondary structure in nanoparticle-protein conjugates. In general, functionalization of NPs with proteins such as antibodies can be achieved by direct chemical covalent conjugation or electrostatic interactions. The Semi-artificial conjugates of nucleic acids and proteins by both covalent coupling chemistry or else through noncovalent biomolecular recognition systems, which include receptor-ligands of complementary nucleic acids for immunological detection assays.

Nanochips

The Nano Chip System by utilizing electronically enhanced hybridization of complementary DNA strands integrates advanced microelectronics and molecular biology into a platform technology with broad commercial applications in the fields of genomic diagnostics and has achieved 100% accuracy in the detection of nanoparticle. This technique helps in investigating DNA sequences or the pairing of separated strands of DNA with complementary DNA strands of the acknowledged collection that act as probes. Currently, DNA chips is known as DNA microarray assays that employs the energy of a digital contemporary that separates DNA probes to unique web sites at the array based on charge and size, then test sample (blood) can be analyzed for identifying DNA sequences via hybridization with these probes.

Microfluidics (Lab-on-a Chip)

Microfluidics is the modern science of fluids on the nanometer scale. The nanodiagnostics involve microfluidic or "lab on a chip" structures, in which the combination of numerous approaches of DNA analysis is mixed on an unmarried chip composed of an unmarried glass and silicon substrate. Academically, it is a subdiscipline of fluid mechanics, as the fundamental equations describing the physics of fluids at larger length scales are identical to the equations underlying microfluidics. In general, the microfluidic biosensor platforms offer numerous advantages compared with other traditional methods such as ultracentrifugation, electrophoresis, nuclear magnetic resonance, chromatographic approaches, etc. for detection of bio-species molecules. This device have capable of measuring aqueous reagent and DNA-containing answers, mixing the solutions together, amplifying or digesting the DNA to shape discrete merchandise, Destiny possibilities for the utility of nanotechnology in healthcare and for the development of personalized remedy seem like amazing.

Nanobiotechnology in Designing of Drug Delivery System

Among the new technologies, nano-biotechnology has evoked considerable interest for application in the pharmaceutical industry. Important applications of nano-biotechnology are in the areas of drug discovery, drug development, and drug delivery, and these are collectively referred to as nano-pharmaceuticals. Nano-biotechnology, particularly the use of nanoparticles, has made significant contributions to drug discovery and development. In addition to the use of nano-biotechnology for drug discovery, some drugs are being developed from nano-materials. Well-known examples of these are dendrimers, fullerenes, and nano-bodies. At present time increasing the use of nano-biotechnology in many pharmaceutical and biotechnology industries is anticipated. In case of drug development from formulate to appropriate dose to administer with optimal delivery systems, nanotechnology is being utilized at all the stages. Nano-biotechnology applications are also involving in diagnosis of diseases. In future it may be possible to computers are connected with nano-biotechnology systems and provide the appropriate and complete knowledge about complete model or an individual cell. This virtual representation might be helpful for researcher or scientist to develop novel drugs with high rate of accuracy and precision without conducting any experiment in living organisms.

The major challenges in front of drug delivery scientists are poor solubility, high molecular size and low bioavailability for clinical candidates. Other challenges in this field are pediatric and geriatric drug administration, protein and peptide drug delivery etc. Today's major demand in drug delivery field is to develop ideal, safe and effective, non-invasive drug delivery methods.

Nano-biotechnology sector play a significant role to overcome the above drug delivery problems. It provides the following solution related drug delivery problems: (a) With the help of this technology particle size of drugs is reduced in nanometer size range, it enhance the surface area and ultimately improve rate of dissolution, (b) Nano-meter size range of drug also useful to improve their solubility, (c) With the help of this technology scientists trying to develop noninvasive routes of drug administration method which can eliminates the use of injectable drugs, (d) developed nanoparticle formulations has better alternate for non-stable and lower shelf lives formulations, (e) nanotechnology based formulations improved the solubility of poorly soluble drug and enhance absorption capability, improved bioavailability and release rate of large molecules, reduced the optimum dose and enhance the safety margin by reducing the side effects, (f) Nano-biotechnology principles help in developing of Sustain and controlled release formulations with better patient compliance.

Nano-Biotechnology in Molecular Diagnostic

Nanotechnology has gained enormous popularity in recent past decades. In lay man language nanotechnology is defined as the use of least possible input to get maximum possible output. Nano word is used in terms of size. This technology is considered as wonderful amalgamation of physics and chemistry. When this technique is applied with biology, it brings about a new field termed as Nano-biotechnology; where biochemical processes are modified to get far better results than that of simple bio technological procedures.

As far as the medical field is concerned, the harness of nano-biotechnology for diagnostic purpose is successfully obtained. It assists to develop more sensitive diagnostic kits than that of existing one. Such instruments are suitable to probe the bodily problems at cellular pores and receptor level. Moreover, it is due to tiny sized diagnostic materials the degree of toxic and adverse effects have been markedly reduced. New dimensions of diagnostic tools have been explored with the aid of this cutting edge technique. Several modifications are still under developmental phase.

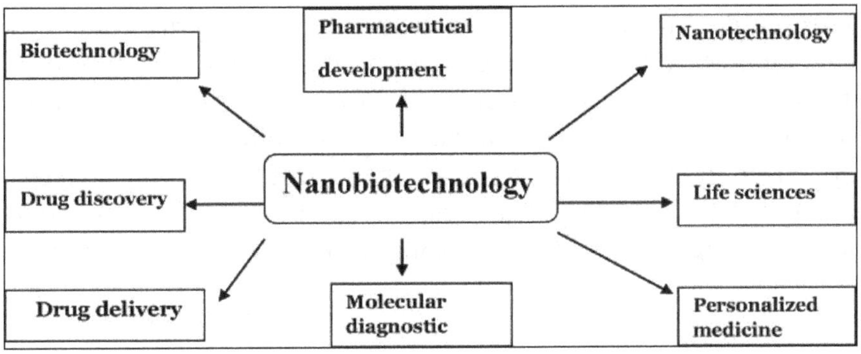

Application of nanobiotechnology for molecular diagnosis.

Nano-particles used in the Diagnosis

Gold nano-particles are also termed as metallic particles. These are produced from gold salts of either organic, aqueous or both origins. A suitable stabilizer is used to get stable particles with good ligand binding capacity. Usable size of particles is ranging between 3 and 100 nm. The major application is due to their electronic, optical, and thermal properties.

Magnetic nanoparticles are formulated from magnetic materials like Fe_3O_4, Fe_2O_3, and many other ferrites. Nanoparticles can be incorporated with bio- marker moieties so that they can be utilized to investigate various biomolecules and help in different processes like separation and purification. The involvement of surface coating materials is significant to ascertain the size and kinetics of these particles. That's why the nature of coating materials should be examined before use.

Quantum dots are semiconductor nano-crystals that are easy to synthesize and have characteristic properties that are between those of bulk semiconductor and discrete molecules. Their diameter ranges from 2 to 10 nm. Their fluorescent property relies on size of the quantum dot.

Carbon nano tubes are composed of graphite. On the basis of number of graphite, these nano tubes are categorized into two classes- tubes with only one layer of graphite are known as mono walled carbon nano tubes on the other hand, tubes consist of multiple layers of graphite are termed as multi walled carbon nano tubes. The main advantage of such tubes is that they can conduct high electricity with less or no heating effect. This happens because of scattering free traveling of electrons throughout the tube.

Liposomes are spherical vesicles where an aqueous core is surrounded by a phospholipid bilayer and cholesterol. The phospholipid consists of a hydrophilic head and two oil-loving tails. The phospholipid that is predominantly used is phosphatidyl choline.

Dendrimers are the emerging and well defined polymeric architectures that are known for their large, complex and well defined structures, versatility in drug delivery and high functionality whose properties resemble with biomolecules. These nanostructured macromolecules have potential abilities to entrap and to conjugate the high molecular weight hydrophilic/hydrophobic drugs by host-guest interactions and covalent bonding (prodrug approach) around an inner core respectively. They have different features like size, shape, branching length, and their surface functionality vary over the design of nanoparticles. For example, Polyamidoamine are usually used dendrimers.

Nano biosensors are applied to investigate biochemical changes with the aid of electronic, optical or magnetic technology. Furthermore, detection and/or quantification of biomolecules like specific base pairs or proteins are also possible. Majority of biosensors work on the principle of key and lock theory or affinity based mechanism; where immobilized tool attaches to target molecule/analyte being sensed in this phenomenon, any change at a localized surface can be examined rather than detecting target in solution. This variation can be sensed by using one of the following methods; viral biosensors, light sensitive biosensors, resonant cantilever and quartz crystal microbalance.

Nanotechnology used in Diagnosis

Nanotechnology-based Biochips/Microarrays

A nanomaterial can be perfectly examined by using same sized probe. This hypothesis is applied in molecular diagnosis with the help of biochips or microarrays. This is because the cell organelles/biological moieties exist in nano size and the biochips also falls under in almost similar tiny sized scale. Nanofluidic arrays and protein nanobiochips are the examples of nanotechnology based biochips. These chips are capable to isolate and analyze living cell molecules such as genetic material DNA. In future this technique will certainly bring about revived investigation tool for cancer. Nano-fluidic technology is expected to have broad applications in systems biology, personalized medicine, pathogen detection, drug development, and clinical research.

Nanotechnology-based Cytogenetic

Cytogenetic has been used mainly to elaborate spatial arrangement of the chromosome and screening of abnormalities related to disease. The use of fluorescent in situ hybridization (FISH) is now reaching its limit. Molecular cyto-genetics is now enhanced by use of biomedical nanotechnology, e.g., use of atomic force microscopy (AFM) and quantum dot (QD) FISH.

Nano-proteomic-based Diagnostics

A comprehensive study including identification, characterization and analysis of a series of bio protein at a specified time is known as proteomics. Proteomics help to recognize the variations which are associated with various pathological conditions. These variations are identified on the basis of spatio-temporal position of protein molecules. A small change in molecular level can bring about such variations as a consequence proteome may be altered; which are further quantified for clinical diagnosis. Gel- based and gel- free are the common proteomics techniques, which work on separation, observation, analysis and marking the altered proteome.

Nanoparticle-based Nucleic Acid Diagnostics

Screening of pathogenic microorganism on the basis of genomic arrangement has become relevant in current clinical determination. It has been well established central dogma of life that each and every living organism has a peculiar set of DNA, which has been transferred from one generation to next without any sort of alteration; this can also be said to be the individualization by nature. In clinical diagnosis, this DNA fragment may help to check the availability of the microorganism in the test sample. This investigation opens the door for molecular detection. This is carried out by using polymerase chain reaction (PCR) or hybridization techniques, known as amplification method and non-amplification method respectively. Therefore, this technique is considered to be more sensitive and precise than that of other techniques.

Nanobiosensors

Biosensors are sophisticated probes, which are used to investigate the existence and/or concentration of a biological analyte, say for example bio molecule, histology of biological material or any microorganism. Biosensors comprise of following parts: amplifier that detects the analyte and generate a signal along with it a signal transducer that converts the signal into electric impulses and a reader to record and investigate the signal.

Safety Issues of Nanoparticles

Nano-biotechnology is the fine amalgamation of multi disciplines of science. In pharmaceutical field, the scientists and researchers are trying to overcome the demerits of existing drug molecules with the aid of this wonderful technology. There are many expected potential benefits in bank. In spite of that safety is the main concern with the in- vivo application of nano engineered medicines yet.

It has been revealed that while production or use of nano medicines, such particles can easily get inside the human body and ultimately blood stream and then in various vital organs; where they exert unintended adverse effects, sometimes these effects are fetal for the user.

An in-vivo study on monkeys and rats have disclosed the aggregation of carbon and manganese nano materials in the olfactory bulb, which has produced severe pathological consequences pertaining to nervous and respiratory system. Several animal studies have shown such sort of abnormal pathological disturbances.

In order to combat with this problem firstly, an efficient probe will have to develop for thorough assessment of risk associated with the use of nano medicine. Second and the most significant challenge would be to develop authentic validation protocol for testing nano medicines in animals so as to get the data of toxic effects in early stage of testing. A restructured, compact and integrated regulatory approach is much needed to look into the expected risks for delivering the promising and safe medicines.

Regulatory Perspective of Nanobiotechnology

In the event that we talk about the most recent decade, nano bio-innovation has generated some new impedance for the research community, industry, and regulators. Applicable techniques have some loopholes regarding fabrication and processing of nano medicines, that's why special provisions would be required to tide-over such complications. FDA and other regulatory authorities experience many unavoidable consequences for the release of such products. This is due to non-authentic validation program in clinical trial phase. A wing of researchers proposes sturdy contentions, revealing both positive and negative aspects pertaining to the advancement of nano-materials. They suggest that vague protocol throttle research and product development, which represses the development and viability of new nano-medicines. Apart from that, a robust administrative body is made so as to push the business advancement by creating certainty for pharmaceutical firms and trust within the ultimate consumers. Currently, the FDA, EMA (European medical agency), and other regulatory agencies examine each new nanoparticle individually. There is commonly an absence of benchmarks in the assessment of nano-medicines as a unique class of therapeutic agent. Afterwards, administrative corporations might imagine of a complete listing of screening and a powerful approbate system that enclose the complete variety of particle characterization, pharmacology, and toxicology troubles.

Future Prospects

In case we consider in trendy approximately the two major capabilities of nanobiotechnology, first is imaging and diagnostics (quantum dots) and second are proteomics. In proteomics especially study and detect about nucleic acids and proteins. Currently, the Thermal Sensitive drug delivery system has been developed for the treatment of cancer. In this, drug is given in the injection form, which reaches inside the body and converts into insoluble form and accumulates in the tumor cells because the temperature of the tumor cell is more than the normal cell. In spite of the tremendous credibility of nanobiotechology, its use has been considerable due to its toxicity and environmental problems. Despite all this, the drug delivery, drug discovery, gene therapy, molecular imaging, biomarkers, and biosensors fields are exploring possibilities in the nanobiotechnology. Nanobiotechnology has brought a bright future for the clinical diagnosis and targeted drug delivery sector. Now in the field of diagnosis, it has become easier to detect diseased cells rapidly, due to which it is possible to prevent diseased cells from spreading in the body. There is a constant development in this field right now and it is possible that some incurable diseases in the future will actually be cured with the resources of nanobiotechnology.

References

- Medical-nanotechnology-and-nanomedicine, anesthesia-analgesia-11000: lww.com, Retrieved 02, March 2020
- Role-of-nanobiotechnology-in-drug-discovery-development-and-molecular-diagnostic, applications-of-nanobiotechnology: intechopen.com, Retrieved 23, July 2020

Chapter 4

Food Nanobiotechnology

Nanotechnology can be used within the food industry for various purposes. A few of them are using nanobiosensors for increasing food safety, nanoformulations and using nanotechnology for food packaging. This chapter closely examines the different applications of food nanobiotechnology to provide an extensive understanding of the subject.

Nanotechnology is a distinct branch for production of sub-micron systems. Due to increase in human population, the global demand of food, feed and energy has increased, so there has been an enlarged attention in using nanotechnologies in foodstuff sector, with spanning applications in this sector. Potential uses of nanotechnology have been identified by scientists and trade stakeholders in practically every segment of the food industry. The applications begin with agriculture where different aspects are covered such as pesticide, fertilizer or vaccine delivery; animal and plant pathogen detection and extend to another track of food processing where encapsulation of flavor or odor enhancers; food textural or quality improvement; new gelation or viscosifying agents. It also has its diverse application in food packaging covering pathogen detection, gas or abuse sensors; anti-counterfeiting devices, providing UV-protection, and films made from different polymers that are impermeable and finally to bioactive supplements which covers nutraceuticals with higher stability and bioavailability. Various nanotechnology tools are being used to improve the food quality as well as to increase the bioavailability of many food components. In food systems, the application of nanotechnology provides new insights for advances in the safety and nutritional significance of food products. The different aspects such as safety, efficiency, bioavailability, and nutritional aspects as well as the molecular synthesis of new products and ingredients are affected at the nanoscale level of food. By the use of nanotechnology even in food sector different new functional materials in a range of microscale and nanoscale can be developed and such design of methods and instrumentation in food sector may also benefit. The applications of nanotechnology in food sector are:

- Food Safety Nanobiosensors.
- Nutraceutical Nanotechnology.
- Food Nanopackages.

Nanobiosensors for Food Safety

Food Safety: Global Public Health Concern

Food safety is a critical and persistent public health issue. The concerns associated with food safety are further intensified by improper hygiene, poor food handling practices, and contaminated food supplies, leading to a financial burden of foodborne disease (FBD). The FBDs are often linked to consumer illness, which bears high medical costs and loss of productivity and sales. To combat the threat of FBDs, an increased and comprehensive awareness of food safety is of paramount importance. The safety of the food enormously influences consumer health. There are several factors that ensure the safety of processed and packaged food commodities from pathogenic microorganisms, such as Listeria monocytogenes, Escherichia coli O157:H7, Toxoplasma gondii, Campylobacter jejuni, Salmonella, Staphylococcus aureus, Campylobacter coli, Bacillus cereus, Norovirus, and numerous others that can deleteriously impact human health. Hence considering the roles that food safety plays in both health and development, relevant actions have been taken by various agencies in different nations to improve the safety of the food supplied to the consumers. In the United States, the US Food and Drug Administration (FDA) and the United States Department of Agriculture (USDA) are key agencies providing food safety guidelines, standards, and policies to US entities and many other nations. The European Food Safety Authority and the Food Standards Agency in the United Kingdom were authorized to survey the quality and safety of foods sold in stores. In addition, they are also assigned the responsibility of performing research in food safety and, therefore, plays crucial roles in uplifting the food safety scenario in the EU, the UK, and across the globe.

Various foodborne diseases cause significant morbidity and mortality worldwide. Historically, efforts to reduce the life-threatening consequences of food contamination have been made by innovation in food preservation. Sun drying and cooking were conceivably the first methods used; later more sophisticated technologies, such as fermentation and canning, came into existence. In recent times, advanced technologies in food preservation and packaging have made food safer. As the global population is increasing, scientists with innovative approaches towards science and technology are working efficiently to provide the best quality and safest foods to consumers. In between these approaches, nanotechnology provides an advanced and powerful platform utilizing unique properties of materials emerging from nanometric size (1–100 nm) that have the prospect of revolutionizing agriculture and food sectors, biomedicine, environment safety, energy conservation, and many other areas.

Food Safety: A Challenging Field for Nanotechnological Innovations

The far-reaching implications of nanotechnology in the twenty-first century confer substantial impact on health, industry, and the global economy. They possess the potential to maximize agricultural productivity and food quality through proper management of processing and packaging within food production facilities. Apart from the food-related issues, this technology can also be implemented in the areas of toxic sensing and prevention by exploiting unique properties of nanomaterial. The integration of nanotechnology in electrochemistry, photosynthesis, and photo-electrochemical cells along with other modern technology to address the growing need for safe and renewable energy resources continues to provide high-impact solutions to these sectors. Nanotechnology plays a major role in all stages of the food production chain starting from the initial agricultural production to food processing in which food ingredients can be encapsulated within nanomaterials to achieve enhanced functionality. In addition, nanotechnology has been explored tremendously in food packaging. For example, the food-packaging sector has seen many transformations in barrier perfection with the advent of various nanoscale fillers, which have also resulted in reduced effects of accelerating factors due to contamination and spoilage. In addition to packaging, preservation of food materials also holds great potential in the application of nanotechnology. All these issues are regulated by different segments of nanotechnology, of which nanosensors often play a crucial role.

There are various factors that affect a country's economic conditions, of which agri-business is one of them. In today's world, many nations encounter pressing challenges, including environmental issues, such as the build-up of fertilizers and pesticides, loss of viable use of natural resources, urbanization, and climate change. The field of nanoscience has the potential to help farmers to maintain their farms with precise control. Biomass conversion and material science technologies can be applied to develop proper handling of fertilizers and chemical pesticides. Through the use of nanomaterial-based nanosensors such as surface plasmon resonance (SPR) consisting of gold nanoparticle, quantum dots (QDs), and clay nanotube, farmers can detect pests, plant viruses, evidence of drought, and levels of soil nutrients, to name a few. Nano-encapsulated fertilizers that are released slowly can be used for the further consumption process and to minimize environmental pollution.

Smart packaging is categorized under an advanced generation of packaging, which incorporates nanosensors that react to the physical or chemical change in food samples within the packaging to delay spoilage or contamination. Various assets, such as mechanical, thermal, and optical features of nanosensor-possessing nanocomposites, are used to enhance the properties of foodstuffs to extend shelf life and maintain freshness and quality by regulating the route of unwanted gases and moisture through the packaging material. Nanosensors in the form of an electronic nose or tongue can also be used for detecting chemicals released during spoilage of food, by placing them in direct contact with the packaging material so as to ensure better quality. These nanosensors are in the form of "buttons" placed upon the packaging materials, which help in detecting the change in color. This technique can also be used to indicate whether or not packaged food is stored under optimum conditions.

Biosensors, which are increasingly integrating nanotechnology tools, is a portable sensor incorporating molecular recognition elements derived from biological materials, such as nucleic acid, antibodies, cells, or enzymes with suitable physicochemical transducing mechanisms. These biological sensing elements are either integrated within or associated with transducers, which transform physicochemical interaction into a decipherable form with the help of a transduction and electromechanical interpretation, depending upon target groups of analytes. A conventional biosensor consists of three components: bioreceptor, transducer, and detector. The first component, bioreceptor, acts as a prototype for the detection of the analyte. The second component is the transducer system, which is an interface that determines the energy that arises with the reaction at the bioreceptor and then transforms that energy into the detectable electrical output. There are various elements of the transducer system, mostly optical, electrochemical, thermometric, piezoelectric, magnetic, and other ones. The third component, the detector system, amplifies signals and analyses them in a microprocessor. Data are transformed into concentration units, which are then transferred to a data storage device.

Biosensors can be used to identify a broad array of targets, ranging from protein molecules to large pathogens. Compared to the traditional methods, biosensors can provide results more rapidly than other culture-based methods, making them vital to the lab and real-world applications, promoting robustness, sensitivity, accuracy, real-time assay, and ease of use. The innovation of sensors is essential to the agrofood sector, as any incorrect result can lead to an extensive economic loss along with the loss of consumer confidence.

Nanosensors

The increasing application of sensors in different sectors, including medical diagnostics and therapeutics, food, air, and water quality monitoring, is evident. The word "nanosensors" is an integration of two segments: nano, which suggests a size range of 10^{-9} unit of the scale, and sensors, which is derived from a Greek word Sentire meaning to perceive. Broadly, it is a device that ultimately converts a physical stimulus such as a thermal, luminescent, electric, or magnetic effect into an electrical signal. Sensitivity, selectivity, resolution, stability, and calibration characteristics are some of the crucial parameters to be considered during their functional operation in any field of the response.

General classification of nanosensors can be done in two major groups. One group has nanoscale dimensions that are further subgrouped into two types that is sensors whose three-dimensional properties are in nanometric proportions, like QDs resembling fluorescent nano-sized semiconductor crystals and thin films of any metal oxides for gas sensing having one nanoscale dimension. The other group is of nanoscale measurements but does not have any particular nanoscale dimensions. Further classification of nanosensors can be as active nanosensors that require a source of energy such as a thermistor (a thermally sensitive resister) and passive nanosensors that do not require the energy source like a piezoelectric sensor (a sensor that uses the piezoelectric effect to measure the strain, force and pressure). Another broad classification can be according to the signal detected in the form of energy, such as a physical, chemical, or biological effect. Physical nanosensors comprise mechanical, optical, and electromagnetic nanosensors, which are used for measuring properties such as force, pressure, density, viscosity, absorbance, fluorescence, luminescence, dielectric constants, and permeability. Although chemical nanosensors are meant to identify any foreign chemical particle, concentrations of certain substances, and pH, biological nanosensors (biosensors) have other essential purposes.

Optical Nanosensors

Optical nanosensors use optical signals to transduce biological or chemical sequences for analytical purposes of all particles with dimensions less than 1000 nm. They include the option of measuring in small volume, high specificity, low toxicity, and other interferences similar to water on the sensors properties. The first optical nanosensor was based on fluorescein entrapped in a polyacrylamide nanoparticle, which was used for pH sensing. The fluorescent molecules were composed of at least one photoactive element and a substrate unit. In the process, a light of particular wavelength was absorbed, which was followed by the emission of quanta of light possessing an energy equivalent to the difference between the excited state and the ground state, also known as luminescence phenomenon. This phenomenon is preferred for nanoscale chemical analysis and biological or environmental applications. An optical receptor or a sensing element is integrated into the optical nanosensor consisting of an inert matrix for detecting the properties of the assigned analyte. Currently, fluorescence is the only transduction process due to the high relative ease of measurement and sensitivity.

Fiber-optic Nanosensors

Optical nanosensors usually consist of a chemically or biologically sensitive layer, which can either be made up of chemical recognition element or biological recognition elements covalently attached to the optical transducer. This transduction mechanism holds a very attractive platform for chemical sensing application, among which minimal invasiveness is the most substantial technique for cell monitoring. The small size, light weight, exemption to electromagnetic interference (also called radiofrequency interference), and easy configuration of the optical fibers used in the nanosensors are some unique characteristics that enable sensing of biological and physiological processes in the corresponding fields. The interaction between the surface of a target analyte and the receptor is such that a physicochemical perturbation can be converted to a measurable electrical signal that is received by the optical probe, and then the signal is transmitted to the database. The sensing strategy for fiber optic nanosensors is based upon two methods; one is a direct method in which intrinsic optical properties of target analytes are measured. The other is an indirect method in which optically detectable bioprobes are monitored. It is also classified into two broad categories that fiber-optic chemical nanosensors and fiber optic biosensors.

The first fiber-optic sensor in the field of food science was designed for detection of E. coli O157:H7 from the complex matrix of assigned food material. Analyte 2000 is a core type fiber-optic sensor designed by Research International (Monroe, WA, USA) and is widely used for the detection of various pathogens in a sandwich form using fluorophore-labeled antibody including E. coli O157:H7, L. monocytogenes, Salmonella enteritidis, and staphylococcal enterotoxin B. A slightly different fluoroimmunoassay utilizing optical fiber probes consisting of polystyrene was applied to detect E. coli O157:H7. During food packaging, a sensitive coating of syndiotactic polystyrene, a new crystalline engineering thermoplastic with a boiling point of 270 C high heat and chemical resistance with good dielectric properties, is used along with the fiberoptic nanosensors. Furthermore, sol–gel-based fiber-optic nanosensors along with carbon dioxide, a sensing element that employs dual luminophore internal referencing, was developed. Sensing of pH remains of great interest even though cross-sensitivity to ionic strength is the central problem for optical sensors. An emerging application of pH sensing inside cavities, such as microtiter plates and micro-bioreactors, can help avoid microbial contamination in foods due to noncontact measurement. These nanosensors can also be used to detect benzo[a]pyrene inside a single cell that is found to be in slightly elevated in foods that are cooked well on the barbecue, particularly in steaks, chicken with skin, and hamburgers.

Probes Encapsulated by Biologically Localized Embedding

PEBBLES is the acronym for "Probes encapsulated by biologically localized embedding." It is a spherically optical nano-sized sensor typically ranging in size between 20 and 200 nm in diameter and transduces biological or chemical events into a measurable optical signal. Limited analysis has been done using optical PEBBLE nanosensors; however, it has been used to detect various analytes in food additives in which Mg^{2+} PEBBLE sensor is one of them. It was designed by encapsulating a hydrophilic dye (cyanine which has an affinity for water molecules) and a commercial reference dye (Texas red which is a red fluorescent dye used for staining cell specimen) in polyacrylamide nanoparticle (a white polyamide related to acrylic acid which possesses a hydrophilic property). This sensor was designed to determine the role of Mg^{2+} inside human macrophage cells in the presence of attacking Salmonella (a Gram-negative rod-shaped bacteria which is a major cause for foodborne illness transmitted to individuals through ingestion of contaminated food of animal origin i.e., meat, poultry, mainly eggs, and milk).

Recent Developments in Optical Nanosensors

1. Surface-enhanced Raman scattering: In contrast to the detection strategy based on refractive index profile, surfaceenhanced Raman scattering (SERS) is a unique vibrational spectroscopic method that generates vibrational signatures for small molecular analytes to distinguish between molecules such as structural isomers of fructose and glucose possessing certain similarities. This was first observed on a silver surface electrode in 1974. The benefit of using Raman spectroscopy is its ability to display the molecular structure of a sample by transducing the data in the form of signal. With the progress being made in the fabrication of metal nanoscale and nanomaterial surfaces and the advancing technique of Raman scattering, SERS has become a delicate method, relevant for examining foods, and drugs.

SERS can be used for detecting quantitatively by picking an appropriate calibration data set, as its intensity varies between samples due to the substrate division. Real-time response, low detection limit, and quantitative and qualitative analytical capabilities are some of the advantages of this mechanism, which makes it applicable for identification and characterization of bacteria, pharmaceuticals, and other molecular species.

The pH sensor based on SERS is developed using a silver nanoparticle with a gold-silica core nanoparticle. The sensor is made up of a 50 to 80 nm diameter of silver nanoparticles mixed with para-mercaptobenzoic acid, which shows a characteristic SERS spectrum dependent on the pH of the adjoining solution. It is made sensitive to pH changes in the range between 6.0 and 8.0 and has been successfully examined for the rapid detection of B. subtilis spore, which is a nonpathogenic surrogate for B. anthracis. Ideally, the limit of detection (LOD) should be less than the threshold dose of a pathogen. The sensor was found to have an LOD of 2.1×10^{-14} M (2.6×10^{3} spore in 0.2 µL, 0.02 M HNO_3) for B. subtilis, which was found to be less than anthrax infection dose of 10^{4} spore. Conventional Raman spectroscopy based sensors have also been used to examine the antifungal movement of ZNO nanoparticles against Penicillium expansum and Botrytis cinerea, as they play a major role in the economic loss during postharvest handling of fruits. In various food industries, these nanosensors are used to monitor and detect chemical fungicide that is, thiabendazole (TBZ) against mold and blight formed on bananas and citrus fruits. With a portable Raman spectrometer, the estimated amount of TBZ in a 5 g of citrus sample (peel) is found to be 78 mg kg^{-1}, which is almost higher than the maximum allowed range of presentday regulations. Another crucial development has emerged for fungicide detection at ultralow levels in solutions that is by using a dogbone-shaped gold nanoparticles in SERS sensors. Nanorods (aspect ratio of 2.19 and length of 38 nm) were used for the detection of dithiocarbamate fungicide Thiram in acetonitrilewater solutions in which LOD and limit of quantitation were estimated to be around 11 and 34.4 nM, respectively, which is much below the Environmental Protection.

Agent tolerance value of 16.6 mM. But when dogbone-shaped gold nanoparticles were used, the LOD value further decreased to 44 nM. Conventional high performance liquid chromatography (HPLC) analysis techniques can be integrated with the SERS-based detection system for rapid screening of food sample to eliminate false results. Along with the SERS mechanism, gold nanoparticles or nanorods can be used as a nanomaterial for efficient detection of perchlorate (a food pollutant) at the nanometric level in contaminated water samples. Silver substrates along with the SERS technique can rapidly screen L. monocytogenes, E. coli, and Salmonella. Gold substrates have previously been used as a nanomaterial to detect viruses such as adenovirus, parvovirus, coronavirus, herpesvirus, norovirus, simian rotavirus, and Sendai virus with detection limit of 100 particles. Application of magnetic nanoparticles in SERS-based detection of microbes renders a particular advantage. In this approach, first the magnetic nanoparticles that are covered with capture molecules (such as antibodies) participate in the initial capture and separation of microbes from sample matrices takes place. This is followed by direct detection of the captured species by Raman tags attached to the same or different nanoparticles in the same system.

2. Surface plasmon resonance: SPR sensors utilize free electron oscillation at the interface of a metal surface with dielectric constants of opposite charges during the coupling of excited light of a specific wavelength. In 1982, this phenomenon was first implemented for gas detection and in biological sensors; since then this technology has been extensively used for the detection of chemical and biological analytes in a solution.

SPR nanosensors consist of an optical system and a transducing medium that correlates the chemical (or biological) properties with optical outputs along with an electronic database. In this sensor, the optoelectronic components allow the processing of data through the transducing medium. These two components contribute in determining the capabilities of this sensor, that is, sensor stability, sensitivity, and resolution, and completely depend on the properties of the optical and transducing systems. This sensor produces a surface wave that moves parallel to the dielectric axis and the surface of the metal in between them while producing a field strength that demonstrates evanescent characteristics. The strength reduces rapidly as a function of distance from the surface, usually up to several hundred nanometers. When the target analyte is captured by any biorecognition molecule like antibodies, enzymes, or receptors, a change occurs in the reflection intensity, which is then converted into the refractive index. Usually, SPR is treated as a label-free detection system, as it does not require any distinguishing fluorophore or reagents at the sample preparation phase. However, for the specificity of the target analyte, biorecognition molecules or antibodies are occasionally predeposited onto the SPR surface.

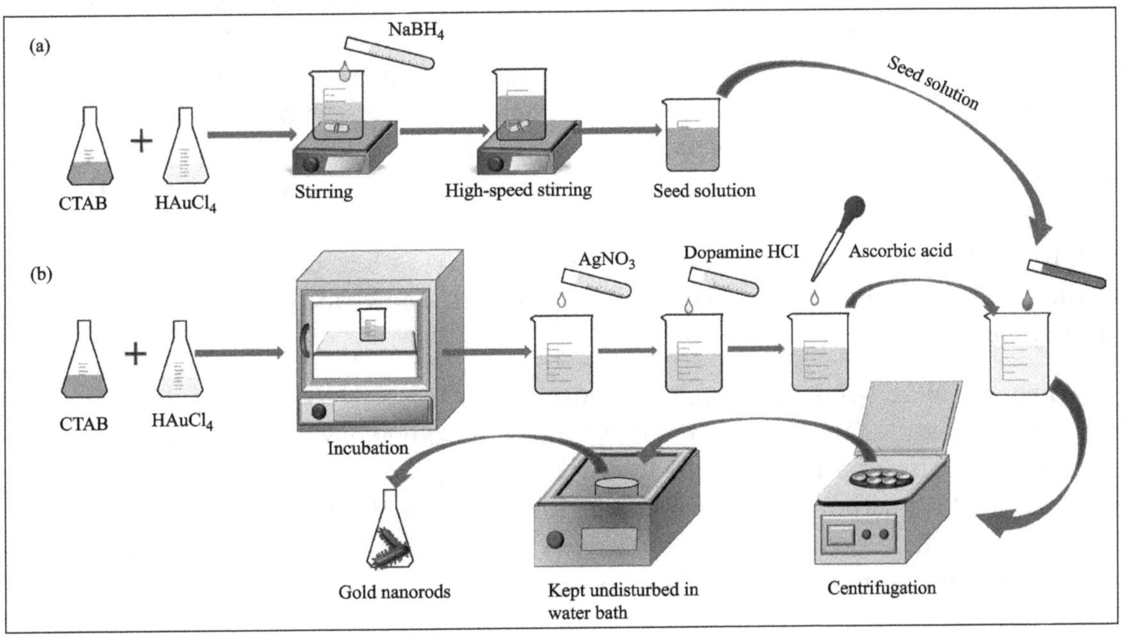

Seed mediated synthesis of gold nanorods: (a) Process denotes the preparation of seed solution, (b) process denotes the preparation of growth solution. After the growth solution is prepared, seed solution is mixed in it with high speed and then kept undisturbed for accumulation of the precipitate. Finally the gold nanorods are observed through localized surface plasmon resonance technique.

This sensor has been used to detect damaged and live cells of L. monocytogenes at 10^6 cell mL^{-1} using rabbit antibodies or phage displayed single-chain fragment variable antibodies on a sensor platform that is covalently restrained Detection of Salmonella enterica serovar Typhimurium, at a concentration of 10^2 CFU Ml^{-1} using protein G molecules immobilized with anti-Salmonella antibodies on the gold surface. SPR biosensors can also be used effectively for the detection of Salmonella in milk at 1.25×10^5 CFU mL^{-1} concentration, enterotoxins from Staphylococcus, mycotoxins, and E. coli O157:H7.

Chemical Nanosensors

These nanosensors involve a broad spectrum of transducing processes performed optically, calorimetrically, gravimetrically, or in various other ways which transform data acquired from chemical analytes (ranging from the concentration of a single specific component to a mixed component) into a detectable signal. The chemical selectivity and analytical performance of this nanosensors are dependent on the transduction process and the coupling of chemical recognition elements with the physical part of the sensor.

Nanomaterials like nanotubes or nanowires are compatible with the properties of chemical nanosensors; hence, they are used as sensing components. Electrochemical-based nanosensors possess another standard technique by which nanomaterial-based chemical sensors with application in food industries can be extensively compared to colorimetric or fluorimetric methods, whereas the former approach would be more useful for food matrices as the light absorption and scattering from the various surface of the food can be avoided. These nanosensors mainly function by binding selective antibodies to conductive nanomaterials like carbon nanotubes (CNTs) and then applying changes to the material conductivity as soon as the target analyte binds to the antibodies. Other electrochemical systems based on nanomaterials includes an immunosensor, which is composed of a CeO_2 nanoparticle and chitosan (partially or fully deacetylated chitin) nanocomposite to detect a vital foodborne fungal contaminant that is, ochratoxin-A. It could also detect staphylococcal enterotoxin B using silicon nanowire transistor and Cholera-toxin using CNTs. As analytes are not restricted to harmful substances, a study revealed that CNTs-based electrochemical nanosensor in microfluidic devices can be used to quantify vitamin content, antioxidant, and flavor in vanilla beans and apples.

Apart from chemical analytes, nanomaterials also aid in the electrochemical detection of microorganisms. A conductive TiO^2 nanowire coated with antibodies has been developed for screening L. monocytogenes. This method was able to detect about 4.7×10^2 CFU mL^{-1} of L. monocytogenes pathogen in almost 1 h without any unusual interference from other foodborne pathogens. Change in resistance or conductance across the circuits of these nanosensors has also been done for the detection of Bacillus, Salmonella, E. coli as well as viruses.

Biosensors and Biological Nanosensors

The invention of biosensors has proved to be a promising tool in food safety, medicine, environmental protection, and many other fields. The safety and quality of food have been the main concern for food technologists and environmental scientists. Biosensors provide a rapid and efficient way to assess safety and quality of food and related products. Currently, industries need quick and economical methods that enable development of products of best quality. High selectivity, specificity, quick response, and cost affordability make biosensors as one of the most efficient tools available to the food industry. A biosensor is defined as a device integrated with a biological sensing element of receptor along with one or more transducers capable of relaying selective or semiquantitative information to an output device. The integration of a diverse group of cross-disciplinary expertise, such as biochemistry, physical science, molecular engineering, biotechnology, and material sciences, enables the possibility of detecting changes at nanolevel, makes biosensors the most versatile analytical tools ever used.

In food industries, a major area of biosensor use is in the detection of pathogens, pesticides, and toxins present in various food samples and packaging materials. Traditional methods of detection of contaminants include physiochemical and biochemical tests that take days or weeks to yield results. However, the results often lack sufficient selectivity and sensitivity with a significant investment of money and time for preparation of samples. Biosensors with all its advantages of selectivity, sensitivity, simplicity, low cost, and rapid processing time are found to be a great nondestructive alternative approach. The method implemented in its construction also allows its usage in determining the composition of raw and processed food samples and on-line control of fermentation process. Biosensors are expected to improve the quality of life in upcoming years.

Regarding the conceptual and fundamental mode of operation, there are three components of a typical biosensor: bioreceptor/biorecognition element, transducer, and detector. The main sensing elements of biosensors are enzymes, microorganisms, antibodies, biocatalysts, antigens, nucleic acids, whole cells, etc. The organic element acts together selectively with the target analytes (pathogens, microorganism, pesticides, or toxins), providing the discriminatory power of the sensors. Speed and selectivity of biorecognition elements are the most important and crucial aspect of biosensors. Enzymes and antibodies are most widely used biological recognition elements. The second key component is a transducer, which converts the response resulting from the chemical interaction between bioanalyte and bio receptor into an electrical signal. The name itself defines the ability to convert one form of energy into another as the word "trans" means change and "ducer" means energy. The transducers may convert biochemical energy into electrical energy. Some of the interactions are temperature-dependent with heat output or electrochemical reactions with the electronic output. Some other mode of signal transduction may include, light output or light absorbance between reactant and product which are based on the mass of the reactant, absorption or emission of electromagnetic radiation, mass and micro-viscosity alteration of wave propagation, etc.

The third component is the detector, which receives electrical energy and processes it. The detector mainly consists of a microprocessor or other electrical or optical processing tools. In many cases, the signal produced by the transducer is so weak that it cannot be detected by the processor. In that scenario,another component, an amplifier, can also be installed in the detector to amplify the signal received for further improvements of processing of the signal. The specificity of the sensor for a specific target analyte is assured by biorecognition elements; whereas, the accuracy and sensitivity is influenced by the type of transducer employed. In addition to these separate components, a critical structural feature of biosensors is governed by the conjugation chemistry to immobilize the bio receptor on the sensor platform. Proper immobilization is important to ensure the accuracy and speed of the biosensor. However, the process is affected by temperature, pH, types of contaminants, and other physiochemical changes.

Nanotechnology-based Biosensors

In past decades, with the progress and successes in nanotechnology, most of the nanomaterials and their properties have been studied extensively paving the way to their applications for the development of biosensors. The use of diverse groups of nanomaterials, such as nanoparticles, nanorods, nanotubes, and nanowires, has enabled faster detection with enhanced reproducibility of the target analytes. The main advantage of nanomaterials is their dimension. Nanomaterials with their high surface area allow more efficient and better use of the surface-to-volume ratio as compared to other materials on a sensor platform. Nano biosensors are biosensors that use nanomaterials as an integral part of their design.

Nanomaterials-based Biosensors

Carbon Nanotube

Nanotubes are long and narrow tubes that can be assayed onto a strong, stable, and malleable honeycomb-like structure. Nanotube-based biosensors are popular for their large surface area and ability to immobilize large numbers of biomolecules. They usually have a tubular or porous membrane with very active mechanical, electrical, chemical, and optical properties. Nanotubes can be reinforced into composite structures with high fracture and thermal stability. Among nanotubes made out of different materials, CNTs are the most popular and have a wide range of applications. They have applicability in many fields due to their graphite structure, highly effective surface chemistry, electronic potential, and typical nanostructure. The CNTs have various applications in biomedicine, food processing, chemistry, bioanalysis, and biosensing. For example, polymeric CNTs have good mechanical properties and high electrical conductivities, which makes them ultrasensitive and effective electrochemical sensor. The interesting properties of CNTs in electrochemical measurement are their abilities to provide large active surface area at electrodes of relatively smaller magnitudes. The high rate of electron transmission and sorption capacity, and with their potential to be used as effective electrode material for immobilization of substrate on the surface, makes CNT as one of the most popular materials used in biosensors for detection of pesticide and food analysis.

CNTs can be classified as single CNTs or multi CNTs. Single CNTs have a single layer of graphite sheet enclosed in a cylindrical tube, whereas, multi CNTs consist of an array of single nanotubes nested in a concentric manner. They can be coupled with microchips as a novel electrochemical material for analysis and detection of food additives, sugars, flavors, vitamins, and isoflavones. These biosensors have the possibility to supplement and even replace the old method of pesticide detection through a simple, fast, and cost-effective method of enzymatic biosensors. Enzymatic biosensors used to screen for carbamate pesticides through the enzymatic inhibition of acetylcholinesterase and their substrate, acetylcholine. The carbamate inhibits the catalytic activity through binding to the active stearic site of the enzyme and blocking the serine residue in the catalytic triad of AChE by phosphorylation or carbamylation. For detection of carbamate pesticide in food and vegetables, the core–shell structure of CNTs has a compact layer of polyaniline, and detection principle is based on immobilization of acetylcholinesterase on it is surface so that it can be used to quantify methomyl and carbamyl in fruit and vegetable samples through chronoamperometry.

Quantum Dots

Food contaminants such as, added pesticides, and bacterial toxins (e.g., botulinum toxin, enterotoxin produced by S. aureus, and E. coli) can be detected by the use of water-soluble bi-conjugated QDs. The advantage of these aqueous synthesized QDs includes photo stability for longer periods of time, broad absorption, stability, and a very specific emission spectrum with very high compatibility. They can be combined with various biomolecules to develop an integrated and hybrid form of biosensors with the combination of unique optical and magnetic properties having specific and sensitive detection capability. These characteristics make QDs an attractive fluorescent probe for quantitative and qualitative analysis. QDs can be arranged in an assembly for detection of toxin compound (paraoxon) produced by the organophosphorus insecticide parathion by covering it with layers of chitosan, organophosphorus hydrolase, and thioglycolic acid. This detection regimen is based on the principle of inhibition of acetylcholinesterase, resulting in the accumulation of acetylcholine in cholinergic synapses. The unique ability of QDs makes it an essential tool for bioanalysis and food safety.

Nanofibers

Nanofibers are electrospunfibers with diameters of around 100 nm. These nanofibers with large surface-to-volume ratios offer a great prospecting detection scenario in which high porosity is required. A nanofiber converted into a porous structure can be a dynamic source for detection in which pore size and shape can be changed depending upon requirement. Biosensors made from nanofibers similar to nanofiber mats and arrays have the advantage of an extremely large surface area, which increases the availability of immobilization sites for biorecognition elements in biosensors. In addition to the physical properties, the chemical characteristics of the sensor can be modified by blending with different polymers, nanocomposites made of organic or inorganic materials, or some bioactive ingredients. Its novel properties can be utilized for sample preparation and analyte detection.

Other Sensors Utilizing Nanoscale Materials

Electronic Nose

An electronic nose is a device that consists of groups of chemical sensors with a high degree of electivity, sensitivity, and the ability to recognize and analyze simple and complex odors arising from test samples. Ideally, this sensor can replace human olfactory system. The versatility of the electronic nose to recognize and differentiate between different types of odor with accurate results makes it a very popular tool in food, clinical, and biomedical industries. Electronic noses can be defined as "an intelligent system of chemical-array sensors that mimic the human olfactory system". The concept of the electronic nose has been introduced in food for quality inspection, shelf life investigation, freshness evaluation, and process monitoring. The smaller aspect ratio structures of nanoparticles, nano spheres, and nanocubes are typically used as a film for sensing element, whereas high aspect ratio structures can be utilized as a single or in multiple films for fabrication of an electronic nose. Considerable work has been done to develop these sensors and tested on food samples including fish, vegetables, meats, cheese, mushroom, wine, etc.

The fabrication of electronic nose devices for detection and discrimination between food samples and odor consists of an array of broadly tuned sensors which are treated with odor-sensitive biological or chemical materials. The chemical sensors that are the key components of the system comprise of inorganic crystalline materials like semiconductors, organic materials, polymers, and biologically derived materials. The sensing element depends on the type of analyte to be detected. The odor stimulus generates a signal or fingerprint from the sensor array. The signals are then captured by data acquisition system for processing through a processor or computer software. The control and data acquisition system can be integrated into a single device called microcontroller. Specific signals from known analyte are used to develop a database and a pattern recognition system is created to enable classification, identification, and quantification of the odor based on the data stored in the database. The system is produced to have long-term usage with repeatability and reproducibility. The implementation of nanotechnology in its construction offers the advantage of simple operation, ease of fabrication, low cost, compatibility with microelectronics processing, and low power consumption.

Electronic noses have several applications in food processing and quality assurance of raw and manufactured food products. The determination of the quality of harvested fruits and vegetables is an example in which electronic nose system is used. One of the main concerns of the food industry currently is to monitor and control the quality variation of fresh harvest fruits and vegetables. Electronic noses have the specificity and precision to differentiate between numerous types of fruits and to calculate the extent of damage (such as spoilage) and the duration of storage. The measurement is done by aroma transformation, changes in defect caused by skin cuts or over-ripening, changes in starch or sugar composition, etc. Application of the electronic nose greatly depends on upon the type of sensor array used and gives an idea about the contribution of the sensor array in the analysis of food.

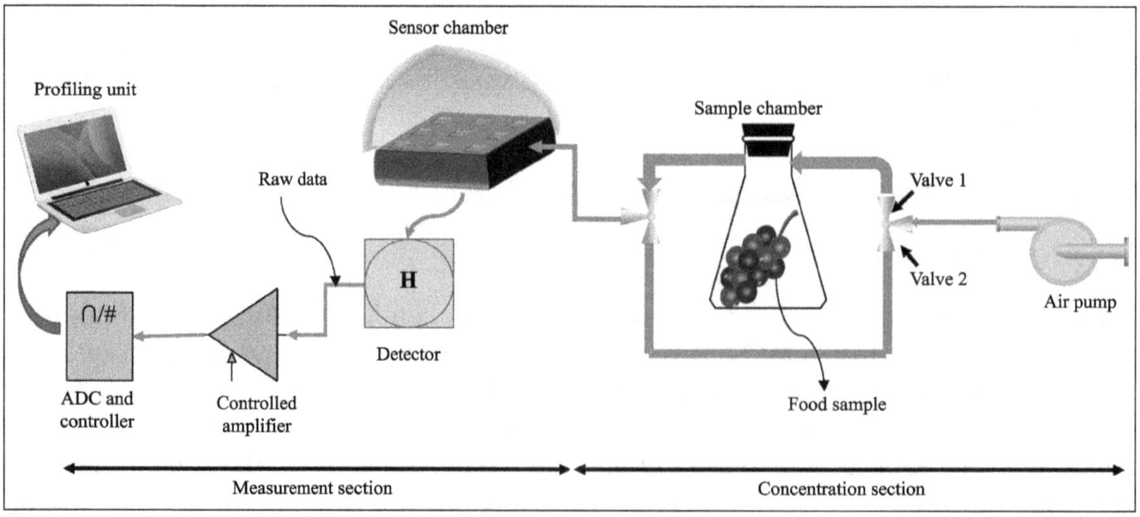

Electronic noses composed of quartz crystal microbalance coated by modified metal porphyrins and related compounds can be used to understand the level of spoilage in red wine and pulped tomato. The spoilage in tomato pulp was detected by estimating the amount of acetic acid present in it. Experiments show that the LOD of an electronic nose is much lower than that of human sensory perceptions. Therefore, these devices may yield very sensitive results for fresh produce quality control. In the case of red wine, spoilage can be detected by an alteration of aroma as a result of mixing with the air. In this case, a correlation between sensor response with time and expected wine taste is reported which rendered it as a powerful tool for quick quality assessment of food products. Moreover, electronic noses can be used as a nondestructive method for altering the ripening process in fruits through a SnO-based chemical array sensor and neural pattern recognition system. These systems sense the aromatic volatile compounds emitted by fruits using the electronic olfactory system, which is capable of classifying fruits into three different states of ripeness (green, ripe, and over-ripe) with great accuracy.

Schematic of an electronic nose. The system comprises two sections. In the concentration section, the food sample is kept in a closed chamber and left for certain time so that aroma containing volatile compounds get accumulated in the headspace. Meanwhile, synthetic gas from the pump is used to carry the concentrated flavor to the sensor chamber. In the measurement section, the sensor creates a specified fingerprint for aroma that is detected by the detector and forwarded for amplification. At the end, the analog signal after amplification is converted into digital and used for further analysis.

Array Sensors

Array-based sensors consisting of nanoparticles for detection of bacterial pathogens have been developed by different groups to improve microbial diagnostics. In such sensing strategies, array of polymer constructs (such as nanoparticle-fluorescent polymer) or (poly(para-phenyleneethynylene)) with conjugated gold-nanoparticles (AuNP) were utilized for bacterial detection. In this sensor, arrays of analyte receptors conjugated with AuNPs are assembled as "chemical noses" to discriminate and detect analytes based on their unique response diagrams. The hydrophobic surfaces of bacteria are recognized by arrays consisting of AuNPs (diameter 1.6 nm) functionalized with the polymer (poly-lysine), following a complementary electrostatic interactions of AuNPs with live bacteria resulting in the self-assembly of the arrays with captured bacterial cells.

Another "Array Biosensor" prototype is based on the total internal reflection fluorescence (TIRF). This biosensor utilizes the sandwich immunoassay format for analysis and detection. In the process of TIRF, fluorophores that are either attached to or in proximity to the surface of the waveguide are selectively excited via an evanescent wave. The infiltration depth of the evanescent wave allows measurement of the binding event on the surface of the waveguide without significant interference with an excess of fluorophores, particulate, and other compounds in the bulk fluid. Background contribution from endogenous fluorophores is minimized with the more widespread use of fluorescent dyes, such as Cy5 (far red fluorescent cyanine dye having excitation wavelength in between 633 and 647 nm), which are excited at much longer wavelength than naturally occurring fluorophores. Array biosensors have been utilized to detect and analyze toxins, large pathogens, and bacteria from food samples.

Sandwich and fluoroimmunoassay have been developed to identify low or high molecular weight toxins in complex food samples. NRL array biosensor was employed to interrogate up to 12 different samples for the presence of multiple contaminants. Among other major applications, the array sensors were utilized for detection of Shigella and Campylobacter at a very low concentration. The biosensor was also successfully employed in the detection of deoxynivalenol (DON) (a mycotoxin produced by various Fusarium species), which is a common contaminant of cereal grains, such as barley, corns, oats, and wheat. The traditional method uses chromatography for its detection, but the method is time-consuming and not appropriate for large-scale detection. Hence, the application of the array biosensor increases the speed and eliminates highly complex method for quantification and identification of DON. The analysis does not require any sample clean-up and sample preconcentration; therefore, it minimizes the cost and time.

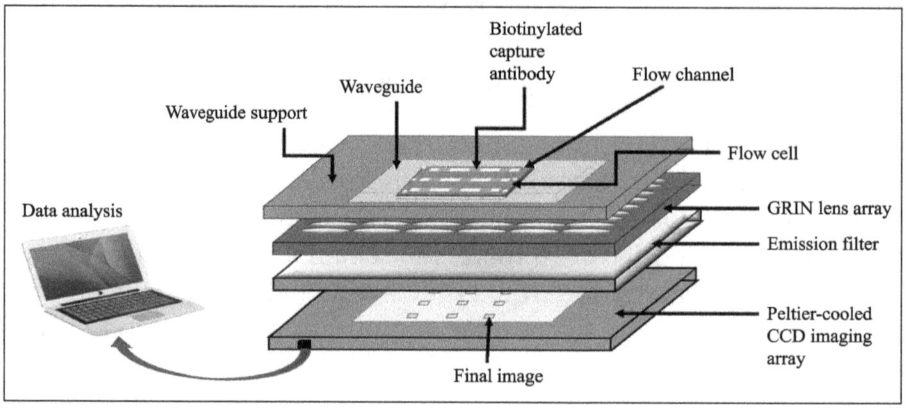

Schematic of an array sensor. Physically isolated patterning technique is used for printing the antibody sample on the waveguide surface (made up of borosilicate glass slides). Flow cell (with polydimethylsiloxane in its flow channels capable of molding and forming 3D structures) is placed perpendicular to the columns of the recognition antibodies so that assays could be analyzed independently. A diode laser is directed toward the waveguide and its excitation light gets coupled in the flow cell (by reflective cladding). The fluorescence emission from the cell automatically focuses toward the graded index (GRIN) lens which then passes through the optical filter onto a Peltier-cooled charge-coupled device (CCD) imaging array. Data are extracted from the image using appropriate software.

Nutraceuticals

It is well known that natural plant-based foods, including fruits, vegetables and whole grains promote health and prevent diseases. The interest in nutraceuticals, functional foods and dietary ingredients has increased because of their safety potential as they naturally occur in food and possess high therapeutic value with a great potential to treat chronic diseases. Their health benefits have grabbed the attention of the world and now the researchers are showing great interest to explore their full potential to treat chronic diseases such as cancer. Increasing awareness among people nowadays regarding the beneficial effects of nutraceuticals and functional foods reminds of what Hippocrates said – Let food be your medicine and medicine be your food. Dr. Stephen DeFelice in 1989 coined the term "Nutraceutical" by combining the terms "Nutrition" and "Pharmaceutical", both being the key contributors to human wellness. Nutraceuticals can be defined as food or part of the food that provides physiological health benefit and helps in curing or treating a chronic disease. Sometimes the terms functional foods and nutraceuticals are used interchangeably, the main difference being in the mode of consumption. Nutraceuticals are generally consumed in the form of tablets, pills, capsules whereas functional foods are consumed in the form of food only. India is already a home to a large number of medicinal herbs and spices and they have been used for centuries in the traditional Indian Ayurvedic Medicine system. Presently over 470 nutraceutical products are available with documented health benefits.

Fruits and vegetables are an important source for the daily intake of healthy constituents like minerals (calcium, phosphorous, magnesium and other minor minerals), water-soluble vitamins (B and C), fat-soluble vitamins (A, E and K) and a wide variety of phytochemicals. Among phytochemicals, there are bioactive molecules capable of protecting against diseases and producing a physiological health benefit. The most important phytochemicals from fruits and vegetables are polyphenols, carotenoids, organo - sulfur and seleno - compounds. Nutraceuticals, their sources and their potential benefits have been listed in Table.

Bioactives	Source	Benefits
Carotenoids (Isoprenoids)		Antioxidant activity, protects against cancer.
Lycopene	Tomato, carrot, guava, papaya, watermelon	Anticancerous, protects against eye related macular.
Lutein	Corn, avocado, egg yolk, spinach	
β- carotene α- cryptoxanthin Zeaxanthin	Carrots Oranges Corn, avocado	Antioxidant activity, neutralizes free radicals, anticancerous.
Dietary fibres		
Soluble fibres	Legumes, oats, barley	Aids in digestion, anticancerous.
Insoluble fibres	Whole grain foods, bran, nuts	Anti-cancerous, aids in digestion.
Polyphenolic compounds		
Flavonones	Citrus fruits	Antioxidants, anti-cancerous.
Flavonols		Antioxidants, anti-cancerous.
Anthocyanins	Blueberries, blackberries	Lowers blood sugar level, counteracts inflammation, anti cancerous.
Phenolic acids	Berries, legumes	Reduce oxidation of LDL cholesterol.
Curcumin	Turmeric	Wide range of therapeutic activities including Strong anti-inflammatory, antioxidant, anti-cancerous activity.

Phyto-chemicals are naturally occurring chemical compounds in plants. They provide plants with color, odor and flavor. There are a wide variety of phytochemicals including carotenoids, flavonoids, indoles, polyphenols, terpenes etc.

Terpenoids

The class terpenes may be identified as hydrocarbons that come from isoprene which is the major type of hydrocarbon synthesized by plants. Terpenoids are believed to possess antioxidant properties and associated with reduced risk of heart diseases and cancer. This large class of phyto-chemicals mainly includes limonoids and carotenoids that protect cells from becoming cancerous or slow the growth of cancer cells and can be isolated from peels, seeds, pulp, pressed oil, juice or whole fruit of many citrus plant.

Limonoids are important phyto-chemicals with less known nutritional bonus present in citrus fruits and plants of families Rutaceae and Meliacae. Citrus limonoids are secondary metabolites that occur in the category of terpenoids that are found mainly in seeds, fruits and peel tissues of citrus fruits including lemons, pumellos,lime, oranges, grapefruits and mandarins. Chemically, they are a group of highly oxygenated triterpenoids as they contain high numbers of oxygen atoms of up to 7-11. They are classed as tetranorterpenoidsand represent an example of triterpenes in nature. Citrus limonoids confer alkaloid taste to citrus fruits after processing (delayed bitterness). This delayed bitterness develops in juice of citrus fruit suffering physical damage which initiates biochemical transformation of a tasteless limonoid, limonate A-ring lactone to a bitter limonoidaglycon i.e. limonin.

They possess moderate polarity, insoluble in water and hexane but soluble in hydrocarbons, alcohol and ketones. They are mostly bitter in taste. The most commonly found and most studied limonoids in citrus fruits are limonin and nomilin. Citrus limonoids have prototypical structure either containing or derived from the precursor with 4, 4, 8-trimethyl-17 furanyl-steroid skeleton. The skeleton undergoes a variety of reactions such as cyclization, aromatization, rearrangements and oxygenation to give various limonoids. The reference compound consists of four rings condensed in steroid like manner and designated as A, B, C and D. All naturally occurring limonoids contain a furan ring attached to ring D at C-17 as well as oxygen containing functional groups at C-3, 4, 7, 10 and C-17. Structurally, they are classified into two classes, aglycons and glucosides. Among them, aglycons are neutral, insoluble, bitter in taste and without sugar moiety. They are mainly responsible for delayed bitterness of citrus juices. Citrus seeds and peels constitute the natural repositories of citrus limonoid aglycons. They contain aglycons upto 1% of their fresh weight. Glucosides are monoglucosylated derivatives of D- ring limonoid aglycons. They are acidic, soluble and mainly tasteless. Reaction of aglycons with beta-D-glucose results in formation of glucosides. The limonoid glucoside content of citrus peel and flesh solids is approximately 500ppm.

Limonoids are very important as they perform variety of functions. Aglycons (mainly limonin) and glucosides of citrus limonoids possess various anticancer, antimicrobial, insecticidal properties and some other pharmacological activities. Limonoids present in citrus fruits and their juices have cancer protective property. They inhibit the growth of estrogen receptor-negative and positive human breast cancer cells in culture. They target and stop neuroblastoma cells. Limonin 17-beta-Dglucopyranoside present in orange juice decreases colon tumor genesis. Limonin is active inhibitor of promotional phase of carcinogenesis. Obacunone, limonin, nomilin and their glucosides inhibit chemically induced carcinogenesis. They also have the ability to induce specific carcinogen-metabolizing enzymes such as glutathione-S-transferase and quinine reductase in liver and mucosa of small intestine to detoxify chemical carcinogenesis.

Other Health Promoting Activites

Limonin and nomilin have been shown to inhibit the replication of HIV-1in a number of cellular systems. A novel limonoid isolated from Clausenaexcavata have shown HIV-1 inhibitory activity. Limonoids from various plants belonging to Meliaceae as well as Rutaceae were reported to have antifungal and antibacterial activity. Radical scavenging and anti-oxidant activity is also demonstrated by some limonoids from various citrus fruits. Several other citrus limonoids and limonin derivatives includingnomilin, obacunone and limonindiosphenol have insect controlling activities as has been reported in coloardo potato beetle. They work as toxins, feeding deterrents (antifeedant), insect repellent, growth disrupter and reproduction inhibitor. They are also known to possess some pharmacological activites. They have cardioprotective effect as they cause reduction in low density lipo-protein and hence cause reduction in cholesterol. Some limonoids have concentration dependent and nonreversible spasmogenic, uterotonic, anti-inflammatory, anti-arthritic, antipyretic, hypoglycemic, anti-gastric ulcer, spermicidal and diuretic activities.

As stated previously, citrus limonoids are mainly classified into limonoid aglycons and limonoid glucosides. Among them, limonoid aglycons, mainly limonin, are water insoluble in nature due to the absence of glyconic moiety which is responsible for conferring aqueous solublilty to limonin. Therefore, to utilize their nutraceutical properties it is necessary to confer aqueous solubility characteristic to limonoids. Moreso, the metabolic pathway of limonin glucosides indicates that upon ingestion these undergo phase I metabolism and covert to the corresponding aglycon. Therefore it is important to develop lower molecular weight and water soluble derivatives of limonoid aglycons which can enhance their delivery and bioavailability in humans.

Carotenoids are tetraterpenes which belong to a large family of isoprenoids and form one of the most important classes of plant pigments. Carotenoids impart red, orange, and yellow hues to plant leaves, fruits, and flowers, as well as the colors of some birds, insects, fishes, and crustaceans. They play a very important role in photosynthetic processes and carry out a protective function against damage by light and oxygen as they have high antioxidant activity. Animals are unable to synthesize carotenoids de novo, and rely upon the diet as a source of these highly bioactive compounds with remarkable antioxidant activity. Carotenoids are of two types based on their structure:

- Hydrocarbon carotenoids: These are known as carotenes, example β-carotene.
- Oxygenated carotenoids: These are derivatives of hydrocarbons known as xanthophylls, examples are zeaxanthin and lutein (hydroxy), spirilloxanthin (methoxy), echinenone (oxo) and antheraxanthin (epoxy).

Xanthophylls are present either in a free form or as esters with fatty acids. In some foods like corn, spinach, broccoli or other green leafy vegetables, xanthophylls are present in free form, unesterified form. In other fruits and vegetables such as pepper, wolfberry, sea buckthorn, apple, xanthophylls are present in esterified form.

The major provitamin A carotenoids are β carotene, α carotene and β cryptoxanthin. These carotenoids contribute to vitamin A supply and may prevent vitamin A deficiency. Carotenoids are also beneficial in preventing age related macular degeneration (AMD), which causes irreversible blindness. Carotenoids function as antioxidants based on their ability to scavenge free radicals. Moreover, carotenoids also boost immune system by enhancing T and B lymphocyte proliferation. Many carotenoid supplements work as sun protectants, the protective effects are thought to be related to the antioxidant properties of carotenoids. Carotenoids intake has also been associated with reduced cancer risk.

Lycopene is the main carotenoid that is found in ripe tomato and is the main pigment that provides red color to it. Many studies have shown that Lycopene protects against prostate cancer, breast cancer, atherosclerosis, and associated coronary artery disease. It also reduces low-density lipoprotein oxidation and helps in reducing cholesterol levels in the blood. Structurally, lycopene is an unsaturated acyclic hydrocarbon and contains 13 double bonds, out of which 11 are conjugated. Lycopene present in red tomato typically contains 94-96% of all-trans-lycopene. However, lycopene undergoes trans-to-cis isomerization during tomato processing which results in the increased proportion of cis isomers. The structural formula of trans-lycopene is shown in figure.

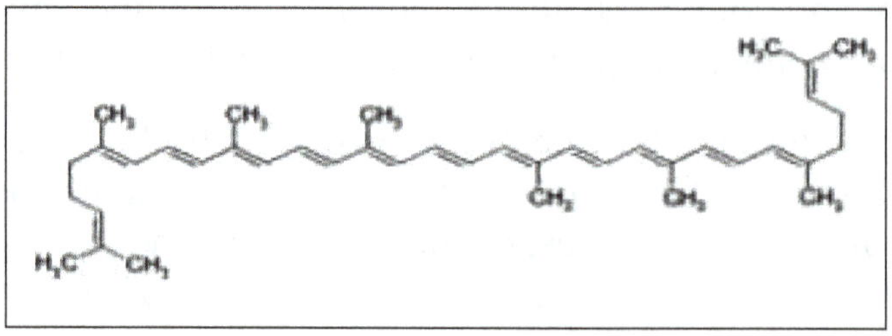

Structure of trans –lycopene.

Carotenoids are lipid-soluble and hence possess poor aqueous solubility. Because of this they have low bioavailability. Their absorption in the body is shown in the Figure. After consumption of carotenoid-containing foods, carotenoids are released from their food matrix and incorporated into mixed micelles in gastrointestinal tract, which has bile acids, free fatty acids, monoglycerides, and phospholipids. Small intestines absorbs carotenoids by mucosa via passive diffusion and get packaged into triacylglycerol- rich chylomicrons. Then secreted into the lymph get into bloodstream and degraded by lipoprotein lipase. The resulting chylomicrons are rapidly taken up by the liver. The liver secretes carotenoids associated with hepatic very low density lipoprotein.

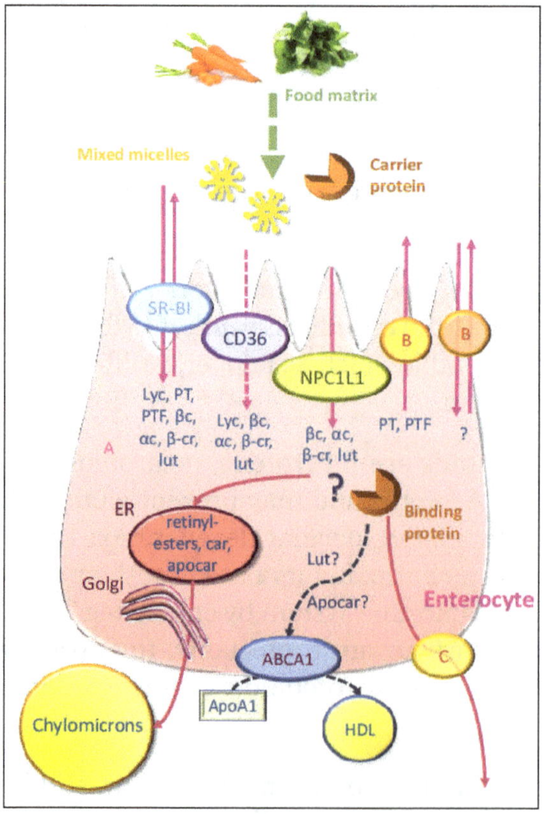

Absorption of carotenoids.

Polyphenols

Polyphenols are naturally occurring compounds that are present as micronutrients in our diet which can be categorized primarily into flavonoids and non-flavonoids. Polyphenols can be defined chemically as substances which possess an aromatic ring bearing one or more hydroxyl substituent, including functional derivatives (esters, methyl ethers, glycosides etc). Flavonoids are the largest group of polyphenols, widely distributed in plant based foods. They are formed in plants as secondary metabolites to show multiple functions for plant growth, development and defence.

Flavonoids

Flavonoids are present in variety of fruits and vegetables such as citrus fruits, apples, berries, onions, banana, grapes, red wine, green tea and are known to show promising health promoting benefits in humans such as antioxidant, hypocholesterolemic, anti-inflammatory effects as well as ability to modulate cell signalling and gene expression related disease development. Structurally, flavonoids are low molecular mass compounds with a basic carbon based skeleton of C6-C3-C6. These consist of two benzene rings linked by a 3-carbon chain that forms a closed pyran ring with benzene ring. Based on the variation in the heterocyclic "C" ring, there are six common food flavonoid classes as Flavonols that are present mostly in food sources such as apples, berries, red grapes etc. Some examples of compounds under this class are kaempferol and quercetin. Another class named Flavanols exist in two forms – monomeric flavan-3-ols and condensed tannins. Their common sources are berries, cherries, green tea etc.

Also, these compounds are known to possess high antioxidant properties. Flavones is another significant class of flavonoids under which polymethoxyflavones a unique class is also present whose sources are green leafy herbs and some citrus fruits. The examples of compounds are apigenin, luteolin, tangeretin and nobiletin. Isoflavones are found in leguminous plants such as soy and its products. These have property to bind to the estrogen receptors. Flavanones is a class abundant in citrus fruits and naringenin and hesperetin are the major flavanones found in citrus whereas anthocyanins are the natural pigments found in cherries, pomegranate, red grapes and cyanidin and malvidin are the classes of anthocyanins.

Although the phenolic phytochemicals have many positive effects on human health but their absorption rate in our body is very low. There are some reasons responsible for lower absorption in the body. The major limitation is their low bioavailability. Bioavailability can be defined as the fraction of the nutrient that is available for the human body for its physiochemical functions. On the basis of structure, low bioavailability of flavonoids is due to its two aromatic rings bound together by three carbon atoms oxygen that form oxygenated heterocyclic ring which renders them insoluble in water. Polymethoxyflavones are the least soluble due to stearic effects of methoxy groups. This makes flavonoids less bioavailable. Another aspect of low bioavailability are low stability of flavonoids: Flavonoids are stable at acidic pH. The pH of our stomach is in the range of 2-4 which favor the stability of flavonoids. As digesta passes from the stomach to the small intestine the pH usually increases to approximately 7. This leads to the degradation of flavonoids. Another reason is the low permeation into small intestine: Low permeation is attributed to the fact that there are no specific receptors for flavonoids to pass through epithelial cells. All flavonoids undergo an active efflux processes by which majority of them are pumped back to the lumen. Flavonoids that manage to enter the epithelial cells through passive diffusion undergo phase-I metabolism to generate aglycone residues, which further undergo phase-II metabolism into flavonoid methylates, falvonoid sulfates and flavonoid glucuronoids, which are transported to the liver to be circulated thereof. Hence they cannot be absorbed by our body.

Curcumin: Golden Nutraceutic

Turmeric known as "spice of life" and "golden spice" belongs to family Gingiberaceae and has been used as a traditional medicine in India. It contains phenolic compounds known as curcuminoids and there are three curcuminoids present in the turmeric namely curcumin being the principal and other two as demethoxycurcumin and bisdemethoxycurcumin. Curcumin is polyphenolic, golden yellow compound and has a structural formula of 1,7-bis-(4-hydroxy-3-methoxyphenyl)-hepta-1,6- diene-3,5-dione as shown in figure.

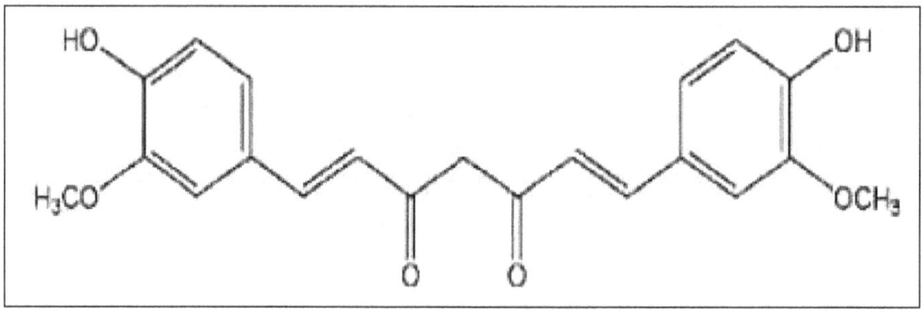

Structure of Curcumin.

Curcumin is known to exhibit beneficial and pharmacological activities including anticancer, antioxidant, anti-inflammatory. Various studies have shown that curcumin is non-toxic to humans. It act as anti-inflammatory by inhibiting a number of molecules responsible for inflammation. It also prevents atherosclerosis by reducing the formation of blood clumps thus acting as cardioprotective in nature. It also helps in inhibiting the growth of Heliobacter pylori which causes gastric ulcers. Due to its numerous health benefits against various diseases such as cancer, neurological diseases, liver diseases, cardiovascular diseases, and inflammatory diseases as shown in figure. and its non-toxic nature has increased the interest of the researchers in this wonderful molecule.

Besides many health benefits of curcumin, it has a limitation of poor aqueous solubility at intestinal pH thus contributing to its low bioavailability and intestinal absorption. The phenyl rings, methoxy groups and two unsaturated carbonyl groups provide hydrophobicity to these curcuminoids which further limits its absorption and bioavailability in humans. Furthermore, its instability in the GI tract and low membrane permeability also limit its use as a nutraceutic. Figure below describes the metabolism of curcumin in humans. Absorption of curcumin through luminal epithelial cells occurs by passive transport which undergo phase-I metabolism, that is, reduction to generate dihydrocurcumin, tetrahydrocurcumin, and hexahydrocurcumin which further undergo phase-II metabolism, that is, conjugation into curcumin glucuronoids, curcumin sulphates, and curcumin glucuronoids/ sulfates mixed conjugates. These curcumin conjugates are actively transported out by multidrug resistance related proteins MRP1 and MRP2 present at the basolateral side of the epithelial cells, which further attribute to their non-bioavailability.

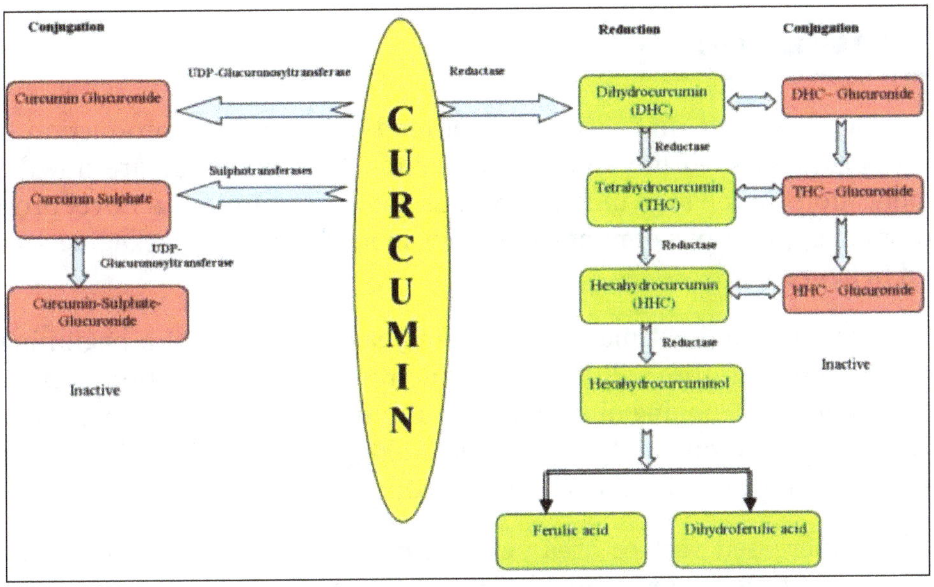

Curcumin metabolism.

To overcome these three limitations, viz. (a) poor aqueous solubility (b) low permeability (c) instability in GI tract, the following approaches are known: (i) Use of adjuvants: Piperine is the most commonly used adjuvant for enhancing the bioavailability of curcumin. It inhibits hepatic and intestinal phase-II metabolism, that is, glucuronidation and thus inhibits the formation of non-bioactive conjugates. A twenty fold increase in curcumin bioavailability by the use of piperine has been reported. However, piperine is also known to enhance the biotransformation of bioactive curcumin into phenolic acids, thereby limiting its use. (ii) Inhibition of P-glycoprotein: On the apical side of epithelial cells are ATP binding cassette transporters, one of which is P-glycoprotein that is responsible for efflux process. Verapamil, a P-gp blocker inhibits the P- glycoprotein and blocking the efflux of curcumin, thus leading to enhanced bioavailability of curcumin. Since verapamil is a drug therefore it cannot be used in preparation of functional foods. (iii) Use of curcumin-phospholipid complex: Phospholipid complexes form an efficient delivery system for lipophilic compounds. Curcumin-phospholipid complexes have been known to provide stability to the molecule in the GI tract by protecting it from enzymatic degradation and are therefore, a favoured concept for enhancing curcumin bioavailability. (iv) Nanoformulations: For the enhanced efficacy and bioavailability of curcumin various nanoformulations have been prepared. These nanoformulations through use of nanotechnology attain reduced size by which colloidal solubility increases. Moreso, the presence of specific clathrin and caveolae receptors on apical surface of epithelial cells play a role in the internalization of nanoformulations. Further, these nanoformulations provide a protective layer to curcumin and prevent its enzymatic degradation and biotransformation into non-bioactive metabolites. Various curcumin nanoformulations such as nanoemulsions, liposomes, polymeric nanoparticles, solid lipid nanoparticles have been prepared which impart colloidal dispersibility, enhanced permeation and greater stability to the molecule. These nanoformulations have been reported to show a significant increase in the bioavailabity of curcumin of upto 9-55 folds. From the above it is clear that nanoformulations show a double thronged action of protective delivery of curcumin to GI tract followed by enhanced intestinal absorption. Coupled with phospholipid complexes, which also confer stability to the molecule, nanoformulations offer a favourable approach for development of turmeric based functional foods.

Nanoformulations

Nanotechnology has an impact on every aspect of food industry from production to processing, packaging, transportation, shelf life and bioavailability. Nanomaterials are now being used for commercial applications in the food industry because of their unique and novel properties. There is diverse plethora of applications of nanotechnology in food and food packaging sector which has been illustrated in figure.

Nanotechnology has found a great market in recent years but it is just a beginning wherein research and development in future will lead to a wide range of different products on the shelves. For instance, one could find cooking oils fortified with nutraceutical components encapsulated in nanocarriers. But still the acceptability of nanotechnology based food products needs awareness and assurance of safety for the sake of health of the consumers.

Nanoparticles have increased surface area per unit mass, therefore, they are expected to show maximum absorption rate as demonstrated in Figure. Bioactive compounds are found naturally in certain foods and possess therapeutic properties which in turn provide physiological benefit to the body by preventing chronic diseases. But most of the bioactive compounds are lipid soluble and hence exhibit poor aqueous solubility. This limitation results into poor absorption in the gastro-intestinal tract and bloodstream rendering very low bioavailability.

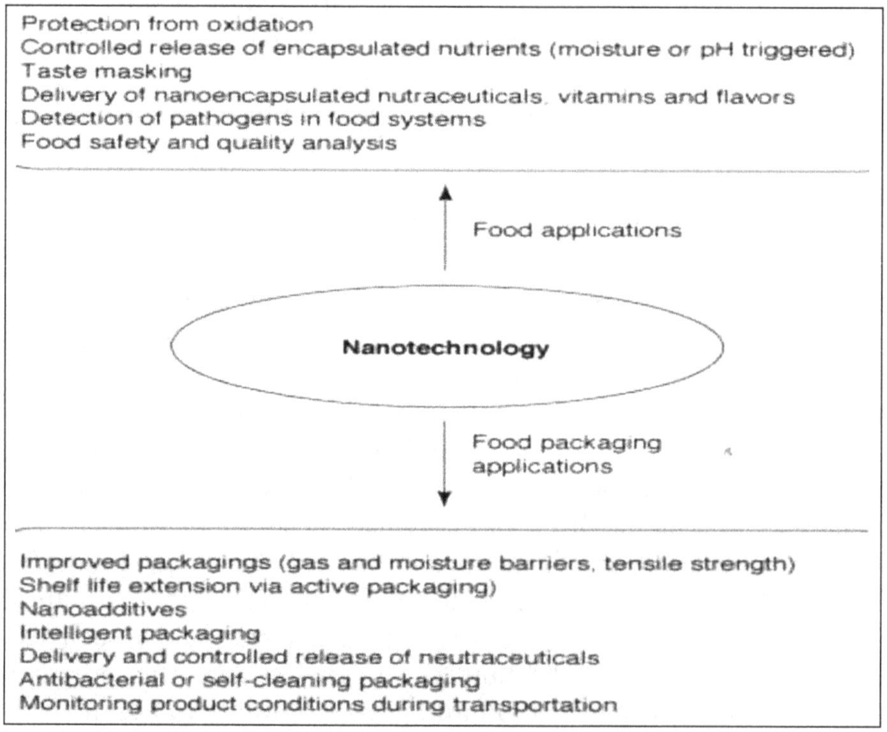

Potential applications of nanotechnology in the food industry.

Nanotechnology here can play a major role by contributing to improve the functional properties of these bioactive compounds. This can be achieved by increasing their aqueous solubility and hence absorption through cells which will eventually lead to increased bioavailability. Different types of nanoformulations that are generally used in the food industry include polymer nanoparticles, solid lipid nanoparticles, and nanogels as shown in figure. Apart from these, nanoemulsions and liposomes are also widely used for different applications.

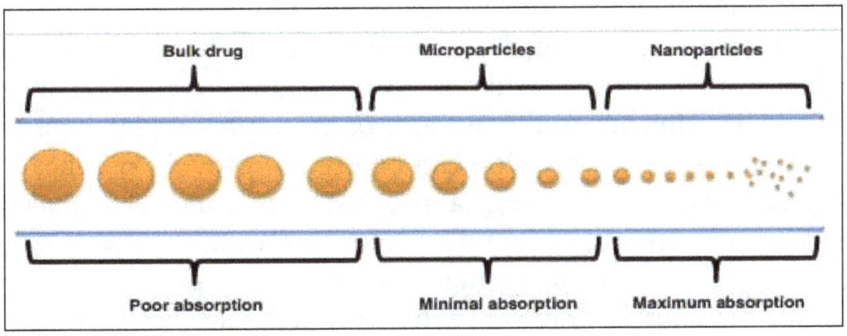

Size dependent absorption characteristics of particles.

Nanocarriers are designed such that they encapsulate the particle inside them and are carried in the gastrointestinal tract. In this mechanism the bioactive particle is prevented from the different harsh conditions in the body and is available for absorption by the body). A number of nanoformulations have been prepared for poorly hydrophilic and lipophilic compounds to enhance their bioavailability. Polymethoxyflavones have been incorporated into protein and polysaccharide nanoparticles and exhibited improved functional characteristics. Besides, many other hydrophobic compounds have been prepared into nanoformulations such as nanoemulsions, nanosuspensions, solid lipid nanoparticles, liposomes, etc. and have shown significantly improved functionality.

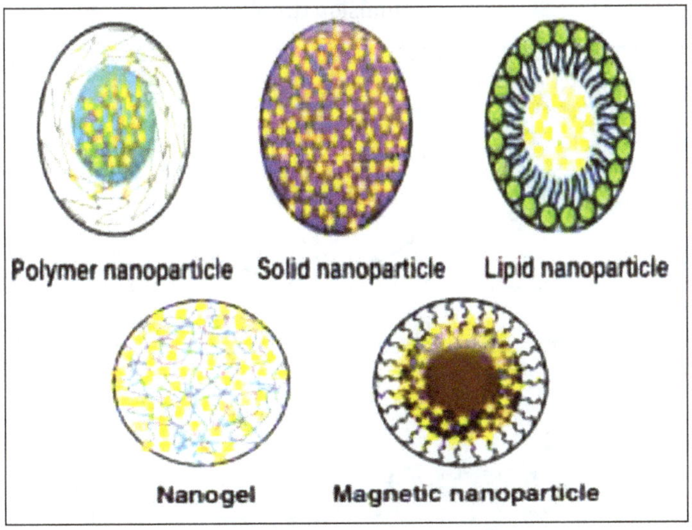

Types of nanoformulations.

There are different types of nanoformulations that can be prepared for incorporation into food matrices.

Nanoemulsions and Nanodispersions

Nanoemulsions are liquid–in –liquid dispersions which are kinetically stable. A nanoemulsion droplet is shown in figure. The reduction in the size leads to increase in surface area per unit volume and imparts stability. Small size also renders enhanced rate of absorption and optical transparent appearance. Therefore, nanoemulsions are the liquid formulations in which the bioactive compounds get entrapped inside the oil carrier. Various oils can be used based on the compatibility with the nutraceutical. Generally nutraceuticals are dissolved in those oils in which they are soluble to a maximum extent to increase the loading as well as the stability of the emulsion. Nanoemulsions can be prepared by using two kinds of approaches such as high energy and low energy approaches as shown in figure. There are many conventional as well as novel methods that can be used to engineer the nanoemulsions by using any of the two approaches. Among high energy approaches, high pressure homogenization and ultrasonication can be used to provide significant energy to reduce the droplet size whereas in case of low energy approaches one can opt for phase inversion method and emulsion inversion point. In case of nanodispersions, solvents are used as carriers, as these efficiently dissolve bioactives and on the later stages solvents are evaporated. But, as these bioactives are for human wellness, therefore solvents are avoided as these are not permitted in food.

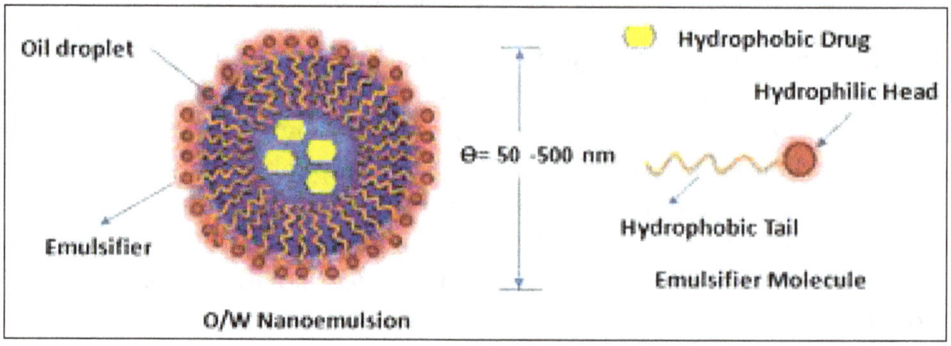

Nanoemulsion droplet.

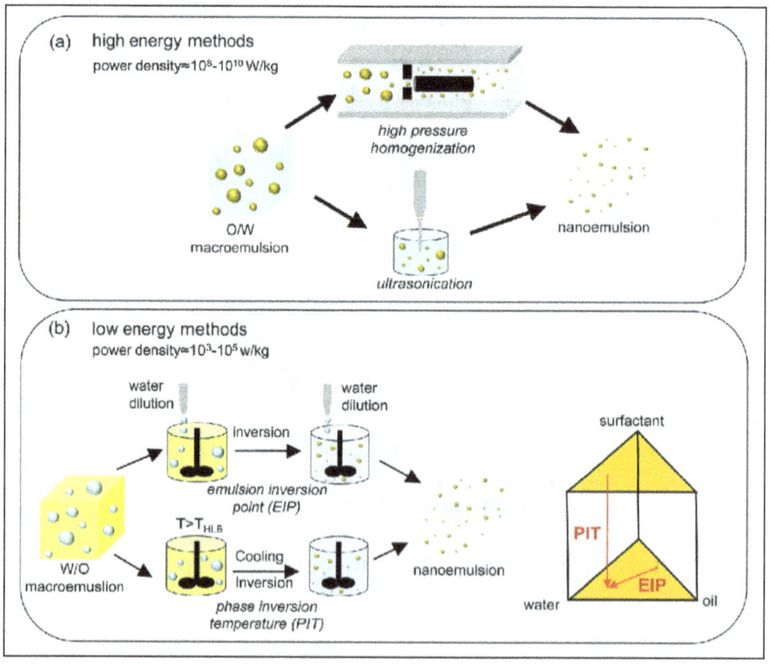

Methods for the formation of the nanoemulsion.

Solid Lipid Nanoparticles

In comparison to emulsions, liposomes and polymeric nanoparticles, solid lipid nanoparticles are also an efficient carrier system. These nanocarriers have attracted major attention for applications in intravenous delivery. Structurally, these particle size ranges between 50 to 1000 nm contributing to their colloidal nature. These are generally composed of lipid in the core dispersed in water or in aqueous surfactant solution. The two phases interact at the interface and there is replacement of liquid lipid with solid lipid. They are highly biocompatible and non-toxic for safe delivery of the drugs.

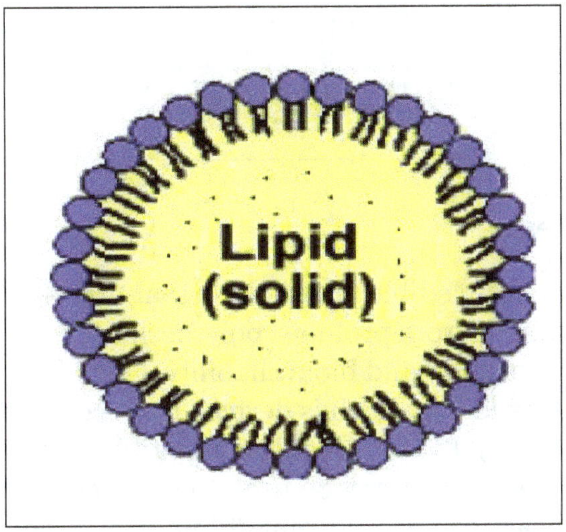

While comparing solid lipids and liquid lipids, solid lipids have an added advantage as they demonstrate controlled release of encapsulated compounds in addition to the stability of incorporated bioactive compounds. There are diverse potential beneficial effects of these carriers as they possess many advantageous physicochemical characteristics of the lipid phase. Firstly, they efficiently lower the rate of mobility and chemical degradation of bioactive agents in the matrix. Secondly, they also prevent phase separations, thus making them highly stable. Thirdly, the absorption of bioactive compounds which are otherwise poorly absorbed can be increased by incorporation into solid lipid nanoparticles.

Liposomes

Liposomes can be defined as vesicles that are spherical in shape and prepared using cholesterol and phospholipids. These are efficient vehicles for delivery of important functional molecules due to their size as well as the presence of both hydrophobic and hydrophilic character. The surface charge, lipid composition and the methodology used for the preparation influences the properties of liposomes to a great extent. Generally, these can be defined as spherical vesicles having a particle size range between 30 nanometers to several micrometers. Structurally, they are composed of more than one lipid bilayers, where the polar head groups are arranged in a way that they align in the pathway of the periphery and external layer of aqueous phases. These are broadly used as carriers for numerous bioactives. Liposomes can offer various advantages such as biocompatibility, biodegradability and low toxicity. Additionally, food industries have extensively used liposomes for encapsulation of unstable compounds such as antioxidants, flavors and bioactive elements to enhance their functionality as well as bioavalaibility. In the context of applications of liposomes in terms of hydrophobic and hydrophilic components, both can be efficiently entrapped. Furthermore, its decomposition can be avoided and even the targeted and controlled release of the encapsulated compound can be achieved.

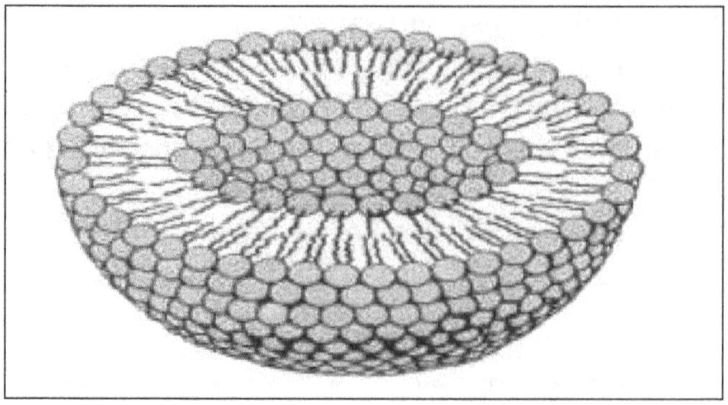

Polymeric Nanoparticles

Nanocarriers are usually made from edible materials so that they don't have any harmful effect on the body. These are generally made from proteins, polysaccharides and lipids which can be ingested thereby increasing their absorption and bioavailability. One of the best biomaterial for the delivery of bioactive compounds is PLGA which is derived from copolymerization of poly lactic acid (PLA) and poly glycolic acid (PGA). PLGA is highly compatible and efficient as it can be processed into any shape and size and therefore it can be used to encapsulate bioactive compounds of any size. It has very good mechanical strength, biocompatibility and biodegradability. PLGA has better interaction with the bioactive compound and prevents it from degradation in the harsh conditions of gastrointestinal tract. Since PLGA is hydrolyzed in the body into its metabolite monomers, it is FDA approved for use in humans.

Food Nanopackaging

Global food industry is under rising pressure to meet consumers demand for safe, healthy and fresh food, along with a challenge to meet updated strict food safety regulations. Awareness through easy access to digital media have motivated consumer toward demand for fresh, minimally processed, nutritious, safe and ready-to-eat food products with well-defined labels. In-order to make certain the safety and authenticity of food stuffs throughout the food supply chain, food manufacturers, traders, buyers and food regulatory authorities seeks a novel, cost-effective, fast and consistent tool to monitor the packaged food quality; more effectively to the traditional passive barrier concept, as packaging is a crucial component of every segment of food industry. Nano-technological interventions to food packaging chiefly explore three possibilities *viz.* direct incorporation into food products, incorporation in food packaging material, and application in food processing. The commercialization, successful execution and responses to diverse applications of nanotechnology are determined by the consumer's outlook toward and acceptance of newly introduced technologies and their applications. Outcome from majority of the studies (apart from those of sample population) indicates wide consumer acceptance of nanoparticles as packaging materials and also when they are used during processing activities; as compared to their direct incorporation into agri-food products.

Nanotechnology mainly comprises of fabrication, characterization and manipulation of nano-range

(<100 nm) molecules. The application of nanotechnology in polymers involve the design, manufacturing, processing and application of polymer materials filled with nano-particles and/or devices of nano range. The enormous potential of this promising intervention has gained attention of researchers from multi-disciplinary areas i.e., biological sciences, chemistry, engineering and physics. Owing to high global interest, nanotechnology has been proposed to impact the global economy by around $3 trillion by 2020, generating a requirement of approximately 6 million professionals in different inter-related sectors. As predicted by the Institute for Health and Consumer Protection (IHPC), the nanoparticle based market will touch $20 billion mark by 2020. It can be anticipated that nanotechnology will create a major thrust for the development of advanced packaging systems for the sake of consumers. Differently from the materials at macroscale, nanomaterials display specific and improved physicochemical properties. By virtue of their small size, nanoparticles hold a huge surface-to-volume ratio and surface activity. When affixed to desirable polymers, nanomaterials results in improved mechanical strength, electrical conductivity and thermal stability etc. Nanomaterials thus improve the mechanical and barrier properties of food packages; along with offering active and intelligent packaging systems.

In order to meet effective food packaging requirements, advanced nanomaterial augmented polymers will help to amplify the benefits associated with existing polymers, with enhanced safety, besides addressing environmental concerns. The developed packaging material will contribute in reducing any serious interaction between packaging and food matrices, impact over consumer's health, drop in quantity of waste material, improved biodegradability and barrier shielding to gases and light, and reduced CO_2 emissions. The ongoing advancement(s) in the area of polymer nanomaterials for food packaging (PNFP) faces a challenge ahead to overcome the safety and regulatory barriers. Upon qualifying the safety parameters and production at a competitive ratio between cost and performance, the new PNFP will be an ideal candidate for widespread application in food packaging.

The introduction of polymer nanotechnology in food packaging aims to improve the principal features of traditional packaging systems i.e., *containment* (ease of transportation and handling), *convenience* (being consumer friendly), *protection and preservation* (avoids leakage or break-up and protects against microbial contaminants, offering longer shelf life), *marketing and communication* (real-time information about the quality of enclosed food stuffs, besides the nutritional constituents and preparatory guidelines.

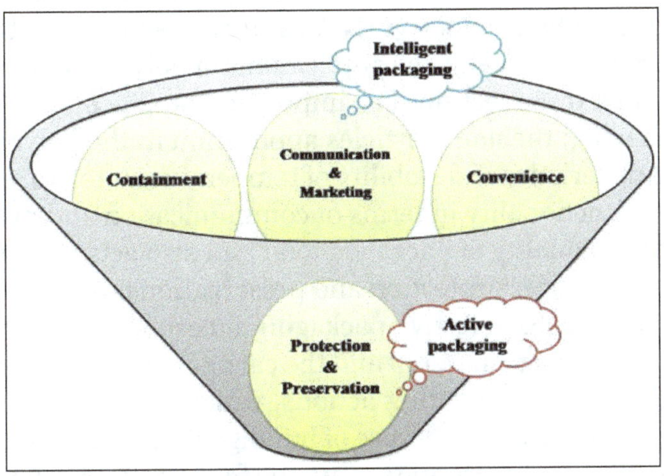

Concept behind and mode of action of active packaging and intelligent packaging.

In the twenty-first century, food producers/processors/vendors and consumers, seeks novel and resourceful food packaging systems, with committed food safety, quality and traceability. This demand needs innovative tools/technology that can be assembled together for application in food packaging. The current interventions and developments in food packaging to be commercially feasible and effectively acceptable, must meet regulatory guidelines along with a justified outcome that outweighs the associated expenses of added novel technology. Figure reflects the growing interest in this field as can be substantiated with the growing number of publications since 2000.

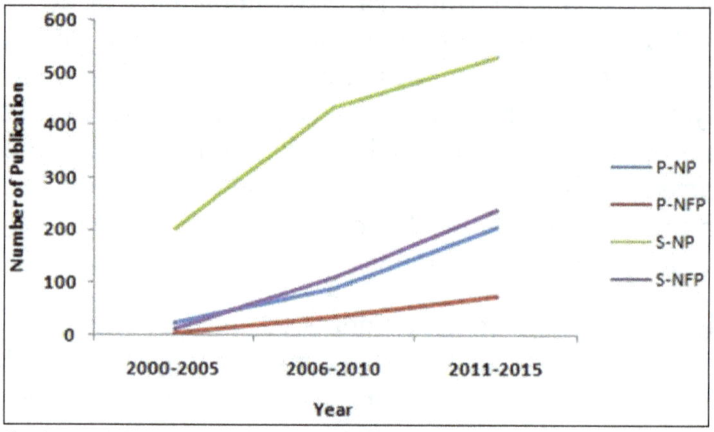

The global trend of research interest in nano packaging (NP) and Nano food packaging (NFP), as estimated by the number of Pubmed (P) and Scopus (S) publications hits from 2000 to 2015.

The major drawback associated with food packaging material(s) is their permeable nature. None of the packaging material offers complete resistance to the atmospheric gases, water-vapors and food and packaging material presently. Among the available options organic polymer materials *viz.* polypropylene, polyethylene, polyethylene terephthalate, polystyrene and polyvinyl chloride, remain the main choice material in food packaging industry, favored due to their lesser cost, simple processing and light weight. However, the key problem rests in their inherent permeability to gases and other small molecules.

A sole polymer fails to offer all the desirable properties needed for efficient food packaging; so blending of polymer(s) or complex multilayer films is generally preferred. This lacuna can be sorted out by using polymer nano-composites as food packaging materials. Nanomaterials are commonly classified into nanoparticles (NPs), nanoclays (NCs) and nanoemulsions (NEs) which have numerous applications in food sector. Basically, the applications of polymer nanomaterials for food packaging will be discussed under three sections i.e., improved, intelligent/smart and active packaging. In case of improved packaging; the nanoparticles application in the polymer matrix improves the elasticity, gas barrier characteristics and stability of temperature/moisture parameters. Intelligent packaging offers superior functionality in terms of communication and marketing. It provides dynamic feedback on the actual quality of packaged food and also acts as a guard against fraudulent imitation. Active packaging assures protection and preservation based on mechanism(s) activated by intrinsic and/or extrinsic factors. Active packaging incorporates the active constituent using nanotechnology into a food package material and the carrier component can interact with internal and/or external factors, thereby stimulating actions, which increases the shelf-life, food quality and safety of the food product. However, in case of intelligent packaging systems, the nanotechnology based indicator/sensor(s) are incorporated into the food package, where they can interact with internal (food components and headspace) and external environmental factors.

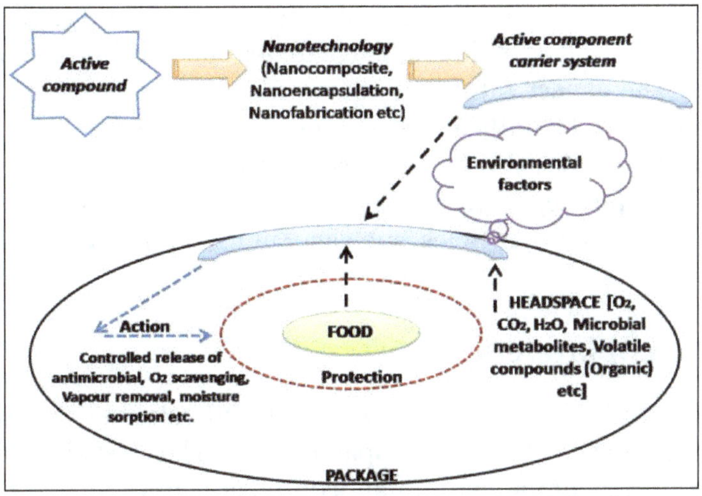

Active packaging and its association with nanotechnology.

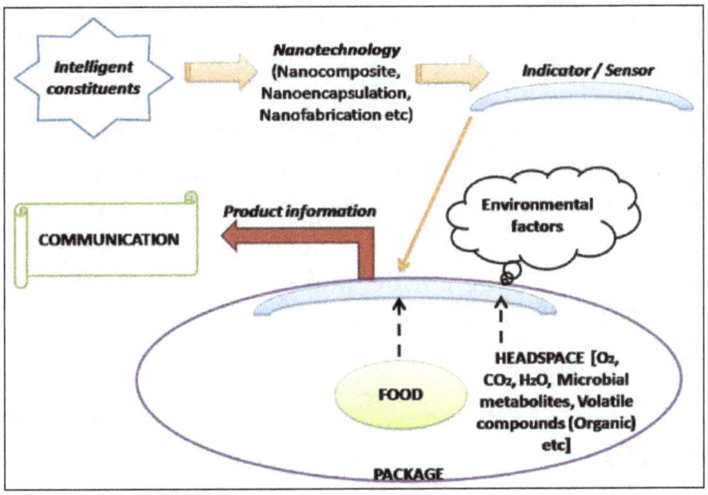

Demonstration of intelligent packaging concept and its association with nanotechnology.

Improved Packaging through Nanocomposites

Nanocomposites, a fusion of traditional food packaging material with nanoparticles are gaining active interest in food packaging sector. In addition to its remarkable antimicrobial spectrum, it displays great mechanical performance and tough resistant characteristics. Nanocomposites are usually made up of a polymer matrix in a continuous or discontinuous phase. It is a multiphase material resulting from the amalgamation of matrix (continuous phase) and a nano-dimensional material (discontinuous phase). Based on the nano-material, the nano-dimensional phase is generally characterized into nanospheres or nanoparticles, nanowhiskers or nanorods, nanotubes and nanosheets, or nanoplatelets. Nano-sized phases augment the mechanical properties of polymer, where the elastic strain is transferred to nano-reinforced material. Owing to this property, nanocomposite has been recognized as a gold standard for improvising the mechanical and barrier characteristics of polymers. Besides improving the mechanical and barrier characteristics, nanoparticles also append active or smart properties to the packaging system. The application of nanotechnology in polymer science can open new avenues for improving the characteristic features and cost-price-competence of packaging materials.

Polymer nanocomposites (PNCs), the mixtures of polymers with inorganic or organic fillers with particular geometries (fibers, flakes, spheres, particulates) have been recently introduced as novel packaging materials. The aspect ratio (the ratio of largest to smallest dimension of filler) of packaging filler material plays a significant role. Fillers having higher aspect ratios possess more specific surface area, with associated high reinforcing properties. Various nanomaterials such as silica, clay, organo-clay, graphene, polysaccharide nanocrystals, carbon nanotubes, chitosan, cellulose-based, and other metal nanoparticles, such as, ZnO_2, colloidal Cu, or Ti are being extensively explored as fillers.

Clay and Silicate Nanoplatelets

Clay and silicates, owing to their availability, low cost and relatively simple processibility have attracted focus of researchers as potential nanoparticles. The layered silicates, about 1 nm thick to a few microns long are usually employed in nanocomposites in their two-dimensional structure. The combination of silicates and polymers impart excellent barrier properties. This combination enhances the diffusive path for an infiltrate molecule. In addition to the typical tactoid structure of microcomposites, interaction among layered silicates and polymers may result in intercalated or exfoliated nanocomposites. The intercalated nanocomposites represent a multilayerd structure with alternating polymer/inorganic layers lying apart by few nanometers. Such structures results through the penetration of polymers chains into the interlayer region of clay lead. The exfoliated nanocomposites comprises of extensive polymer penetration with random dispersion of clay layers. Application of montmorillonite clay as a nanocomponent in extensive variety of polymers such as, polyethylene, nylon, polyvinyl chloride, and starch dates back to 1990s Montmorillonite (MMT) $[Mx(Al_4\text{-}xMgx)\ Si_8O_{20}(OH)_4]$ is the most common clay filler.

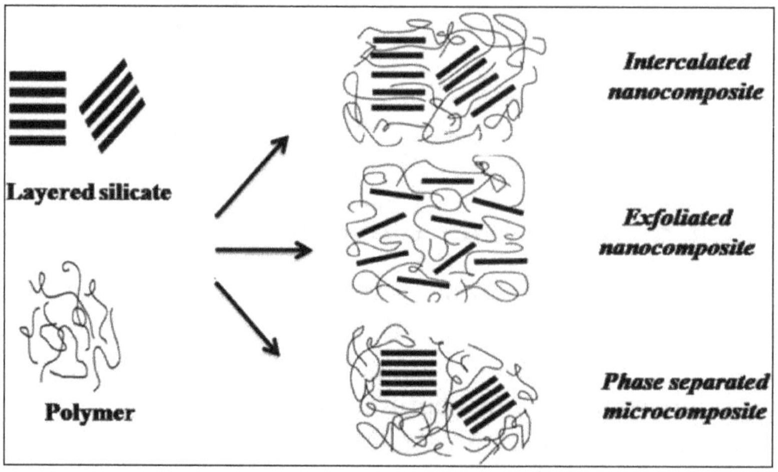

Types of composites obtained from the interaction between layered silicate and polymers.

It represents an octahedral sheet of Al $(OH)_3$ between silica tetrahedral bi-layers, linked together by weak electrostatic forces. The imbalance between the surface negative charges is compensated by the presence of exchangeable cations, Na^+ and Ca^{2+}. The clay layers offers resistance to gases and water vapor permeability. Further, addition of clays improves the mechanical strength of biopolymers. The application of 5% (w/w) of clays in thermoplastic starch (TPS)/clay nanocomposites, improves the mechanical properties with decreased water vapor permeability of starch biopolymer. Clay nanoparticles are further known to impact the glass transition and thermal degradation temperatures. Nano-clays also have prospects in active and intelligent food nano-packaging. Recently, proposed a nano-packaging system developed by inserting blueberry extract between the silicate interlayer spaces of clay. Blueberries possess anthocyanins which changes color with pH, attributed to a shift between quinoidal and flavylium forms. Incorporation of blueberry extract could modify these clays into active and intelligent nanocomposites.

Cellulose-based Nanofibers or Nanowhiskers

Cellulose, the building material of long fibrous cells, is a highly strong natural polymer. It is ubiquitous, cost-effective, environmentally friendly and easy to recycle. Cellulose is also explored as a supporting material for many nanomaterials. Application of cellulose increases the surface area of nanoparticles associated with their enhanced activity. These additive features make cellulose nanofibers an attractive class of nanomaterials. Basically two types of nano reinforcements *viz.* microfibrils and whiskers can be derived from cellulose. The cellulose chains appear as microfibrils (or nanofibers), which are bundles of molecules that are elongated and stabilized through hydrogen bonding. Each microfibril is further formed by aggregation of elementary fibrils, which contain crystalline and amorphous parts. The crystalline parts can be isolated by acid hydrolysis treatments; and are referred to as whiskers, nanocrystals, nanorods or rod like cellulose microcrystals. The dimensions of whiskers, after hydrolysis mainly rely on the percentage of amorphous regions in the bulk fibrils, which differ between different organisms. Microcrystalline cellulose (MCC) consists of huge quantity of cellulose microcrystals associated with amorphous areas.

Cellulose based nano reinforcements improve the strength, thermal characteristics and modulus of polymers, with restricted elongation. The assimilation of cellulose nanowhiskers and starch systems improve their thermo-mechanical characteristics, along with reduced water sensitivity and intact biodegradability. Cellulose nano reinforcements also improvised the moisture barrier properties of polymer films. Earlier, researcher prepared an all-cellulose nanocomposite (ACNC) film consisting of sugarcane bagasse nanofiber and N, N-dimethylacetamide/lithium chloride as solvent. The study utilized a very low-value agricultural waste product for preparing a high performance nanocomposite with a tensile strength of around 140 MPa.

Carbon Nanotubes

Carbon nanotubes (CNTs) are available as either single wall nanotube (SWNT) or multiwalled nanotubes (MWNT). SWNT is generally one atom thick, whereas, MWNT comprises of several concentric tubes with very high aspect ratios and elastic modulus. The tensile strength/modulus of several polymers such as polyethylene naphtalate, polyvinyl alcohol, polyamide and polypropylene have been improvised by incorporating carbon nanotubes and polyamides. CNTs possess elastic modulus of upto1TPa and tensile strength of 200GPa. CNTs exhibit antibacterial properties, attributed to their direct penetration through microbial cells. Researcher documented that CNTs in combination with allyl isothiocyanate could inhibit *Salmonella choleraesuis* for over a period of 40 days of storage. Researcher reported that CNTs infused with polyethylene films used for the packaging of Mazafati dates, prevented fungal invasion for up to 90 days. A single-walled carbon nanotube with cobalt meso-arylporphyrin complexes has also been explored for developing a chemiresistive detector, which detects amines generated during spoilage of meat. However, in spite of having wide application, concerns associated with their processing and dispersion aspects, along with high cost, limits their incorporation in nanocomposites.

Starch Nanocrystals

Starch has been extensively explored over decades as a choice material for food packaging applications. Several associated benefits *viz.* abundance, biocompatibility, non-toxicity, low cost, biodegradability, easy availability, and stable in air further enhance its possible applications. The incorporation of inorganic materials and synthetic polymers further improves its water resistance properties. Native starch granules can be submitted to an extended-time hydrolysis at temperatures below the gelatinization temperature when the amorphous regions are hydrolyzed allowing separation of crystalline lamellae, which are more resistant to hydrolysis. The starch crystalline particles show platelet morphology with thicknesses of 6–8 nm, improve tensile strength and modulus of pullulan films, with decreased elongation property. It has been proposed that the positive charge present over any antimicrobial agent contributes to its antimicrobial activity. As a result, the antimicrobial spectrum of metals incorporated/adsorbed on to polysaccharides surface increases, owing to enhanced surface area.

Chitin/Chitosan Nanoparticles

Chitosan, a heteropolysaccharide is known for its biocompatibility, biodegradability, along with metal complexation. The polycationic nature of chitosan is mainly accountable for its wide antimicrobial activity. Chitosan nanoparticles are formed through ionic gelation, where the positively charged amino groups of chitosan electrostatically interacts with the polyanions engaged as cross-linkers. Earlier, researcher observed that the combination of chitin whiskers with soyprotein isolate (SPI) thermoplastics significantly enhanced the tensile characteristics of the matrix and their resistance to water. Later, Burdock proposed hydroxypropyl methylcellulose (HPMC) act as a potential material for edible packaging films. Researcher added that the chitosan based nanocomposites in HPMC helped in the improvement of the mechanical and barrier characteristics. Researcher developed a chitosan-based antimicrobial film comprising of chitosan and polyvinyl alcohol. The developed film could effectively extend the shelf-life of tomato, displaying antibacterial activity against *E. coli, S. aureus*, and *B. subtilis*. Edible films and coatings prepared from chitosan improvised physicochemical and microbiological quality of fresh-cut vegetables and fruits. The active antibacterial bags consisting of chitosan/polyethylene have displayed efficacy in inhibiting total aerobic mesophilic bacteria, coliforms, molds, and yeasts in chicken drumsticks; besides maintaining pH, color, and hardness of samples. Chitosan/silver, chitosan/gold and chitosan/cinnamaldehyde nanocomposite films have demonstrated antimicrobial activity against *E. coli, S. aureus, P. aeruginosa, Aspergillus niger*, and *Candida albicans*.

Active Packaging

Unlike conventional food packaging, an active packaging is intentionally designed packaging system that incorporates components that would release (antimicrobial or antioxidant agents) or absorb (oxygen or water vapor) material into or from the packaged food or the food environment. Amalgamation of active compounds *viz.* antimicrobial agents, preservatives, O_2 and water vapor absorbers, ethylene removers etc. with polymer renders it more effective for boosting the shelf life and quality of food product.

Inorganic Nanoparticles

Various metals [Silver (Ag), Gold (Au), Zinc (Zn)] and metal oxide [Titanium dioxide (TiO_2), zinc oxide (ZnO), silicon oxide (SiO_2) and magnesium oxide (MgO)] derived nanomaterials have been explored in diverse active packaging applications. These particles either function on direct contact or they can migrate slowly and react preferentially with organics present in the food. Nanoparticles antimicrobial activity might be due to one of these mechanisms: direct interaction with the microbial cells (interrupting trans-membrane electron transfer, disrupting/penetrating the cell envelope); oxidizing cell components; and production of secondary products (e.g., reactive oxygen species - ROS or dissolved heavy metal ions), leading to cell damage.

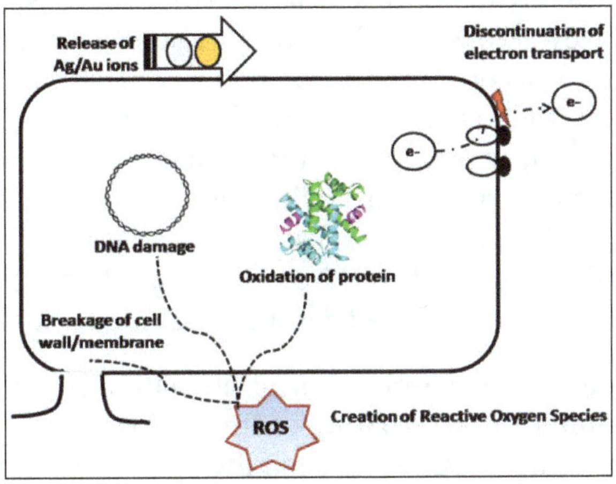

Different mechanisms of antimicrobial activity exhibited by nanoparticles.

Silver Nanoparticle

Silver nanoparticles are among the most explored nanoparticles, owing to their established antimicrobial potential against multiple commensal and pathogenic strain. Besides bacterial strains, silver nanoparticles are known to be inhibitory against multiple fungi and also several viruses. Silver targets bacterial metabolism by binding to its DNA proteins and enzymes; resulting into bacteriostatic effects. Silver nanoparticles destabilize and disrupt both the outer and cytoplasmic membranes. Silver nanoparticles also inhibit the respiratory chain enzymes and can also stimulate the production of reactive oxygen species (ROS). Earlier, researcher evaluated the effectiveness of polyamide 6 and silver micro and nanoparticles incorporated films against *E. coli*. The developed nano-particles based film completely inhibited the *E. coli* growth. However films with silver micro particles killed 80% of the bacteria. Researcher reported noteworthy antimicrobial activity of silver nanoparticles incorporated agar films against *Listeria monocytogenes* and *E. coli* O157:H7. In an earlier study, researcher documented strong antibacterial activity of PLA/silver-OMMT nanocomposite against *Salmonella* spp. Antibacterial activity of nanocomposite systems such as PVA dispersed with cellulose nanocrystals and silver nanoparticles has been displayed against *E. coli* and *Staphylococcus aureus* (Researcher analyzed the antimicrobial effect of cellulose/chitosan-silver and cellulose/chitosan nanocomposite films and reported better activity in films with silver nanoparticles. Researcher studied biosynthesized silver nanoparticles incorporated into sodium alginate films in food packaging, and showed noteworthy antibacterial effect against *E. coli* and *S. aureus*. In another study conducted by researcher, silver nanoparticles deposited over multilayered linear low density polyethylene (LLDPE) showed 70% reduction of *Aspergillus niger*.

Recently, researcher observed the UV/ozone treated commercial low-density polyethylene (LDPE) films coated by layer-by-layer (LbL) method by alternating the deposition of polyethyleneimine (PEI) and poly(acrylic acid) (PAA) polymer solutions and antimicrobial silver. They demonstrated that the resultant films containing antimicrobial Ag NPs could be explored in antimicrobial packaging. Researcher reported that size of nanoparticles also affects the antimicrobial efficacy as they evaluated the antibacterial properties of hydroxypropyl methylcellulose films containing silver nanoparticles with diameters of 41 and 100 nm against *E. coli* and *S. aureus* and observed that silver nanoparticles with smaller size (i.e., 41 nm) had greater antibacterial properties than the larger one.

Silver zeolites are also used to create antibacterial polymer composites. AgNP-based nanocomposites are stable and offer slow release of silver ion into stored foods, resulting in persistent antimicrobial activity. In a study where AgNP/SiO_2 nanocomposite material was compared with that of Ag zeolite and $AgNO_3$/SiO_2 composite, effective antimicrobial activity was displayed by both the materials; however, longer period of activity was displayed by the nancomposite; while a better immediate effect was observed with zeolite-based material. Nanocomposite based silver nanoparticles may find application in food packages requiring longer transportation or storage. To explore the antimicrobial potential of AgNP on the shelf life of food, various AgNP/polymer nanocomposite materials have been inspected within actual food systems.

Researcher proposed an edible antibacterial film of alginate and AgNPs over sterilized carrots and pears. Researcher documented that coating of fresh asparagus spears with AgNP/polyvinyl pyrrolidone nanocomposite films, extended its shelf life by 25 days at refrigerated storage. When the Chinese jujube fruit was stored in bags made up of AgNP/nanoparticulate TiO2/polyethylene films, lesser decaying rate along with delayed ripening over a period of 12 days was observed. Orange juice stored at 4°C in LDPE films having P105 (TiO_2 and 10 nm nanosilver mixture) displayed considerable reduction in *Lactobacillus plantarum* growth over the storage period of 112 days. Recently, researcher prepared binary blend films comprising of agar and banana powder (A/B) composite films reinforced with silver nanoparticles (A/B/AgNPs). These films displayed strong antibacterial activity against *E. coli* and *L. monocytogenes*. With the increase in the concentration of banana powder, characteristics such as UV light absorption, water vapor barrier and antioxidant activity of A/B blend also amplified whereas their mechanical properties decreased. Researcher observed another kind of application i.e., AgNPs containing cellulose pads, which reduces the levels of microbial exudates of meat stored in modified atmosphere packaging. In a follow up study, antimicrobial activities of AgNP-containing cellulose pads were displayed in fresh foods, along with slower ripening rates with extended shelf lives. Further, cellulose pads containing silver nanoparticles have also been successfully applied for beef coating, wherein significant reduction in microbial load was reported.

Other Antimicrobial Nanoparticles

Copper nanoparticles were shown to inhibit the growth of *Saccharomyces cerevisiae, E. coli, S. aureus* and *L. monocytogenes* on a polymer composite after 4 h exposure. Researcher also revealed good antibacterial effect of copper nanoparticles against *E.coli* and *B. subtilis* in polyurethane nanofibers containing copper nanoparticles. Copper nanoparticles leads to multiple toxic effects such as, generation of reactive oxygen species (ROS), lipid peroxidation, protein oxidation and

DNA degradation; which might be responsible for its antimicrobial activity. Zinc nanocrystals have also been used as an antimicrobial and antifungal agent, when incorporated with plastic matrix. Different nanoparticles oxide(s) such as titanium dioxide (TiO_2), zinc oxide (ZnO), silicon oxide (SiO_2), and magnesium oxide (MgO) do found application in food packaging, due to their ability to act as UV blockers and photo-catalytic disinfecting agents. Among all, TiO_2 particles have been most promising. The antimicrobial activity of TiO_2 nanoparticles is photocatalyzed and these antimicrobial particles are active only in the presence of UV light. TiO_2 nanoparticles displayed good activity against *S. choleraesuis, Vibrio parahaemolyticus*, and *L. monocytogenes* under UV illumination but not in the dark.

Titanium Dioxide (TiO_2)

Naturally, titanium dioxide exists in three primary phases i.e., anatase, rutile, and brookite; having varied crystal sizes. TiO_2 possess photocatalytic abilities and at nanoscale TiO_2 shows surface reactivity, which connects it with biological molecules (phosphorylated proteins and peptides) and DNA. The surface energy of TiO_2 nanoparticles amplifies with size and is the significant factor in polymer/filler interaction. The surface energies of rutile particles are higher than those of anatase particles of similar size. TiO_2 is being highly explored in preparing several nanomaterials *viz.* nanoparticles, nanorods, nanowires, nanotubes, mesoporous and nanoporous TiO_2 containing materials. The antibacterial properties of TiO_2 is well known. However, the antibacterial capacity of nano-TiO_2 particles confined to the exposure of UV irradiation. Although, the exact mechanism of biocidal activity of TiO_2 is unclear, it may be attributed to its initial oxidative attack over the outer/inner bacterial cell membrane, alterations of Coenzyme A-dependent enzyme activity, and DNA damage through hydroxyl radicals.

Researcher examined the photo-activated biocidal properties of TiO_2 nanoparticles based EVOH films against nine microorganisms (*B. stearothermophilus, S. aureus, E. coli, P. fluorescens, Bacillus* sp., *L. plantarum, E. caratovora, P. jadinii, Z. rouxii*). The TiO_2 nano-particles were uniformly dispersed ultrasonically. They reported over 5 log reduction for *B. stearothermophilus, Bacillus* sp., *L. plantarum* and *P. jadinii* after 30 min of irradiation in the presence of the TiO_2/EVOH materials. Chawengkijwanich and Hayata observed 3 log reduction in *E. coli* count over fresh cut lettuce treated with TiO_2 nanoparticles coated oriented-polypropylene (OPP) film after 3 h of illumination. In contrary, uncoated films could reduce the *E. coli* count by only 1 log under similar conditions. In another study, TiO_2 nanoparticle coated plastic films were evaluated against *Penicillium expansum* spoilage in apple, tomato and lemons. Results indicated that the growth of *P. expansum* was suppressed due to the photocatalytic properties of TiO_2 particles upon exposure to light. Combination of TiO_2 nanoparticles with silver has been shown to enhance the antimicrobial properties to a significant level.

Silicon Oxide (SiO_2)

Silica nanoparticles ($nSiO_2$) also have potential to improve the mechanical and/or barrier properties of various polymer matrices. Researcher examined the tensile characteristics (i.e., strength, modulus and elongation) of $nSiO_2$ incorporated polypropylene (PP) matrix. Addition of $nSiO_2$ into starch matrix could improve the tensile properties along with decreased water absorption by starch (Researcher integrated $nSiO_2$ in an isotactic polypropylene (iPP) matrix using maleic anhydride grafted polypropylene (PP-g-MA) and this $nSiO_2$ enhanced the storage modulus of iPP and making the material stiffer with improvised O_2 barrier capacity of matrix. Researcher generated polyvinyl alcohol and SiO_2 nanocomposites through radical copolymerization of vinyl silica nanoparticles and vinyl acetate. The resultant nanocomposites exhibited enhanced thermal and mechanical properties, in comparison to pure polyvinyl alcohol. This feature may be due to strong covalent interactions between $nSiO_2$ and the polymer matrix. In another study, Tang and Liu fabricated starch/PVOH/$nSiO_2$ biodegradable films and documented that the tensile and water resistance properties of films enhanced with an increase in their $nSiO_2$ content. Additionally, increased intermolecular H bonding and formation of C–O–Si groups between $nSiO_2$/starch and $nSiO_2$/PVOH were observed. Such bonding resulted in improvised miscibility and compatibility between the film components.

Food packaging applications of nano-silica has been explored as food contact surface materials. Bayer Polymers, Germany made silicate nanoparticles enriched packaging film which retards the entrance of O_2 and other gases and loss of moisture; preventing food spoilage. Nanocor Inc, Chicago, IL, USA, developed a clay nanoparticle based nanocomposite. Researcher reported that nanocomposites comprising of 5% silica nanoparticles lead to the improvement of mechanical and physical properties. Recently, Farhoodi showed that the application of SiO_2 nanoparticles as fillers in food packaging materials leaves a twisting pathway for gases. Researcher modified paper to form a lotus like super hydrophobic surface by coating with R812S silica nanoparticles and Polydimethylsiloxane (PDMS) silicone oil. The coated paper displayed strong water repellent characteristic.

Zinc Oxide (ZnO)

Zinc oxide particles display good antibacterial activity, which further enhances with decrease in particle size. ZnO requires visible light for stimulation. Its direct contact with microbial cell wall may result in destruction of bacterial cell integrity, liberation of antimicrobial ions i.e., Zn^{2+} ions, and generation of ROS. Sawai observed that ZnO was the most effective antibacterial agent for *S. aureus* growth inhibition when compared to MgO and CaO. Sevinc and Hanley explored that ZnO lessen the growth of bacterial biofilms (*Streptococcus sobrinus*) nearly 80% in dental materials. In an another study, researcher observed different approaches i.e. powder, film, PVP capped and coating; for the application of nano-ZnO in food systems and concluded that nano-ZnO displayed antibacterial effects against *L. monocytogenes* and *Salmonella enteritidis* in liquid egg white and in culture media. Researcher compared antimicrobial activity of ZnO powder and ZnO nanoparticles against food borne pathogens (*Bacillus cereus, E. coli, S. aureus, S. enteridis*) and observed that ZnO nanoparticles displayed better antibacterial activity than non-nano powder against all the tested bacteria. Petchwattana and Naknaen prepared PBS (polybutylene succinate)/thymol film and evaluated its antimicrobial activity against *E. coli* and *S. aureus*. Recently, researcher displayed significant antibacterial activity of poly butylenes succinate (PBS)/zinc oxide composite films against *E. coli* and *S. aureus*. Researcher analyzed the inhibitory activity of carvacrol oil against *S. aureus* and *E. coli*. Essential oils act can as a promising antimicrobial agent in PBS film but their unpleasant odors affected its acceptance. To overcome its odor, when ZnO is added to PBS, then the resultant product displayed better antibacterial activity devoid of any odor from volatiles.

Oxygen Scavengers

Access to O_2 in the food package reduces the shelf-life of product owing to diverse degradation processes such as browning, rancidity, growth of aerobic microflora, lessening of vitamins, off-odors, off-flavors and undesirable color changes. Along with it, during storage, fresh fruits and vegetables produces some undesirable compounds i.e., production of ethylene vapor in climacteric fruits, which accelerates the postharvest ripening and shortening the shelf-life. In turn to circumvent these detrimental effects, the entry of O_2 should be eradicated within the package or lessened to acceptable level according to the product. The content of O_2 in the headspace of packaged food can be reduced through vacuum sealing or by generating inert gas atmosphere in the packaging (N_2, CO_2) or both, and through the use of oxygen absorbent materials. Various active packaging perceptions based on nanocomposite films have been explored for oxygen and ethylene scavenging or limit the path of oxygen diffusion. A varied number of polymers having nanostructured oxygen scavengers can be explored as active packaging material for sliced meat, poultry, beverages, cooked pasta, ready-to-eat snacks, fish etc. Researcher developed O_2 scavenger films and incorporated TiO_2 NPs to different polymers and this nanocomposite material was further explored as packaging films in O_2 sensitive food stuffs.

Later on, other researchers various researchers modified the high density polyethylene (HDPE) films by integrating them with iron containing kaolinite to produce O_2 scavenging packaging films. At nanoscale, TiO_2 can be photoinduced by UV radiation and electrons got accumulated on the TiO_2 surface. In photocatalytic reactions, the transfer of electron to oxygen in the air is the rate-determining step and the photogenerated holes (h+) have strong oxidation potential, which can react with water and other compounds. Researcher revealed that the deposition of nanocrystalline TiO_2 on glass and acetate films formed a deoxygenated closed environment and the photocatalytic properties of TiO_2 utilized for the removal of ethylene vapor to delay the ripening of climacteric fruits. Maneerat and Hayata developed TiO_2 coated polypropylene films for the removal of ethylene vapor in packaged horticultural products. Researcher explored the photoelectrocatalytic degradation of ethylene using TiO_2 supported on activated carbon; the membrane obtained is capable of degrading ethylene vapor when irradiated with UV radiation. This strategy can be explored for delaying the ripening of climacteric fruits by constantly removing the ethylene vapor during long-distance transportation.

Enzyme Immobilization

Enzymes possesses multifunctional role in food industry but possess few limitations i.e., sensitivity to processing atmosphere (temperature, pH) and/or to enzyme inhibitors, sometimes restricts their applicability in food system but the concept of immobilization and incorporation of enzymes in packaging materials provides an alternative to the direct use of enzymes in food matrix. Immobilization at nanoscale increases the surface area and boosts the performance by enhancing the stability to pH, temperature and resistance to proteases; besides helping controlled release of enzymes into food system. According to researchers, immobilization is an efficient way to improve the stability of enzyme toward pH and temperature, resistance to proteases and other denaturing components and provides an ample environment for their repetitive use or controlled discharge in food system. Researcher observed that the enzymes (lactase or cholesterol reductase) inclusion to packaging substances might enhance the food product values and match the consumer requirement with enzyme deficiencies. Rhim and Ng explored the approach of enzyme adsorption into nanoclays incorporated to polymers. It is further supported by Gopinath and Sugunan that nanoclay has a high affinity for protein adsorption and could be utilized as a competent enzyme carrier. Researcher reported that conductive polymers may be explored as immobilizing matrices for biomolecules. However, researcher also validated the same by immobilizing the glucose oxidase onto films of poly (aniline-co-fluoroaniline). Researcher modified SiO_2 nanoparticles to immobilize glutamate dehydrogenase and lactate dehydrogenase, which displayed better enzyme activity after immobilization.

Intelligent/Smart Packaging System

Intelligent or smart packaging systems boost the communication aspect of a package and such type of advanced packaging recognize any characteristics of the packaged food and apply different mechanisms to register and convey information regarding the existing quality or condition of the food with regard to its safety and digestibility. Such system uses different innovative communication methods i.e., Nano-sensors, time-temperature indicators, oxygen sensors, freshness indicators etc. The inclusion of nanosensors in food packaging systems helps in detecting the spoilage associated changes, pathogens and chemical contaminants, and thus providing the exact scenario of food products freshness. Nanosensors are nanotechnology enabled sensors characterized by a range of variations. Generally, nanosensors can be applied as labels or coatings to add an intelligent function to food packaging in terms of ensuring the integrity of the package through the detection of leaks (for foodstuffs packed in a vacuum or inert atmosphere), indications of time–temperature variations (e.g., freeze–thaw–refreeze), or microbial safety (the deterioration of foodstuffs). Gas sensors are basically used for revealing the gaseous analyte in the package. Optical O_2 sensors works on the principle of luminescence quenching or absorbance changes caused by direct contact with the analyte; whereas optochemical sensors are used to check the quality of products by sensing gas analytes such as, hydrogen sulfide, CO_2 and volatile amines. The unique chemical and electro-optical properties of nanoscale particles respond to environmental changes (e.g., temperature or humidity in storage rooms, levels of oxygen exposure), product degradation or microbial contamination. Therefore, such technology would apparently benefit consumers, industry stakeholders and food regulators.

O_2 Sensors

Food packaging systems with restricted oxygen availability are preferred over those giving free oxygen accesses. In order to achieve this packaging needs to be carried out under vacuum or nitrogen gas along with incorporation of irreversible O_2 sensors. In case of modified atmosphere packaging (MAP), the headspace O_2 concentration is either deliberately reduced to an optimal level or entirely eliminated as per food product requirements. Researcher developed a UV based colorimetric O_2 indictor that utilizes TiO_2 nanoparticles for photosensitization of triethanolamine induced reduction of methylene blue in a polymer encapsulation medium. UV irradiation leads to bleaching of sensor, which remains colorless until exposed to O_2. The pace of change in color is relative to the level of O_2 exposure. Mills and Hazafy used nanocrystalline SnO_2 as photosensitizer and a colorimetric O_2 indicator consisting of glycerol (as electron donor), methylene blue (a redox dye) and hydroxyethyl cellulose (as encapsulating polymer). Upon exposure to UV-B light, indicator gets bleached and photoreduction of dye takes place by SnO_2 nanoparticles. Another study done by Mihindukulasuriya and Lim generated UV based activated O_2 indicator membrane through electrospinning method. Encapsulating of the active components *viz.* TiO_2 nanoparticles, glycerol and methylene blue within electrospun poly fibers, enhanced the oxygen sensitivity of the membrane.

CO_2 Sensor

To monitor freshness or quality of packaged foods during their storage condition, researcher formed chitosan-based CO_2 indicator. For the enhancement of signal strength of indicator, 2-amino-2-methyl-1-propanol (AMP) was used as an additive to the chitosan solution. The indicator solution of chitosan and AMP can be easily packed into sachets, which are permeable to gaseous CO_2 inside food packages. Researcher develop an optochemical CO_2 sensor which uses a phosphorescent reporter dye phosphorescent Pt-porphyrin PtTFPP and a colourimetric pH indicator (naphtholphthalein) integrated in plastic matrix together with a phase transfer agent (tetraoctyl or cetyltrimethyl ammonium hydroxide). The sensor material was optimized for food packaging applications and underwent complete characterization with respect to its CO_2 sensitivity and cross-sensitivity to O_2. Researcher concluded that the optical CO_2 gas sensors, particularly, dry optical sensors having pH-sensitive dye indicators can be explored for food packaging applications.

References

- Nanosensors-in-food-safety-318339488: researchgate.net, Retrieved 17, February 2020
- Functional-Foods-and-Nanotechnological-Interventions-I-Nutraceuticals-and-their-limitations: inflibnet.ac.in, Retrieved 06, May 2020

Chapter 5

Understanding Biogenic and Stealth Nanoparticles

Biogenic nanoparticles are the particles which are synthesized through the use of biological organisms are known as biogenic nanoparticles. Nanoparticles which are able to avoid recognition by the immune system is known as a stealth particle. This chapter has been carefully written to provide an easy understanding of the varied facets of biogenic and stealth nanoparticles.

Biogenic Nanoparticles

Nanoparticles synthesized using biological systems (eg. Microbes, fishes, plants etc.) are referred to as biogenic nanoparticles. The process of synthesis of the nanostructures by organisms is also known as biomineralization. As the process involves no harsh chemicals and solvent systems, this method of nanoparticle synthesis is classed as a 'Green synthesis'. These nanoparticles possess the advantage of having uniform size and shape. Moreover, they have also been found to possess better stability owing to stabilization by proteins and other biomolecules from the organism. The biogenic nanoparticles may be sequestered in separate intracellular compartments inside the cells and are referred to as intracellular biogenic particles. Separation of these nanoparticles requires disruption of the cells. On the other hand, some organisms synthesize these nanoparticles outside the cell or send them outside post-synthesis. Such nanoparticles are referred to as extracellular biogenic nanoparticles.

Why do organisms synthesize nanoparticles? One of the major reasons involved in such synthesis is detoxification. The major route to shield the cell(s) from the toxicity of certain soluble ionic species such as metal ions is to convert them into insoluble forms through reduction or precipitation.

In general, four major detoxification strategies have been identified in organisms:

(a) Modifications in the cellular transport mechanisms to restrict entry of the toxic ions into the cell.

(b) Sequestration of the toxic species within the cell (intracellular sequestration) or outside the cell (extracellular sequestration).

(c) Activation of energy-dependent efflux pathways to eliminate the toxic species.

(d) Enzyme catalyzed oxidation or reduction of the toxic species to a less toxic form.

In certain cases, the synthesis of nanoparticles is initiated to meet the cell's requirements for a functional component such as iron oxide or silica. Such synthesis occurs via oxidation and condensation-association routes. Some of these nanostructures have exquisite and complex morphology that is too difficult to be recreated using chemical routes of synthesis. Moreover, large-scale production of these biogenic nanoparticles can be easily accomplished and hence this area has been among the most widely explored areas in nanoscience and technology.

Synthesis of Biogenic Metal Nanoparticles

The synthesis of metal nanoparticles by biological systems is mainly a strategy employed to protect the system against the toxic effects of the soluble metal ions. Several species of microbes and plants exhibit tolerance to metal ions through this strategy. Both bacterial and fungal species have been employed for synthesizing metal nanoparticles. The most widely synthesized metal nanoparticles are silver and gold due to their widespread applications in many fields of science and technology. The metal nanoparticles synthesized using fungal species is referred to as 'mycogenic nanoparticles' while those synthesized using bacterial species is known as 'bacterioform nanoparticles'. Even higher plants have been shown to be effective in synthesizing silver and gold nanoparticles and the amount of reports available in this topic is too large to be summarized.

> Did you know?
>
> The huge number of fungal species that have been used to synthesize different nanoparticles with excellent stability and size control has led to the genesis of a new field known as "Myconanotechnology". The cost-effective large-scale production of metallic, semiconductor and metal oxide nanoparticles is being developed for various applications in this rapidly expanding field.

Some of the major organisms that have been employed to synthesize nanoparticles are listed in the following Table.

Table: Biological systems used for synthesizing nanoparticles.

Name of the biological species	Nanoparticle	Size (nm)
Acalypha indica	Silver	20-30
Actinobacter spp.	Magnetite	
Aspergillus clavatus	Silver	10-25
Aspergillus flavus	Silver	8.92±1.61
Aspergillus fumigatus	Silver	5-25
Aspergillus niger	Silver	20
Apiin extracted from henna leaves	Silver & Gold	39; 7.5-65 (respectively)
Avena sativa (oat)	Gold	5-20 (pH 3-4) & 25-85 (pH 2)
Azadirachta indica	Silver	50
Brassica juncea (mustard)	Silver	2-35
Bacillus licheniformis	Silver	50
Bacillus licheniformis (culture supernatant)	Silver	50

Bacillus megaterium	Silver	46.9
Bacillus sp.	Silver	5-15
Bacillus subtilis	Silver	5-60
Bacillus subtilis (culture supernatant)	Silver	5-60
Brevibacterium casei	Silver	50
Candida glabrata	CdS	Not available
Carica Papaya	Silver	60-80
Cinnamomum camphora leaf	Silver	55-80
Citrus limon (lemon)	Silver	<50
Cladosporium cladosporioides	Silver	10-100
Clostridium thermoaceticum	CdS	Not available
Cochlospermum gossypium	Silver	3
Coriandrum satiyum leaf extract	Silver	26
Coriandrum sativum	Gold	6.75-57.91
Corynebacterium sp.	Silver	10-15
Cymbopogon flexuosus (lemongrass)	Gold	200-500
Cycas sp. (cycas)	Silver	2-6
Datura metel	Silver	16-40
Desmodium triflorum	Silver	5-20
Eclipta sp.	Silver	2-6
Enhydra fluctuans	Silver	100-400
Escherichia coli	Silver	1-100
Escherichia coli	CdS	Not available
Eucalyptus camaldulensis (river red gum)	Gold	1.25-17.5
Eucalyptus citriodora (neelagiri)	Silver	~20
Eucalyptus hybrida (safeda)	Silver	50-150
Euphorbia hirta	Silver	40-50
Ficus bengalensis (marri)	Silver	~20
Fusarium acuminatum	Silver	5-40
Fusarium oxysporum	Silver	5-50
Fusarium oxysporum and Verticillium sp.	Magnetite	Not available
Fusarium semitectum	Silver	10-60
Fusarium solani	Silver	5-35
Garcinia mangostana (mangosteen)	Silver	35
Geobacter sulfurreducens	Silver	200
Gliricidia sepium	Silver	10-50
Glycine max (soybean) leaf extract	Silver	25-100
Honey	Silver	4
Ipomoea aquatic	Silver	100-400
Jatropha curcas (seed extract)	Silver	15-50
Klebsiella pneumonia (culture supernatant)	Silver	50

Lactic acid bacteria	Silver	11.2
Lactobacillus Strains	Silver	500
Ludwigia adscendeous (ludwigia)	Silver	100-400
Magnetospirillium magneticum	Fe_3O_4, Fe_3S_4	Not available
Mentha piperita (peppermint)	Silver & Gold	5-30, 90; 150
Moringa oleifera	Silver	57
Morganella sp.	Silver	20±5
Murraya koenigii	Silver & Gold	10; 20
Nelumbo nucifera (lotus)	Silver	25-80
Ocimum sanctum (tulsi; root extract)	Silver	10
Ocimum sanctum (tulsi; leaf extract)	Silver	10-20
Penicillium brevicompactum WA 2315	Silver	23-105
Penicillium fellutanum	Silver	1-100
Phanerochaete chrysosporium	Silver	100
Phyllanthus amarus	Silver	18-38
Proteus mirabilis	Silver	10-20
Psidium guajava (guava)	Gold	25-30
Pseudomonas stutzeri AG259	Silver	200
Scutellaria barbata D. Don (Barbated skullcup)	Gold	5-30
Schizosaccharomyces pombe	CdS	Not available
Sesbania drummondii (leguminous)	Gold	6-20
Stapphylococcus aurens	Silver	1-100
Syzygium aromaticum (clove)	Gold	5-100
Syzygium cumini (jambul)	Silver	29-92
Terminalia catappa (almond)	Gold	10-35
Trichoderma asperellum	Silver	13-18
Trichoderma viride	Silver	5-40
Verticillium sp.	Silver	25±12

The differences in the reducing agents, stabilizing moieties as well as templating molecules between different species cause variations in the size, size distribution as well as crystal type of the nanoparticles.

Mechanism Involved in the Synthesis of Biogenic Metal Nanoparticles

Though different biogenic sources have been identified for synthesis of nanoparticles, the mechanism of synthesis is yet to be deciphered clearly. Many theories have been suggested to explain the mechanism of synthesis, especially for metal nanoparticles. As biogenic metal nanoparticles have been found to be stable for three months and more, it is evident that there should be a reducing agent as well as a stabilizing agent in the system that would facilitate formation and stabilization of the metal nanoparticles. The identification of these components from the complex cellular environment is extremely challenging. The nature of the reducing agent and stabilizing agent also varies with the species employed for the synthesis. In general, peptides, proteins, polyols and heterocyclic compounds have been suggested to have an important role in the synthesis of metal nanoparticles through an electron shuttle or charge capping mechanism.

Peptides have been investigated for their role as possible reducing agents as well as stabilizing agents in many microbial species. Interestingly, several reports have indicated that a single amino acid might not be as effective as a polypeptide sequence containing the same amino acid residue for the synthesis. The metal ion is first reduced to form metal nanoparticle, which acts as a nucleus for further growth of the metal crystal. The peptide is initially thought to adsorb to the surface of the metal nanoparticle clusters causing a localized reducing environment that results in the reduction of more metal ions at the interface between the peptide and the metal nuclei. As a result, nanoparticles with large size distributions are formed. Figure depicts a cartoon on the formation of a silver nanocrystal in the presence of peptide.

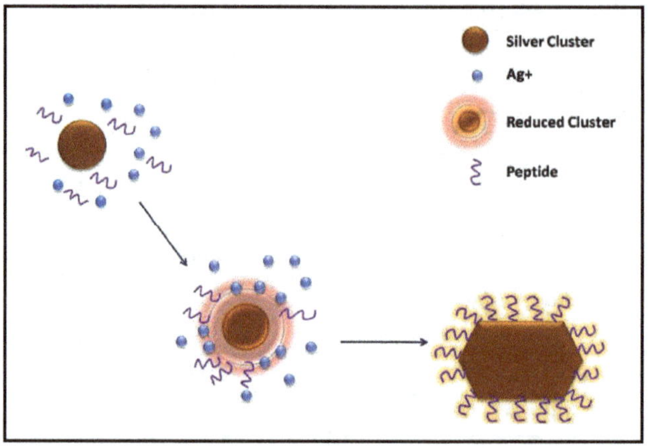

Formation of silver nanocrystals in the presence of peptide.

Formation of twinned crystals could also occur in this process. A twinned crystal forms when two crystals have some similar lattice points in a symmetrical manner. A similar kind of mechanism has been proposed with proteins in plants and is referred to as the 'recognition-reduction-limited nucleation'. The positively charged metal ions interact electrostatically with negatively charged residues in the proteins followed by reduction resulting in nucleation at specific sites. The protein template further directs the growth of the crystals in a specific orientation.

Several amino acid residues have been implicated in the reduction and stabilization of silver and gold nanoparticles. These include arginine, cysteine, lysine, methionine, tyrosine and tryptophan. Tyrosine residue has been shown to reduce silver and gold ions under alkaline conditions. Figure gives the possible reaction involved in the formation of metal nanoparticles in the presence of tyrosine residue.

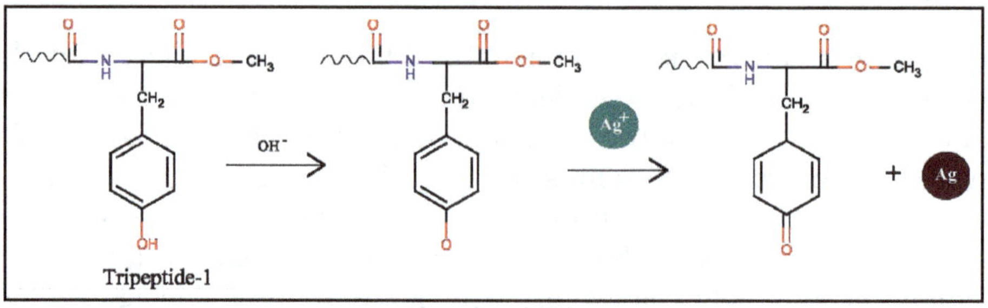

Role of tyrosine residue in the reduction of silver ions to form silver nanoparticles.

The phenolic group of tyrosine gets converted into a semi-quinone during this process. The N-terminus of the peptide will interact with the metal nanoparticle conferring stability. Is there any evidence for this theory? Yes! It was found that an engineered viral protein template developed from the capsid of cowpea chlorotic mottle virus (CCMV) containing tyrosine residue was found to effectively synthesize gold nanoparticles.

Removal of the tyrosine residues resulted in no reduction reaction and hence no gold nanoparticle formation occurred.This observation strongly suggests that the presence of tyrosine residue is a key factor in the formation of metal nanoparticles. Figure shows a schematic representation of a viral protein containing tyrosine involved in the synthesis of metal nanoparticles.

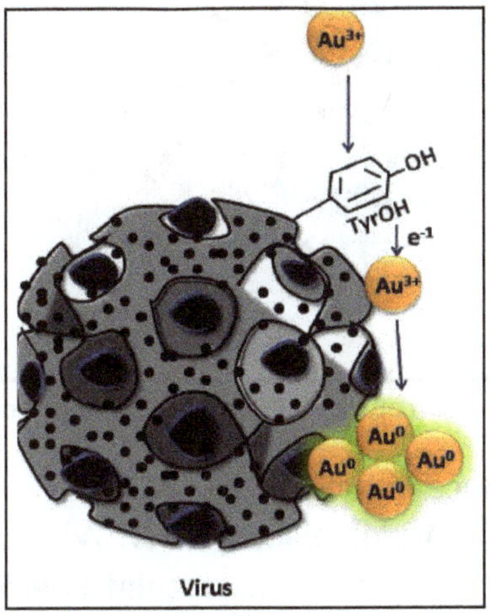

Viral capsid involved in the synthesis of gold nanoparticles.

Tryptophan residue in peptides also has been shown to play a key role in the formation of metal nanoparticles. The tryptophan residue forms a transient tryptophyl radical after donating an electron to the metal ion during the reduction reaction. The highly unstable radical forms ditryptophan, a fluorescent dimer and kynurenine, another fluorescent molecule. This scheme of reactions is represented in Figure.

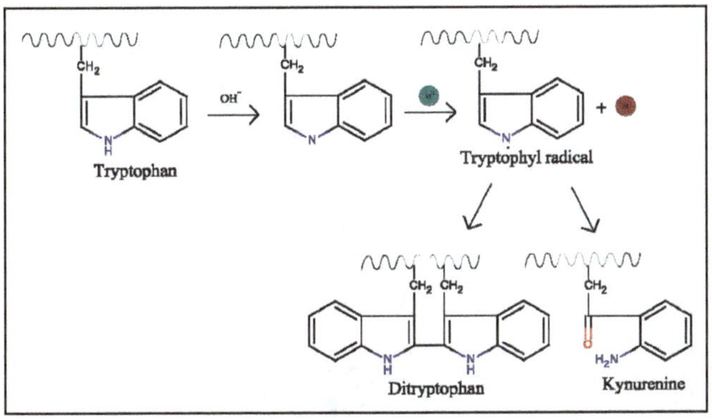

Role of tryptophan residue in the formation of metal nanoparticles.

Biogenic Synthesis of Metal Nanoparticles

It has also been recognized now that in several bacterial species, the presence of the co- enzyme NADH along with the NADH-dependent nitrate reductase enzyme enables the reduction of the metal ions to the respective metal particle. The stability of the resultant particle is ensured by the presence of proteins as capping agents on the surface of the nanoparticle. The NADH-NAD$^+$ redox system acts as an electron shuttle between the enzyme and metal ion. Hence this mechanism is referred to as 'electron shuttle enzymatic metal reduction process'. Figure represents the sequence of events that occur in a NADH-dependent formation of silver nanoparticles.

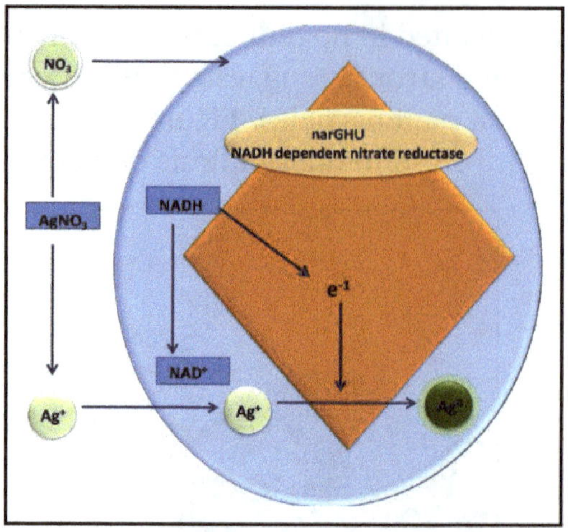

NADH-dependent electron shuttle enzymatic metal reduction process.

The reaction did not occur in the absence of the NADH coenzyme. However, if the NADH was replaced by anthraquinone, the reduction reaction took place readily. Why? Anthraquinone is a redox mediator. In other words it can shuttle electrons from one molecular species to another. Thisimplies that the presence of a redox mediator that could shuttle electrons between the enzyme and metal ion is essential in the reduction process leading to the formation of metal nanoparticles. Figure shows the scheme of events in the reduction process involving an anthraquinone moiety as the redox mediator.

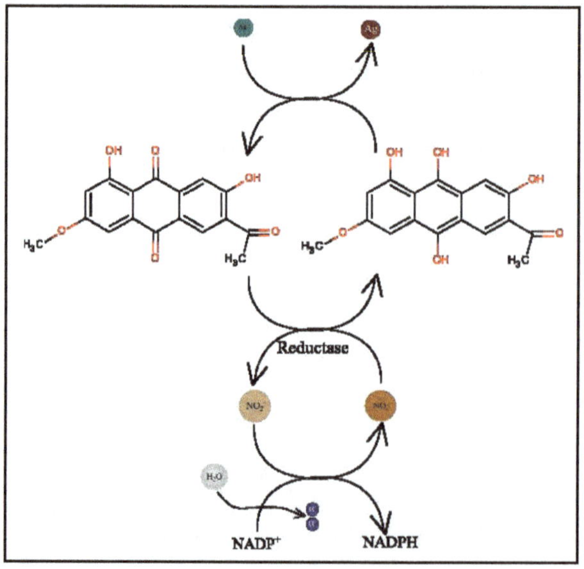

Anthraquinone mediated reduction of silver ions to form silver nanoparticles.

In several species of bacteria utilizing NADPH/NADP+ redox system, charge capping of the resultant metal nanoparticles by the negatively charged phosphate from the coenzyme has been observed indicating an additional stabilizing role for the coenzyme.

Different enzymes have been identified in different species to have been involved in the formation of metal nanoparticles. In Lactobacillussp. mediated synthesis of metal nanoparticles, a NADH dependent lactate dehydrogenase was involved in the reduction reaction along with glutathione and thioredoxin systems.

In a fungus-mediated synthesis of platinum nanoparticles from chloroplatinic acid (H_2PtCl_4), a two stage, two-step reduction mechanism is proposed for the conversion of Pt^{4+} to Pt^{o}. This is because a single four-electron transfer is energetically unfavourable.

The enzyme cytoplasmic hydrogenase that is sensitive to oxygen catalyzes the conversion of tetravalent platinum (Pt^{4+}) to the divalent platinum (Pt^{2+}). An oxygen tolerant periplasmichydrogenase enzyme catalyzes the reduction of Pt^{2+} to the metal. This final step is inhibited by copper ions. Figure depicts the scheme of events involved in the two-step process.

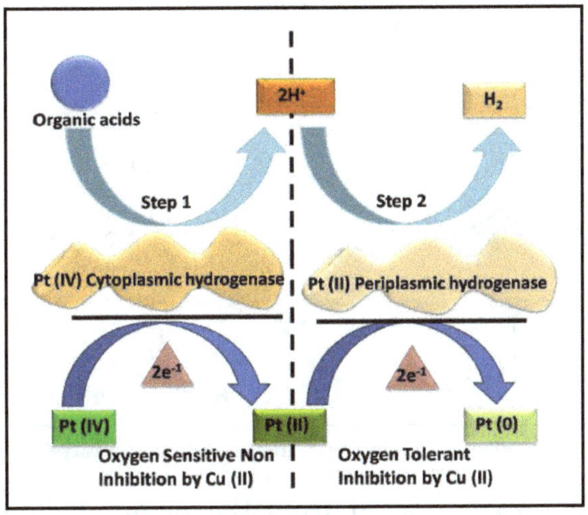

Two-step reduction reaction involved in the formation of platinum nanoparticles.

The pH of the medium is an important factor that determines the efficiency of the reduction process. The type of enzyme dictates the optimum pH at which the catalysis is most facilitated. Higher pH facilitates creation of anionic centres that could act as nucleation sites for the positively charged metal ions by promoting electrostatic interactions.

Reducing sugars such as glucose found in the biological systems also have the ability to reduce metal ions to form metal nanoparticles. However, this reaction is pH dependent. The monosaccharide sugars predominantly exist in the cyclic structure. At alkaline pH, the ring opening occurs leading to the linear form that possesses the free aldehyde functional group. This aldehyde catalyses the reduction of the metal ions while the sugar molecule itself gets converted to the corresponding carboxylic acid. Figure depicts a reduction reaction involving glucose moiety.

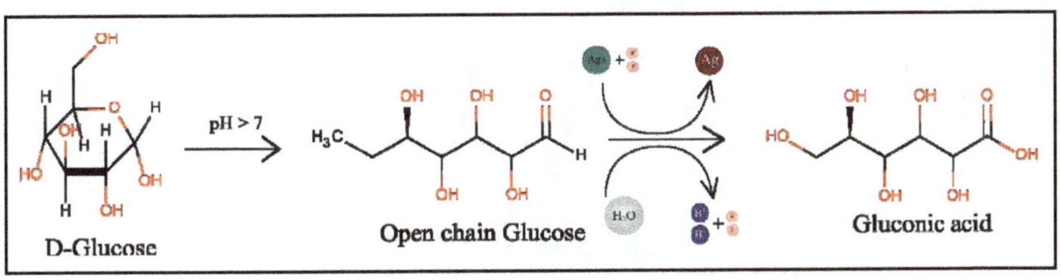

Formation of metal nanoparticles in the presence of glucose.

This reaction also forms the basis of biochemical assays in laboratories to detect the presence of reducing sugars (Eg. Benedict's test, Tollen's silver mirror test etc.), though the particle sizes usually are not in the nanoscale owing to the absence of any stabilizers. Whether this reaction is dominant in biological systems to synthesize metal nanoparticles is still not very clear. It is also probable that in the cell, the proteins may serve as template to restrict the size of the particles formed to the nano-dimensions.

In higher plants, apart from enzymes, other molecules such as polyols, heterocyclic compounds and flavonoids may have a role in the synthesis of nanoparticles. Terpenoids such as citronellol and geraniol (found in rose oils) can reduce silver ions to silver while getting transformed to the corresponding carboxylic acids. Figure depict the transformation of silver ions to silver nanoparticles in the presence of geraniol.

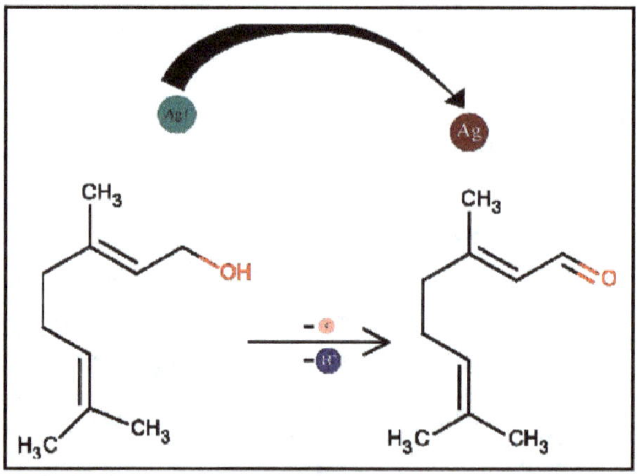

Formation of silver nanoparticles in the presence of a terpenoid, geraniol.

Similarly, phyllanthin, a lignan component from the plant Phyllanthusamarus that possesses anti-microbial and anti-oxidant properties also has been found to convert gold and silver ions to their corresponding metal nanoparticles. The shape of the resultant nanoparticles was influenced by the concentration of phyllanthin. This can happen when the phyllanthin molecules interact with the surface of the nanoparticle clusters and form a cage-like template that dictates the shape of the crystal obtained.

In certain plants, catechol (1,2-dihydroxy benzene) under alkaline conditions converts metal ions to metal nanoparticles while it gets oxidized to protocatechaldehyde and finally protocatechuric acid. This reaction is shown in Figure.

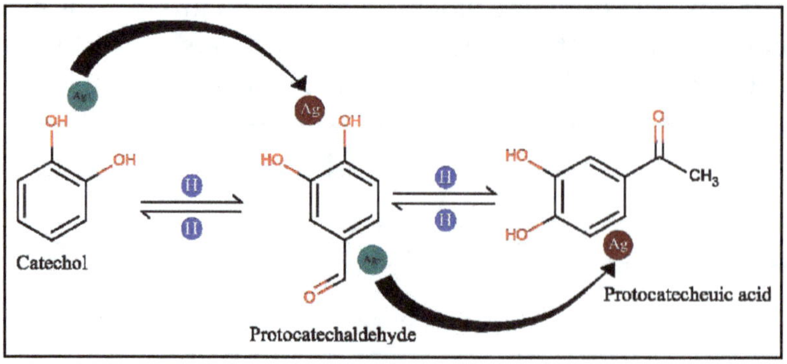

Formation of silver nanoparticles in the presence of catechol.

Similarly, the flavonoid quercetin has been found to be involved in a two-step reduction leading to formation of metal nanoparticles. Ascorbic acid or vitamin C has also been implicated in the reduction of metal ions to metal nanoparticles. The ascorbic acid itself gets transformed into dehydroascorbic acid. However, the factors regulating this transformation are yet to be identified conclusively. Figure shows the reaction involved in the formation of silver nanoparticles by ascorbic acid.

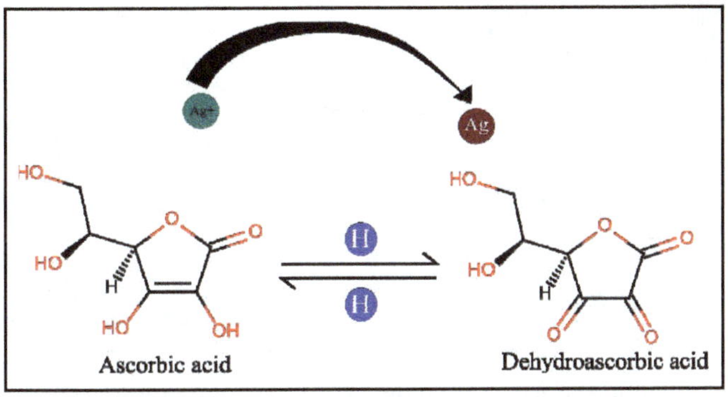

Formation of silver nanoparticles by ascorbic acid.

Anthraquinones such as emodin, cyperquinone, remerin etc. undergo keto-enoltautomerism where the enol form acts as the reducing agent.

The Jatropacurcas plant, more famous for its biodiesel properties also has been found to effectively reduce silver and gold ions to the respective metal nanoparticles. This conversion is facilitated by the presence of two cyclic peptides – curcacyclin A (cyclic octapeptide) and curcacyclin B (cyclic nonapeptide). Figure shows the reaction involving the curcacyclin peptides leading to the formation of silver nanoparticles.

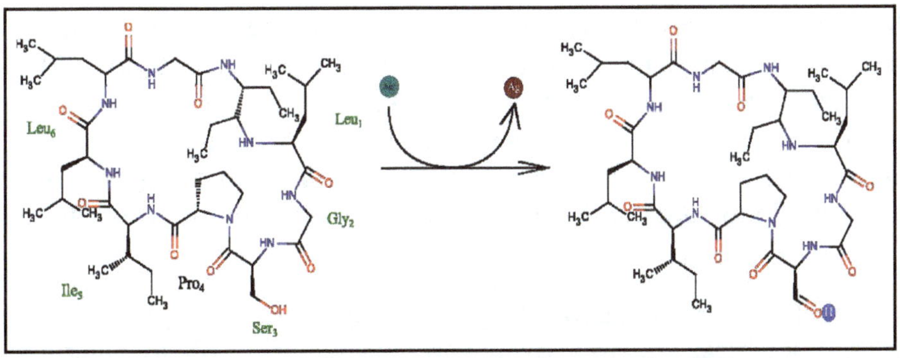

Formation of silver nanoparticles by curcacyclin peptides; Leu: leucine, Gly: glycine, Ser: serine, Ile: isoleucine.

Though the exact mechanism of synthesis is not confirmed, it is proposed that the metal ions are accommodated in the core of the cyclic peptides. The amide groups in the peptide bond stabilize the ions. The reduction reaction is catalyzed by the alcohol group in the threonine residue in curcacyclin A and the serine residue in curcacyclin B.

Thus, it is evident that the biological system abounds with many types of reducing agents that can effectively convert the metal ions to metal nanoparticles. The binding of the metal ions is a key step in the synthesis. The stabilization and size control is effected by templates which are generally peptides or proteins.The amount of the peptide in the system could serve to stabilize the nanoparticles by adsorbing on the surface. However, if a large amount of peptide is present, then they tend to promote aggregation of the metal nanoparticles thereby contributing to increase in size and formation of unique shapes. Figure highlights this aspect using a peptide stabilized gold nanoparticle system as example.

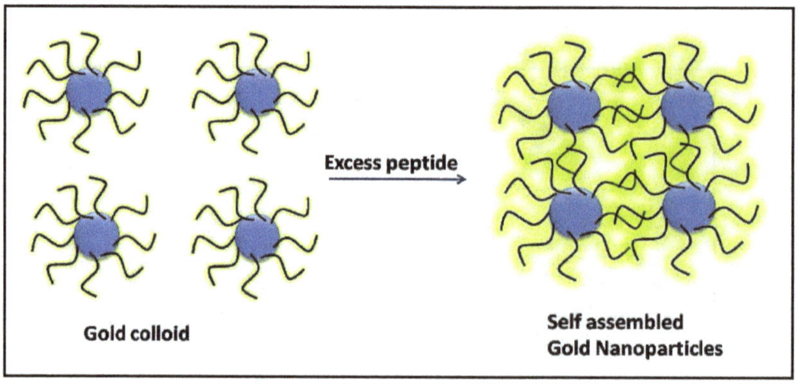

Influence of peptide concentration on size of the gold nano-structures obtained.

Biogenic Synthesis of Quantum Dots

The formation of cadmium sulphide and zinc sulphide nanoparticles in microbes and plants has been mainly attributed to their detoxification attempts as both cadmium and zinc ions are toxic to these organisms in a concentration-dependent manner. Two categories of peptides have been reported to be involved in such detoxification processes. These are phytochelatins and metallothioneins. Though the name 'phytochelatin' suggests a plant origin, this peptide has also been found in animals. The phytochelatins are cysteine-rich oligopeptides that have excellent metal binding properties, especiallywith cadmium. Copper, nickel and zinc also bind with phytochelatins but at higher concentrations. The expression of these peptides is triggered by the presence of the heavy metals. Phytochelatins have repeated (γ-Glu-cys) sequences followed by an amino acid, mostly glycine. Other types containing alanine, serine, glutamate and glutamine as the final residue have also been identified in different species. The (γ-Glu-cys) sequence has an unusual peptide bond formed between the gamma carboxyl group of glutamate and alpha amino group of cysteine as in the case of glutathione. The number of (γ-Glu-cys) repeats in different phytochelatins varies from 2-11.

How does the presence of heavy metal ions trigger the expression of phytochelatins? The presence of metal ions in the biological system activates an enzyme called phytochelatin synthase, which utilizes glutathione from the cells to assemble phytochelatin. The next question that needs to be addressed is how does the phytochelatin aid in the formation of cadmium sulphide nanoparticles? The cadmium ions tend to bind to the thiol groups leading to the formation of a low molecular weight phytochelatin-cadmium complex. This peptide-metal ion complex is most likely transported by ABC (ATP-binding cassette) membrane transport proteins into a vacuole. This might be more common in plant cells. However, as vacuoles are not common in animal cells, the site of sequestration of these complexes in animal cells is yet to be identified.

Once inside the vacuoles, addition of more glutathione units to the complex is continued leading to increase in molecular weight. Simultaneously, sulphide ions are pumped into the vacuole. The source of these sulphide ions still is to be identified. But, it is widely believed thatsulphide generation could arise due to the triggering of specific biochemical pathways by cadmium ions. In certain sulphate reducing bacteria, cysteine sulphinate is utilized to generate sulphide ions. These ions combine with the cadmium ions to form cadmium sulphide. The peptide template directs the size and shape of the resulting cadmium sulphide. Figure shows the schematic representation of the formation of cadmium sulphide using phytochelatin.

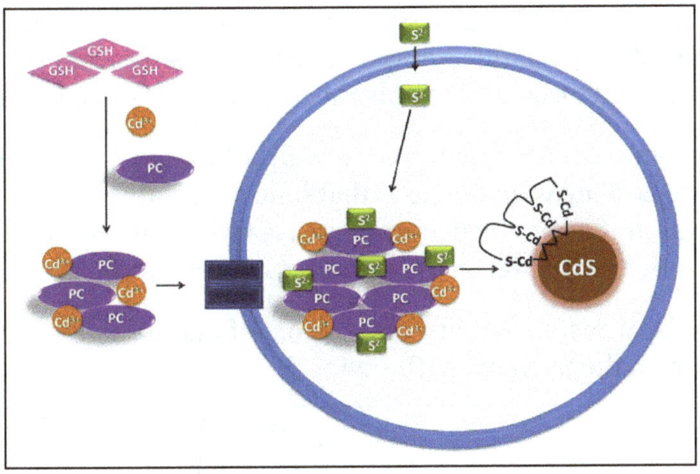

Formation of cadmium sulphide (CdS) by phytochelatin (PC); GSH represents reduced glutathione.

Interestingly, the chelation of the cadmium ions to the phytochelatin units is stable only in weakly alkaline and alkaline pH (pH > 7). The pH in the vacuoles is about 5.5 where the phytochelatin–cadmium chelates loose their stability. Thus the cadmium ions now in the form of cadmium sulphide are dissociated from the phytochelatin and remain in the vacuole while the free phytochelatin units are transported away from the vacuole using specialized transport systems. Some species still retain a peptide coat, presumably an oligomer or monomeric unit from the phytochelatin unit,as a surface coating on the cadmium sulphide that stabilizes the CdS nanoparticles. The size of the nanoparticles depends on the length of the peptide main chain.

Extracellular synthesis of CdS nanoparticles has also been reported in certain fungal species. The cadmium ions present in the extracellular medium will be converted to cadmium sulphide crystals. The sulphide ions are obtained from the reduction of sulphate ions by the sulphatereductase enzymes secreted by the fungus. A co-enzyme siroheme is considered essential for this process.In some cases, the enzyme cysteine desulphydrase is implicated in the generation of sulphide ions from the amino acid cysteine.

Another alternate protein template that has been implicated in the synthesis of CdS nanoparticles is albumin. The globular albumin has a pocket 5- 10 nm in diameter that can accommodate cadmium ions. The cysteine residues found in the albumin sequence can bind with the cadmium ions. This results in monodisperse and highly crystalline particles.

Metallothioneins are another class of cysteine-rich metal chelating proteins that may also have a role in the synthesis of sulphides of cadmium or copper or zinc, most likely in a manner similar to that of phytochelatins. Similarly, glutathione and cysteine are also said to play an important role in the synthesis of zinc sulphide and silver sulphide. However, the co-factors and other biomolecules involved in these processes are yet to be discovered.

Phytochelatin

Phytochelatins are oligomers of glutathione, produced by the enzyme phytochelatin synthase. They are found in plants, fungi, nematodes and all groups of algae including cyanobacteria. Phytochelatins act as chelators, and are important for heavy metal detoxification. They are abbreviated PC2 through PC11.

A mutant Arabidopsis thaliana lacking phytochelatin synthase is very sensitive to cadmium, but it grows just as well as the wild-type plant at normal concentrations of zinc and copper, two essential metal ions, indicating that phytochelatin is only involved in resistance to metal poisoning.

Because phytochelatin synthase uses glutathione with a blocked thiol group in the synthesis of phytochelatin, the presence of heavy metal ions that bind to glutathione causes the enzyme to work faster. Therefore, the amount of phytochelatin increases when the cell needs more phytochelatin to survive in an environment with high concentrations of metal ions.

Phytochelatin seems to be transported into the vacuole of plants, so that the metal ions it carries are stored safely away from the proteins of the cytosol.

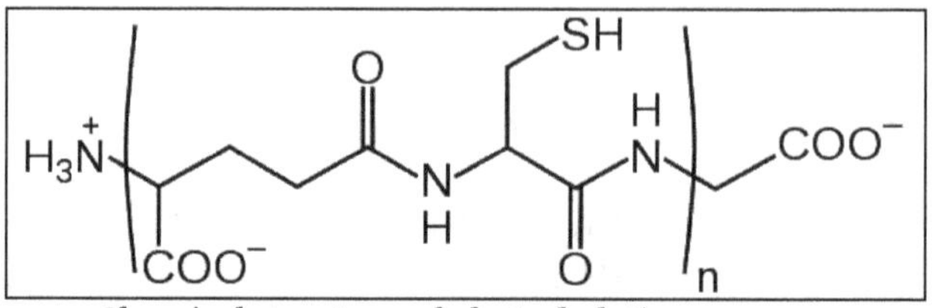

Chemical structure of phytochelatin. n = 2-11.

Related Peptides

There are groups of other peptides with a similar structure to phytochelatin, but where the last amino acid is not glycine:

Phytochelatin-like peptides

Phytochelatin-like peptides			
Type	**Structure**	**Has been found in**	**Precursor**
Phytochelatin	$(\gamma\text{Glu-Cys})_n$-Gly	many organisms	Glutathione
Homophytochelatin	$(\gamma\text{Glu-Cys})_n$-Ala	legumes	Homoglutathione
Desglycine phytochelatin	$(\gamma\text{Glu-Cys})_n$	maize, yeasts	
Hydroxymethyl-phytochelatin	$(\gamma\text{Glu-Cys})_n$-Ser	grasses	Hydroxymethylglutathione
iso-Phytochelatin (Glu)	$(\gamma\text{Glu-Cys})_n$-Glu	maize	Glutamylcysteinylglutamate
iso-Phytochelatin (Gln)	$(\gamma\text{Glu-Cys})_n$-Gln	horseradish	

History

Phytochelatin was first discovered in 1981 in fission yeast, and was named cadystin. It was then found in higher plants in 1985 and was named phytochelatin. In 1989 the biosynthetic enzyme, phytochelatin synthase, was discovered.

Metallothionein

Metallothionein (MT) is a family of cysteine-rich, low molecular weight (MW ranging from 500 to 14000 Da) proteins. They are localized to the membrane of the Golgi apparatus. MTs have the capacity to bind both physiological (such as zinc, copper, selenium) and xenobiotic (such as cadmium, mercury, silver, arsenic) heavy metals through the thiol group of its cysteine residues, which represent nearly 30% of its constituent amino acid residues.

MT was discovered in 1957 by Vallee and Margoshe from purification of a Cd-binding protein from horse (equine) renal cortex. MTs function is not clear, but experimental data suggest MTs may provide protection against metal toxicity, be involved in zinc and copper regulation, and provide protection against oxidative stress. There are four main isoforms expressed in humans: MT1 (subtypes A, B, E, F, G, H, L, M, X), MT2, MT3, and MT4. In the human body, large quantities are synthesised primarily in the liver and kidneys. Their production is dependent on availability of the dietary minerals such as zinc, copper, and selenium, as well as the amino acids histidine and cysteine.

Structure and Classification

MTs are present in a vast range of taxonomic groups, ranging from prokaryotes (such as the cyanobacteria *Synechococcus sp.*), protozoa (such as the ciliate *Tetrahymena* genera), plants (such as *Pisum sativum*, *Triticum durum*, *Zea mays*, or *Quercus suber*), yeast (such as *Saccharomyces cerevisiae* or *Candida albicans*), invertebrates (such as the nematode *Caenorhabditis elegans*, the insect *Drosophila melanogaster*, the mollusc *Mytilus edulis*, or the echinoderm *Strongylocentrotus purpuratus*) and vertebrates (such as the chicken *Gallus gallus*, or the mammalian *Homo sapiens* or *Mus musculus*).

The MTs from this diverse taxonomic range represent a high-heterogeneity sequence (regarding molecular weight and number and distribution of Cys residues) and do not show general homology; in spite of this, homology is found inside some taxonomic groups (such as vertebrate MTs).

From their primary structure, MTs have been classified by different methods. The first one dates from 1987, when Fowler *et al.*, established three classes of MTs: Class I, including the MTs which show homology with horse MT, Class II, including the rest of the MTs with no homology with horse MT, and Class III, which includes phytochelatins, Cys-rich enzymatically synthesised peptides. The second classification was performed by Binz and Kagi in 2001, and takes into account taxonomic parameters and the patterns of distribution of Cys residues along the MT sequence. It results in a classification of 15 families for proteinaceous MTs. Family 15 contains the plant MTs, which in 2002 have been further classified by Cobbet and Goldsbrough into 4 Types (1, 2, 3 and 4) depending on the distribution of their Cys residues and a Cys-devoid regions (called spacers) characteristic of plant MTs.

A table including the principal aspects of the two latter classifications is included.

Family	Name	Sequence pattern	Example
1	Vertebrate	K-x(1,2)-C-C-x-C-C-P-x(2)-C	*Mus musculus* MT1 MDPNCSCTTGGSCACAGSCKCKECKCTSCKKCCSCCPVG-CAKCAQGCVCKGSSEKCRCCA
2	Molluscan	C-x-C-x(3)-C-T-G-x(3)-C-x-C-x(3)-C-x-C-K	*Mytilus edulis* 10MTIV MPAPCNCIETNVCICDTGCSGEGCRCGDACKCSGADCKCS-GCKVVCKCSGSCACEGGCTGPSTCKCAPGCSCK

3	Crustacean	P-[GD]-P-C-C-x(3,4)-C-x-C	*Homarus americanus* MTH MPGPCCKDKCECAEGGCKTGCKCTSCRCAPCEKCTSGCK-CPSKDECAKTCSKPCKCCP
4	Echinoderms	P-D-x-K-C-V-C-C-x(5)-C-x-C-x(4)-C-C-x(4)-C-C-x(4,6)-C-C	*Strongylocentrotus purpuratus* SpMTA MPDVKCVCCKEGKECACFGQDCCKTGECCKDGTCCGICT-NAACKCANGCKCGSGCSCTEGNCAC
5	Diptera	C-G-x(2)-C-x-C-x(2)-Q-x(5)-C-x-C-x(2)D-C-x-C	*Drosophila melanogaster* MTNB MVCKGCGTNCQCSAQKCGDNCACNKDCQCVCKNGPKD-QCCSNK
6	Nematoda	K-C-C-x(3)-C-C	*Caenorhabditis elegans* MT1 MACKCDCKNKQCKCGDKCECSGDKCCEKYCCEEASEKKC-CPAGCKGDCKCANCHCAEQKQCGDKTHQHQGTAAAH
7	Ciliate	x-C-C-C-x ?	*Tetrahymena thermophila* MTT1 MDKVNSCCCGVNAKPCCTDPNSGCCCVSKTD-NCCKSDTKECCTGTGEGCKCVNCKCCKPQA-NCCCGVNAKPCCFDPNSGCCCVSKTNNCCKSD TKECCTGTGEGCKCTSCQCCKPVQQGCCCGDKAKACCTD-PNSGCCCSNKANKCCDATSKQECQTCQCCK
8	Fungal 1	C-G-C-S-x(4)-C-x-C-x(3,4)-C-x-C-S-x-C	*Neurospora crassa* MT MGDCGCSGASSCNCGSGCSCSNCGSK
9	Fungal 2	---	*Candida glabrata* MT2 MANDCKCPNGCSCPNCANGGCQCGDKCECKKQSCHGC-GEQCKCGSHGSSCHGSCGCGDKCECK
10	Fungal 3	---	*Candida glabrata* MT2 MPEQVNCQYDCHCSNCACENTCNCCAKPACACTNSAS-NECSCQTCKCQTCKC
11	Fungal 4	C-X-K-C-x-C-x(2)-C-K-C	*Yarrowia lipolytica* MT3 MEFTTAMLGASLISTTSTQSKHNLVNNCCCSSSTSESSM-PASCACTKCGCKTCKC
12	Fungal 5	---	*Saccharomyces cerevisiae* CUP1 MFSELINFQNEGHECQCQCGSCKNNEQCQKSCSCPT-GCNSDDKCPCGNKSEETKKSCCSGK
13	Fungal 6	---	*Saccharomyces cerevisiae* CRS5 TVKICDCEGECCKDSCHCGSTCLPSCSGGEKCKCDHSTG-SPQCKSCGEKCKCETTCTCEKSKCNCEKC
14	Procaryota	K-C-A-C-x(2)-C-L-C	*Synechococcus sp* SmtA MTTVTQMKCACPHCLCIVSLNDAIMVDGKPYCSEV-CANGTCKENSGCGHAGCGCGSA
15	Plant	[YFH]-x(5,25)-C-[SK-D]-C-[GA]-[SDPAT]-x(0,1)-C-x-[CYF]	
15.1	Plant MTs Type 1	C-X-C-X(3)- C-X-C-X(3)- C-X-C-X(3)-spacer-C-X-C-X(3)-C-X-C-X(3)- C-X-C-X(3)	*Pisum sativum* MT MSGCGCGSSCNCGDSCKCNKRSSGLSYSEMETTETVIL-GVGPAKIQFEGAEMSAASEDGGCKCGDNCTCDPCNCK
15.2	Plant MTs Type 2	C-C-X(3)-C-X-C-X(3)-C-X-C-X(3)- C-X-C-X(3)-spacer- C-X-C-X(3)- C-X-C-X(3)-C-X-C-X(3)	*Lycopersicon esculetum* MT MSCCGGNCGCGSSCKCGNGCGGCKMYPDM-SYTESSTTETLVLGVGPEKTSFGAMEMGESPVAENGCK-CGSDCKCNPCTCSK

15.3	Plant MTs Type 3	---	*Arabidopsis thaliana* MT3 MSSNCGSCDCADKTQCVKKGTSYTFDIVETQESYKEAMIM-DVGAEENNANCKCKCGSSCSCVNCTCCPN
15.4	Plant MTs Type 4 or Ec	C-x(4)-C-X-C-X(3)-C-X(5)-C-X-C-X(9,11)-HTTCGCGEHC- X-C-X(20)-CSC-GAXCNCASC-X(3,5)	*Triticum aestivum* MT MGCNDKCGCAVPCPGGTGCRCTSARSDAAAGEHTTCG-CGEHCGCNPCACGREGTPSGRANRRANCSCGAACNCAS-CGSTTA
99	Phytochelatins and other non-proteinaceous MT-like polypeptides	---	*Schizosaccharomyces pombe* γEC-γEC-γECG

Secondary structure elements have been observed in several MTs SmtA from *Syneccochoccus*, mammalian MT3, Echinoderma SpMTA, fish *Notothenia coriiceps* MT, Crustacean MTH, but until this moment, the content of such structures is considered to be poor in MTs, and its functional influence is not considered.

Tertiary structure of MTs is also highly heterogeneous. While vertebrate, echinoderm and crustacean MTs show a bidominial structure with divalent metals as Zn(II) or Cd(II) (the protein is folded so as to bind metals in two functionally independent domains, with a metallic *cluster* each), yeast and procariotyc MTs show a monodominial structure (one domain with a single metallic *cluster*). Although no structural data is available for molluscan, nematoda and Drosophila MTs, it is commonly assumed that the former are bidominial and the latter monodominial. No conclusive data are available for Plant MTs, but two possible structures have been proposed: 1) a bidominial structure similar to that of vertebrate MTs; 2) a codominial structure, in which two Cys-rich domains interact to form a single metallic cluster.

Quaternary structure has not been broadly considered for MTs. Dimerization and oligomerization processes have been observed and attributed to several molecular mechanisms, including intermolecular disulfide formation, bridging through metals bound by either Cys or His residues on different MTs, or inorganic phosphate-mediated interactions. Dimeric and polymeric MTs have been shown to acquire novel properties upon metal detoxification, but the physiological significance of these processes has been demonstrated only in the case of prokaryotic Synechococcus SmtA. The MT dimer produced by this organism forms structures similar to zinc fingers and has Zn-regulatory activity.

Metallothioneins have diverse metal-binding preferences, which have been associated with functional specificity. As an example, the mammalian *Mus musculus* MT1 preferentially binds divalent metal ions (Zn(II), Cd(II),...), while yeast CUP1 is selective for monovalent metal ions (Cu(I), Ag(I),...). Strictly metal-selective MTs with metal-specific physiological functions were discovered by Dallinger et al. (1997) in pulmonate snails (Gastropoda, Mollusca). The Roman snail (*Helix pomatia*), for example, possesses a Cd-selective (CdMT) and a Cu-selective isoform (CuMT) involved in Cd detoxification and Cu regulation, respectively. While both isoforms contain unvaried numbers and positions of Cys residues responsible for metal ligation, metal selectivity is apparently achieved by sequence modulation of amino acid residues not directly involved in metal binding (Palacios et al. 2011).

A novel functional classification of MTs as Zn- or Cu-thioneins is currently being developed based on these functional preferences.

Yeast

Metallothioneins are characterized by an abundance of cysteine residues and a lack of generic secondary structure motifs. Yeast Metallothionein (MT) are also alternatively named, Copper metallothionein (CUP).

Function

This protein functions in primary metal storage, transport, and detoxification. More specifically, Yeast MT stores copper so therefore protects the cell against copper toxicity by tightly chelating copper ions.

Structure

For the first 40 residues in the protein the polypeptide wraps around the metal by forming two large parallel loops separated by a deep cleft containing the metal cluster.

Examples

Yeast MT can be found in the following:

- *Saccharomyces cerevisiae.*
- *Neurospora crassa.*

Function

Metal Binding

Metallothionein has been documented to bind a wide range of metals including cadmium, zinc, mercury, copper, arsenic, silver, etc. Metallation of MT was previously reported to occur cooperatively but recent reports have provided strong evidence that metal-binding occurs via a sequential, noncooperative mechanism. The observation of partially metallated MT (that is, having some free metal binding capacity) suggest that these species are biologically important.

Metallothioneins likely participate in the uptake, transport, and regulation of zinc in biological systems. Mammalian MT binds three Zn(II) ions in its beta domain and four in the alpha domain. Cysteine is a sulfur-containing amino acid, hence the name "-thionein". However, the participation of inorganic sulfide and chloride ions has been proposed for some MT forms. In some MTs, mostly bacterial, histidine participates in zinc binding. By binding and releasing zinc, metallothioneins (MTs) may regulate zinc levels within the body. Zinc, in turn, is a key element for the activation and binding of certain transcription factors through its participation in the zinc finger region of the protein. Metallothionein also carries zinc ions (signals) from one part of the cell to another. When zinc enters a cell, it can be picked up by thionein (which thus becomes "metallothionein") and carried to another part of the cell where it is released to another organelle or protein. In this way the thionein-metallothionein becomes a key component of the zinc signaling system in cells. This system is particularly important in the brain, where zinc signaling is prominent both between and within nerve cells. It also seems to be important for the regulation of the tumor suppressor protein p53.

Control of Oxidative Stress

Cysteine residues from MTs can capture harmful oxidant radicals like the superoxide and hydroxyl radicals. In this reaction, cysteine is oxidized to cystine, and the metal ions which were bound to cysteine are liberated to the media. As explained in the *Expression and regulation* section, this Zn can activate the synthesis of more MTs. This mechanism has been proposed to be an important mechanism in the control of the oxidative stress by MTs. The role of MTs in oxidative stress has been confirmed by MT Knockout mutants, but some experiments propose also a prooxidant role for MTs.

Expression and Regulation

Metallothionein gene expression is induced by a high variety of stimuli, as metal exposure, oxidative stress, glucocorticoids, Vitamin D, hydric stress, etc. The level of the response to these inducers depends on the MT gene. MT genes present in their promotors specific sequences for the regulation of the expression, elements as metal response elements (MRE), glucocorticoid response elements (GRE), GC-rich boxes, basal level elements (BLE), and thyroid response elements (TRE).

Metallothionein and Disease

Cancer

Because MTs play an important role in transcription factor regulation, defects in MT function or expression may lead to malignant transformation of cells and ultimately cancer. Studies have found increased expression of MTs in some cancers of the breast, colon, kidney, liver, skin (melanoma), lung, nasopharynx, ovary, prostate, mouth, salivary gland, testes, thyroid and urinary bladder; they have also found lower levels of MT expression in hepatocellular carcinoma and liver adenocarcinoma.

There is evidence to suggest that higher levels of MT expression may also lead to resistance to chemotherapeutic drugs.

Autism

Heavy metal toxicity has been proposed as a hypothetical etiology of autism, and dysfunction of MT synthesis and activity may play a role in this. Many heavy metals, including mercury, lead, and arsenic have been linked to symptoms that resemble the neurological symptoms of autism. However, MT dysfunction has not specifically been linked to autistic spectrum disorders. A 2006 study, investigating children exposed to the vaccine preservative thiomersal, found that levels of MT and antibodies to MT in autistic children did not differ significantly from non-autistic children.

A low zinc to copper ratio has been seen as a biomarker for autism and suggested as an indication that the Metallothionein system has been affected.

Further, there is indication that the mother's zinc levels may affect the developing baby's immunological state that may lead to autism and could be again an indication that the Metallothionein system has been affected.

Magnetite Producing Organisms and Magnetosomes

It is now well known that biological systems can produce metal nanoparticles and even cadmium sulphide deposits. Is there any other form of inorganic nanoparticles that can be synthesized by biological organisms? We will explore about a category of magnetic nanoparticles that are biogenic in origin.

An interesting phenomenon observed on the motility of bacteria led to the discovery of magnetic deposits in several bacterial species. The main types of deposits in bacteria are magnetite and greigite. The magnetite (Fe_3O_4) deposits serve as internal compasses and as a result,the bacteria display specific directional movement with respect to the earth's magnetic field. This movement is known as magnetotaxis and such bacteria are called magnetotactic bacteria. The magnetotaxis is influenced by the presence of other ions in the environment. The most investigated magnetotactic bacteria are Magnetospirillium magnetotacticum and Magnetospirillium griphiswaldense.

How is this magnetite produced in bacteria? One of the first steps involved should be assimilation of iron in the form of Fe^{2+} or Fe^{3+} in the bacterial cell. The process is thought to be initiated by protein-assisted invagination of the cell membrane to form a vesicle. This may later mature to form a magnetosome. But this is yet to be confirmed. The transport of iron in the form of Fe^{2+} is aided by a hydrophobic transport protein MagA that is found in the lipid bilayer of the bacterial cell. It has both its N-terminus and C- terminus facing the intracellular direction. For transport it requires an ATPase pump nearby. The addition of ATP fuels the influx of protons into the cell through the ATPase. Once a threshold level of these protons is reached, they are pumped out by the MagA while simultaneously the Fe^{2+} ions are pumped into the system.Once inside the cell, the Fe^{2+} is oxidized to formlow densityhydrous ferric oxide followed by a dehydration step to form high density ferric oxide. Then, a small amount (about one third) of the ferric oxide is reduced and finally magnetite (Fe_3O_4) will be produced. These biomineralization reactions occur in unique lipid bilayer compartments known as 'magnetosomes'.

The lipid composition of magnetosomes shows presence of fatty acids, neutral lipids, sulpholipids, glycolipids, phosphatides, phosphatidyl ethanolamine and phosphatidyl serine. Several proteins that include MAM12, MAM22 and MAM28 have been found to be associated with the magnetosome. These contain charged amino acid residues that serve to establish electrostatic interactions with the cytoplasm and its components. The exact role of these proteins is yet to be identified. Most of the MAM polypeptides have been found to be tightly associated with the magnetosome and exhibit good resistance to detergents and proteases. Some of the MAM proteins show a homology with serine proteases and a few more with heat shock proteins. Several MAM proteins have similar sequences to cation diffusion facilitators implying a potential ion transport role. However, these are yet to confirmed by experimental findings.

It is believed that the magnetosome membrane not only enables transport of Fe^{2+}, it also may aid nucleation, crystallization and regulation of the pH and redox environment in the magnetosome.

Indeed, the formation of magnetite inside the magnetosome enables tight regulation of the nucleation and growth of the crystals. The size of the magnetite crystals can also be directed by the magnetosomes. Typical particle sizes range from 35-120 nm. In the bacterial cell, several crystals orient themselves to form linear chains. The number of chains as well as the number of crystals in each chain varies from species to species as well as their growth environment. Presence of many chains in a single bacterium will result in repulsive forces that push each chain to the corner of the cells resulting in a turgor pressure. Cubo-octahedral, hexagonal prismatic and bullet-shaped morphologies have been observed for magnetite crystals derived from different magnetotactic bacteria. The (111), (110) and (100) crystal faces have been identified in the magnetite crystals with the (111) crystal face being the most common. The alignment of the magnetite crystal chains usually is in the direction of the cytoskeletal filaments. Figure depicts the magnetic deposits inside a bacterial cell.

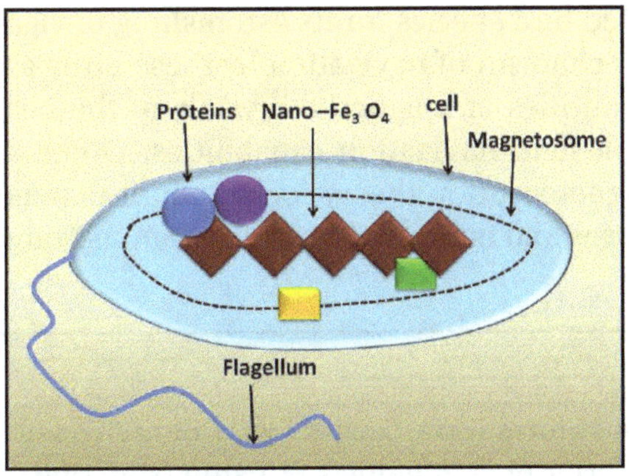

Cartoon of a bacteria containing a magnetosome.

Greigite Synthesis

Several anaerobic bacteria in deep water bodies have been found to contain Fe_3S_4 (Greigite) deposits instead of Fe_3O_4 (magnetite). These bacterial are found well below the OATZ (Oxic-Anoxic Transition Zone). Bacteria found at the OATZ show both magnetite and Greigite deposits indicating that the presence of oxygen is essential for magnetite production. Below the OATZ, hydrogen sulphide content is very high and there is complete absence of molecular oxygen. Hence the microorganisms present in that zone have only Greigite deposits. Rectangular, prismatic as well as cubo-octahedral geometries have been observed in these biomineralized Greigite deposits. The size of each Greigite crystal ranges from 30-100 nm.

Biogenic Nanoparticles from Higher Organisms

Biomineralization of nanoparticles have been also discovered in multi-cellular organisms higher in the evolutionary tree. Deposits of magnetite have been found in the brain of salmon fish (Oncorhynchusnerka), especially in the ethmoid region. The ethmoid region is found in the anterior region of the brain behind the ocular region. Unlike bacterial magnetite, the deposits found in the salmon brain do not contain any amorphous layer of Fe_3O_4 but is highly crystalline. The magnetite crystals have been found to be associated into chains that could be linear or in bundled formation. Each chain is found to consist of 13-45 magnetite crystals. The average crystal size was between 25-60 nm. All the magnetite crystals are found arranged with their (111) face perpendicular to the chain axis. All the magnetite crystals were found to possess cubo-octahedral geometry. What is amazing is that these crystals found in salmon do not exhibit any structural defects! The amount of magnetite in the salmon brain seems to increase with age.

Why does salmon produce magnetite? Well, the most likely explanation is because it uses these magnetite crystals for navigation with respect to the earth's magnetic field as a reference. This phenomenon is known as 'magnetoception' or 'magnetoreception'. The weak magnetic field of the earth can be detected through changes in the magnetic environment around these magnetite crystals present in the brain, which aids the fish to determine its coordinates with respect to the earth's magnetic field. Interestingly, magnetic material other than magnetite have also been found in other tissues such as the eye, skin and olfactory tissues of salmon, but these have no magnetoception role.

A similar mechanism for magnetoception has also been reported for several species of birds. One of the most widely investigated bird species for its astonishing navigation property is the 'homing pigeon'. Though the exact mechanism of navigation by these birds are yet to be agreed upon, the presence of considerable quantities of magnetite crystals in the beak and tissues of the pigeon seems to have a significant role in its navigation capabilities. Other suggestive navigation mechanisms apart from the magnetoception theory include light-dependent navigation and olfactory-based navigation. Honeybees and bats also have been demonstrated to exhibit magnetoception properties.

> Did you know?
>
> In 2009, a South African Information Technology Company sent a 4 GB flash drive containing data through a carrier pigeon named Winston from one of their offices to another office at Durban, 80 km away. At the same time, they also sent the same data to their Durban office through the broadband connection. The pigeon took sixty eight minutes to reach the Durban office and the office staff took about an hour to transfer the data from the flash drive to their systems. The total exercise took two hours six minutes and fifty seven seconds. On comparison, during the same duration, only 4% of the data was downloaded through the broadband connection!

Magnetite deposits have also been discovered in human tissues, especially in the brain. Though there remains some controversy regarding the concentration and localization of these magnetite deposits, it is now an accepted fact that human brain tissues and meninges (membrane layers that surround the brain and spinal cord) contain magnetite deposits. Another category of iron based deposits are ferritin moieties which is paramagnetic as compared to the ferromagnetic magnetite crystals. These magnetite deposits are prismatic as compared to the octahedral morphology commonly observed with geological samples indicating that these crystals have a different route of synthesis in vivo. The formation of these crystals having (111) crystal faces is most likely to resemble formation of a magnetosome in bacteria.

The biogenic magnetite crystalsin human brain have two different size ranges – 10-70 nm and a larger population between 90 and 200 nm. Also, it has been recently discovered that apart from the brain, tumour tissues also have magnetic crystal deposits. Hence it is unlikely that these could be used for magnetoceptive purposes. Then what is the role of such deposits in humans?One of the suggested roles indicates a potential gating molecule role for these crystals in mechanically sensitive ion channels. The opening and closing of ion channels in response to the extremely weak magnetic fields could change signal transduction, membrane polarity and cell response. The channel opening and closing could be triggered due to pressure changes accompanying change in orientation of the magnetic dipole in the crystals associated with the mechanically sensitive ion channel most probably through cytoskeletal proteins.

Another hypothesis is that the magnetite deposits may have a role in cell division and proliferation. However, more in-depth studies are required to probe the role of such biogenic magnetic crystals in humans. Interestingly, a gender-based difference in the amount of magnetite deposition has been also observed. Men show an increased amount of magnetite deposits when compared with women. The reason for this difference would come to light only after understanding the purpose of such deposits in humans!

Applications of Biogenic Nanoparticles

The biogenic nanoparticles have found wide range of applications in diverse fields. The silver and gold nanoparticles synthesized by biological systems exhibit excellent stability. Silver nanoparticles have been employed as anti-microbial agents in textiles, paper, domestic appliances, non-biofouling coatings, catheters, wound dressings etc. Their cytotoxicity is mainly attributed to their ability to alter membrane permeability and interfere with the DNA. They also could disrupt the electron transport components leading to cell death.The noble metal nanoparticles (silver and gold) also find applications as imaging agents. The presence of protein stabilizers around the nanoparticles enables functionalization of the nanoparticles with specific molecules that impart special characteristics to the nanoparticles. For example, attachment of a poly(ethylene glycol) chain confers long circulating characteristics to the particles while attachment of a target specific ligand enables selective internalization of these particles into specific cells. Figure represents a multi-functional magnetite nanoparticle.

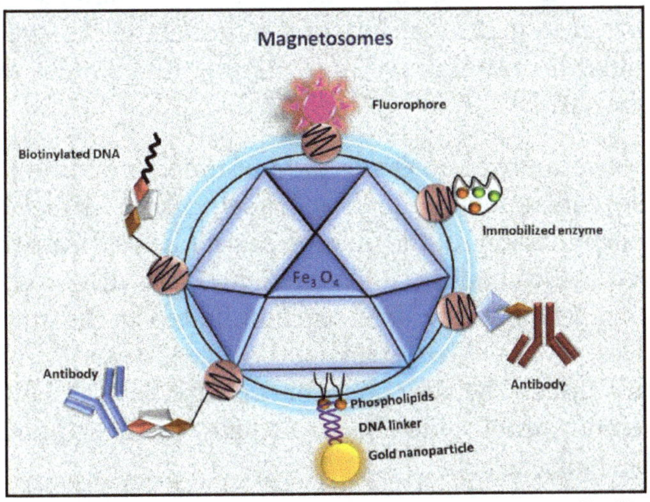

Multi-functional magnetite nanoparticle modified with a fluorescence moiety, antibodies for targeting, immobilized enzyme for catalysis and gold nanoparticle for photosensitivity and a DNA moiety.

The semi-conductor nanoparticles find extensive applications in the electronics industries and for imaging applications. The magnetite nanoparticles have nearly monodisperse size distributions that are in the superparamagnetic regime. This property can be utilized inmagnetic separations, inducing hyperthermia in cancer cells as well as resonance imaging as contrast agents.

Stealth Nanoparticles

One of the major research areas in nanomedicine is in the development of nanocarriers to deliver drugs to specified locations. Though in vitro experiments employing the drug- loaded nanocarriers are successful, the efficiency of these drug carriers in vivo is not encouraging as they fail to reach their target. What could be wrong? There are numerous factors that can influence the efficiency of a drug carrierin vivo. These include the route of administration, preferential uptake by non-target cells (biodistribution) as well as the lifetime of the carrier within the biological system.There are many options for introduction of the nanocarrier into the biological system such as through nasal, transdermal, parenteral, oral, rectal or ocular routes. The efficiency of the system is different in each case. Why? The carrier molecule faces different types of barriers and filters in each case before it can enter the blood. As a result, the amount of drug-loaded carriers that enter into the blood circulation varies with the route of administration. Once it is in the blood, the carrier faces an entirely new challenge! Blood contains an arsenal of components that form part of the body's defense – namely the immune system. The nanocarrier is not native to the biological system and is considered a foreign body by the immune system.Hence, it is either degraded or eliminated from the biological systemas soon as the carrier is recognized by the immune system. As a result, the drug present in the nanocarrier never gets to reach its target! Therefore the carrier needs to escape recognition by the immune system components! The properties of the nanocarrier have a major influence on the response of the immune system as well as its interaction with other non-specific targets in the biological system. These include the particle size, surface charge, surface topography and hydrophilicity. As a result of such non-specific and undesirable interactions with non-specific targets as well as the immune system, the carrier and the drug loaded in the carrier do not get delivered to the target. Therefore, suitable modifications must be made to the nanocarrier to impart the ability to 'hide' from the immune system. The ability of a nanoparticle to evade immune recognition so as to enhance its circulation timein vivo and thereby its chances of reaching the target is known as 'stealth characteristic'.

One of the earliest attempts employed to evade the immune recognition was to flood the system with a placebo carrier before introducing the drug-loaded carrier. A placebo carrier does not contain any drug. The high concentrations of the placebo carrier will activate the immune components, which will then be engaged in eliminating these carriers. If the drug-loaded carrier is introduced at this time, the chances of it getting recognized and eliminated by the immune system is remote as the immune components are involved in destroying the placebo carrier. This diversionary strategy unfortunately comes with a price! The very high concentrations of the placebo carrier introduced into the system can cause undesirable side effects owing to the disruption of the immune system. Hence this strategy has been discarded.

Introduction of immune suppressive agents is another strategy that tends to disable the abilities of the immune system. Though this strategy is used in extreme cases, it is not desirable as the patient becomes vulnerable to other infections during this period. In recent years, a 'Trojan horse concept' has been employed to impart stealth characteristics to the nanocarrier to enable it to escape from immune recognition without causing any impairment to the immune system.

> The origin of the Trojan horse concept...
>
> In Greek mythology, a long war was fought between the kingdoms of Troy and Greece. The Greeks laid siege to the city of Troy for more than 10 years and still no end to the war was in sight. Hence, they decided to win the war by employing a stealth strategy. They constructed a large wooden horse with a secret compartment in which they hid some of their soldiers. Then they pretended to sail away making the Trojans think that the Greeks had given up. The unsuspecting Trojans took the horse inside their city to mark their victory. In the night, the soldiers inside the horse came out and opened the gates of the city to let in the Greek army, which then plundered the city and won the war! In the case of delivering the drugs to selected locations in the biological system, a similar strategy has to be employed to trick the immune system. The carrier becomes the Trojan horse carrying the 'soldiers', namely the drug molecules. The immune system fails to recognize the danger posed and once the carrier enters the 'city of Troy', which in this case is the target cell, the drug molecules come out of the carrier and destroy the cell. Hence the Trojan horse concept is very relevant in nanomedicine to design stealth carrier systems!

In order to devise strategies to impart stealth characteristics to the carrier, one has to understand the pathways leading to immune recognition of foreign bodies. The response of the blood components to the nanocarrier will be of prime importance in deciding the fate of the nanocarrier in the biological system. Apart from the nature of the nanocarrier, the flow parameters of the blood also influence the interactions between the blood and the carrier. These include the kind of stress experienced by the carrier surface owing to the flow of blood – static, laminar or vortex, and the ratio of the blood volume in contact with the nanocarrier surface. Any carrier that does not elicit much response from the blood components is termed as 'hemocompatible'. The search for the ideal hemocompatible material is still on. Apart from the natural vascular endothelial cells lining the inner side of the blood vessel, no material can be classified as 100% hemocompatible!

Nanocarriers

A nanocarrier is nanomaterial being used as a transport module for another substance, such as a drug. Commonly used nanocarriers include micelles, polymers, carbon-based materials, liposomes and other substances. Nanocarriers are currently being studied for their use in drug delivery and their unique characteristics demonstrate potential use in chemotherapy.

Characterization

Nanocarriers range from sizes of diameter 1–1000 nm, however due to the width of microcapillaries being 200 nm, nanomedicine often refers to devices <200 nm. Because of their small size, nanocarriers can deliver drugs to otherwise inaccessible sites around the body. Since nanocarriers are so small, it is oftentimes difficult to provide large drug doses using them. The emulsion techniques used to make nanocarriers also often result in low drug loading and drug encapsulation, providing a difficulty for the clinical use.

Types

Nanocarriers discovered thus far include polymer conjugates, polymeric nanoparticles, lipid-based carriers, dendrimers, carbon nanotubes, and gold nanoparticles. Lipid-based carriers include both liposomes and micelles. Examples of gold nanoparticles are gold nanoshells and nanocages. Different types of nanomaterial being used in nanocarriers allows for hydrophobic and hydrophilic drugs to be delivered throughout the body. Since the human body contains mostly water, the ability to deliver hydrophobic drugs effectively in humans is a major therapeutic benefit of nanocarriers. Micelles are able to contain either hydrophilic or hydrophobic drugs depending on the orientation of the phospholipid molecules. Some nanocarriers contain nanotube arrays allowing them to contain both hydrophobic and hydrophilic drugs.

One potential problem with nanocarriers is unwanted toxicity from the type of nanomaterial being used. Inorganic nanomaterial can also be toxic to the human body if it accumulates in certain cell organelles. New research is being conducted to invent more effective, safer nanocarriers. Protein based nanocarriers show promise for use therapeutically since they occur naturally, and generally demonstrate less cytotoxicity than synthetic molecules.

Targeted Drug Delivery

Nanocarriers are useful in the drug delivery process because they can deliver drugs to site-specific targets, allowing drugs to be delivered in certain organs or cells but not in others. Site-specificity is a major therapeutic benefit since it prevents drugs from being delivered to the wrong places. Nanocarriers show promise for use in chemotherapy because they can help decrease the adverse, broader-scale toxicity of chemotherapy on healthy, fast growing cells around the body. Since chemotherapy drugs can be extremely toxic to human cells, it is important that they are delivered to the tumor without being released into other parts of the body. Four methods in which nanocarriers can deliver drugs include passive targeting, active targeting, pH specificity, and temperature specificity.

Passive Targeting

Passive targeting refers to a nanocarrier's ability to travel down a tumor's vascular system, become trapped, and accumulate in the tumor. This accumulation is caused by the enhanced permeability and retention effect which refers to the poly(ethylene oxide) (PEO) coating on the outside of many nanocarriers. PEO allows nanocarriers to travel through the leaky vasculature of a tumor, where they are unable to escape. The leaky vasculature of a tumor is the network of blood vessels that form in a tumor, which contain many small pores. These pores allow nanocarriers in, but also contain many bends that allow the nanocarriers to become trapped. As more nanocarriers become trapped, the drug accumulates at the tumor site. This accumulation cause large doses of the drug to be delivered directly to the tumor site. PEO may also have some adverse effects on cell-nanocarrier interactions, weakening the effects of the drug, since many nanocarriers must be incorporated into the cells before the drugs can be released.

Active Targeting

Active targeting involves the incorporation of targeting modules such as ligands or antibodies on the surface of nanocarriers that are specific to certain types of cells around the body. Nanocarriers have such a high surface-area to volume ratio allowing for multiple ligands to be incorporated on their surfaces. These targeting modules allow for the nanocarriers to be incorporated directly inside of cells, but also have some drawbacks. Ligands may cause nanocarriers to become slightly more toxic due to non-specific binding, and positive charges on ligands may decrease drug delivery efficiency once inside of cells. Active targeting has been shown to help overcome multi-drug resistance in tumor cells.

pH Specificity

Certain nanocarriers will only release the drugs they contain in specific pH ranges. pH specificity also allows nanocarriers to deliver drugs directly to a tumor site. Tumors are generally more acidic than normal human cells, with a pH around 6.8. Normal tissue has a pH of around 7.4. Nanocarriers that only release drugs at certain pH ranges can therefore be used to release the drug only within acidic tumor environments. High acidic environments cause the drug to be released due to the acidic environment degrading the structure of the nanocarrier. These nanocarriers will not release drugs in neutral or basic environments, effectively targeting the acidic environments of tumors while leaving normal body cells untouched. This pH sensitivity can also be induced in micelle systems by adding copolymer chains to micelles that have been determined to act in a pH independent manor. These micelle-polymer complexes also help to prevent cancer cells from developing multi-drug resistance. The low pH environment triggers a quick release of the micelle polymers, causing a majority of the drug to be released at once, rather than gradually like other drug treatments. This quick release mechanism significantly decreases the time it takes for anticancer drugs to kill a tumor, effectively preventing the tumor from having time to undergo mutations that would render it drug resistant.

Temperature Specificity

Some nanocarriers have also been shown to deliver drugs more effectively at certain temperatures. Since tumor temperatures are generally higher than temperatures throughout the rest of the body, around 40 °C, this temperature gradient helps act as safeguard for tumor-specific site delivery.

Uses

Most of research on nanocarriers is being applied to their potential use in drug delivery, especially in chemotherapy. Since nanocarriers can be used to specifically target the small pores, lower pH's, and higher temperatures of tumors, they have the potential to lower the toxicity of many chemotherapy drugs. Also, since almost 75% of anticancer drugs are hydrophobic, and therefore demonstrate difficulty in delivery inside human cells, the use of micelles to stabilize, and effectively mask the hydrophobic nature of hydrophobic drugs provides new possibilities for hydrophobic anticancer drugs.

In Vitro

In vitro studies are performed with microorganisms, cells, or biological molecules outside their normal biological context. Colloquially called "test-tube experiments", these studies in biology and its subdisciplines are traditionally done in labware such as test tubes, flasks, Petri dishes, and microtiter plates. Studies conducted using components of an organism that have been isolated from their usual biological surroundings permit a more detailed or more convenient analysis than can be done with whole organisms; however, results obtained from *in vitro* experiments may not fully or accurately predict the effects on a whole organism. In contrast to *in vitro* experiments, *in vivo* studies are those conducted in animals, including humans, and whole plants.

Cloned plants in vitro.

Definition

In vitro studies are conducted using components of an organism that have been isolated from their usual biological surroundings, such as microorganisms, cells, or biological molecules. For example, microrganisms or cells can be studied in artificial culture media, and proteins can be examined in solutions. Colloquially called "test-tube experiments", these studies in biology, medicine, and their subdisciplines are traditionally done in test tubes, flasks, Petri dishes, etc. They now involve the full range of techniques used in molecular biology, such as the omics.

In contrast, studies conducted in living beings (microorganisms, animals, humans, or whole plants) are called *in vivo* .

Examples

Examples of *in vitro* studies include: the isolation, growth and identification of cells derived from multicellular organisms in (cell or tissue culture); subcellular components (e.g. mitochondria or ribosomes); cellular or subcellular extracts (e.g. wheat germ or reticulocyte extracts); purified molecules such as proteins, DNA, or RNA); and the commercial production of antibiotics and other pharmaceutical products. Viruses, which only replicate in living cells, are studied in the laboratory in cell or tissue culture, and many animal virologists refer to such work as being *in vitro* to distinguish it from *in vivo* work in whole animals.

- Polymerase chain reaction is a method for selective replication of specific DNA and RNA sequences in the test tube.
- Protein purification involves the isolation of a specific protein of interest from a complex mixture of proteins, often obtained from homogenized cells or tissues.
- *In vitro* fertilization is used to allow spermatozoa to fertilize eggs in a culture dish before implanting the resulting embryo or embryos into the uterus of the prospective mother.
- *In vitro* diagnostics refers to a wide range of medical and veterinary laboratory tests that

are used to diagnose diseases and monitor the clinical status of patients using samples of blood, cells, or other tissues obtained from a patient.

- *In vitro* testing has been used to characterize specific adsorption, distribution, metabolism, and excretion processes of drugs or general chemicals inside a living organism; for example, Caco-2 cell experiments can be performed to estimate the absorption of compounds through the lining of the gastrointestinal tract; The partitioning of the compounds between organs can be determined to study distribution mechanisms; Suspension or plated cultures of primary hepatocytes or hepatocyte-like cell lines (HepG2, HepaRG) can be used to study and quantify metabolism of chemicals. These ADME process parameters can then be integrated into so called "physiologically based pharmacokinetic models" or PBPK.

Advantages

In vitro studies permit a species-specific, simpler, more convenient, and more detailed analysis than can be done with the whole organism. Just as studies in whole animals more and more replace human trials, so are *in vitro* studies replacing studies in whole animals.

Simplicity

Living organisms are extremely complex functional systems that are made up of, at a minimum, many tens of thousands of genes, protein molecules, RNA molecules, small organic compounds, inorganic ions, and complexes in an environment that is spatially organized by membranes, and in the case of multicellular organisms, organ systems. These myriad components interact with each other and with their environment in a way that processes food, removes waste, moves components to the correct location, and is responsive to signalling molecules, other organisms, light, sound, heat, taste, touch, and balance.

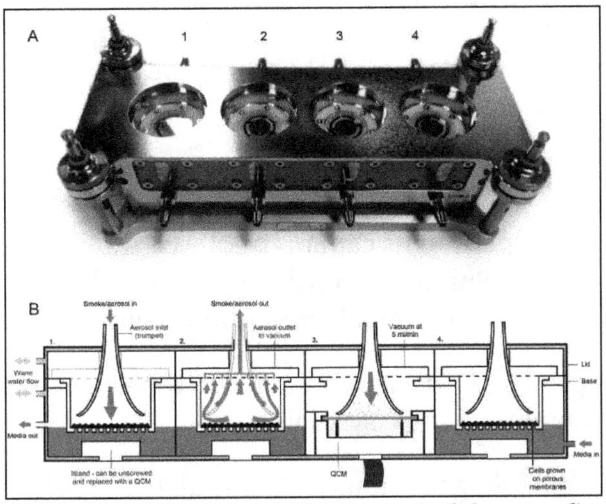

Top view of a Vitrocell mammalian exposure module "smoking robot", (lid removed) view of four separated wells for cell culture inserts to be exposed to tobacco smoke or an aerosol for an in vitro study of the effects.

This complexity makes it difficult to identify the interactions between individual components and to explore their basic biological functions. *In vitro* work simplifies the system under study, so the investigator can focus on a small number of components.

For example, the identity of proteins of the immune system (e.g. antibodies), and the mechanism by which they recognize and bind to foreign antigens would remain very obscure if not for the extensive use of *in vitro* work to isolate the proteins, identify the cells and genes that produce them, study the physical properties of their interaction with antigens, and identify how those interactions lead to cellular signals that activate other components of the immune system.

Species Specificity

Another advantage of *in vitro* methods is that human cells can be studied without "extrapolation" from an experimental animal's cellular response.

Convenience, Automation

In vitro methods can be miniaturized and automated, yielding high-throughput screening methods for testing molecules in pharmacology or toxicology.

Disadvantages

The primary disadvantage of *in vitro* experimental studies is that it may be challenging to extrapolate from the results of *in vitro* work back to the biology of the intact organism. Investigators doing *in vitro* work must be careful to avoid over-interpretation of their results, which can lead to erroneous conclusions about organismal and systems biology.

For example, scientists developing a new viral drug to treat an infection with a pathogenic virus (e.g. HIV-1) may find that a candidate drug functions to prevent viral replication in an *in vitro* setting (typically cell culture). However, before this drug is used in the clinic, it must progress through a series of *in vivo* trials to determine if it is safe and effective in intact organisms (typically small animals, primates, and humans in succession). Typically, most candidate drugs that are effective *in vitro* prove to be ineffective *in vivo* because of issues associated with delivery of the drug to the affected tissues, toxicity towards essential parts of the organism that were not represented in the initial *in vitro* studies, or other issues.

In Vitro to *in Vivo* Extrapolation

Results obtained from *in vitro* experiments cannot usually be transposed, as is, to predict the reaction of an entire organism *in vivo*. Building a consistent and reliable extrapolation procedure from *in vitro* results to *in vivo* is therefore extremely important. Solutions include:

- Increasing the complexity of *in vitro* systems to reproduce tissues and interactions between them (as in "human on chip" systems).
- Using mathematical modeling to numerically simulate the behavior of the complex system, where the *in vitro* data provide model parameter values.

These two approaches are not incompatible; better *in vitro* systems provide better data to mathematical models. However, increasingly sophisticated *in vitro* experiments collect increasingly numerous, complex, and challenging data to integrate. Mathematical models, such as systems biology models, are much needed here.

Extrapolating in Pharmacology

In pharmacology, IVIVE can be used to approximate pharmacokinetics (PK) or pharmacodynamics (PD). Since the timing and intensity of effects on a given target depend on the concentration time course of candidate drug (parent molecule or metabolites) at that target site, *in vivo* tissue and organ sensitivities can be completely different or even inverse of those observed on cells cultured and exposed *in vitro*. That indicates that extrapolating effects observed *in vitro* needs a quantitative model of *in vivo* PK. Physiologically based PK (PBPK) models are generally accepted to be central to the extrapolations.

In the case of early effects or those without intercellular communications, the same cellular exposure concentration is assumed to cause the same effects, both qualitatively and quantitatively, *in vitro* and *in vivo*. In these conditions, developing a simple PD model of the dose–response relationship observed *in vitro*, and transposing it without changes to predict *in vivo* effects is not enough.

Immune Recognition Pathways

One of the first events that occur when a nanocarrierenters the blood stream is 'opsonization'. Blood consists of a huge number of proteins and ions. When the nanocarrier comes into contact with these components, adsorption of the blood proteins on the surface of the nanocarrier occurs. This is a rapid process and is diffusion- controlled in the initial phases. This means that the first interacting components will be the hydrated ions followed by the more mobile proteins. Among the ions, it has been found that the divalent calcium and magnesium ions are the most active. The later stages of opsonization are affinity-controlled as the larger proteins that have greater affinity to the nanocarrier surface displace the weakly adsorbed species. The protein adsorption process over the nanocarrier is thus a competitive process and the sequence of proteins that adsorb on the surface and then on each other is described by 'Vroman effect'. The major blood proteins involved in the adsorption process are albumin, globulin, fibrinogen, fibronectin, Factor XII and high molecular weight kininogen. These proteins are referred to as the 'Opsonins'.

What is the consequence of this protein adsorption? Well, the opsonin proteins can interact with the receptors in the cells of the mononuclear phagocytic system (MPS) that form part of the cell-mediated immune response. This interaction leads to the activation of the monocytes and macrophages,which results in the uptake the nanocarrier by the macrophages or activationof the other components of the immune system. The nanocarrier is taken up by the macrophage by forming a vesicle termed as phagosome. This then fuses with the lysosome containing an array of hydrolytic and degradative enzymes that destroys the nanocarrier. This process of uptake of the carrier by the macrophages or other components of the MPS is termed as 'phagocytosis'.

Why is this process termed 'opsonization'? The term 'opsonin' is derived from the Greek language and means 'seasoning' or 'sauce'. Similar to the process of seasoning food items with herbs, spices or other ingredients to improve its flavor and edibility, the nanoparticle surface is coated or 'seasoned' with proteins to improve its uptake and digestion by the cells of the MPS! Hence this sequence of events is referred to as opsonization. Figure shows a cartoon depicting the sequence of events occurring during opsonization.

Why should opsonization drive phagocytosis? Can't the macrophage phagocytose a foreign body without opsonization? The immune system has evolved mainly to combat and eliminate pathogens. Most pathogens have an outer membrane that is predominantly negatively charged. Unfortunately, the macrophages and the other cells of the MPS also possess a negatively charged cell membrane. Hence uptake of the pathogen by the macrophages is retarded due to the existence of greater electrostatic repulsive forces. Hence, the immune system has evolved an effective strategy of coating the surface of the pathogen with the opsonin proteins that will mask the surface charges of the pathogen thereby facilitating uptake and subsequent degradation by the macrophages.Though it is not necessary that all nanoparticle surfaces will be negatively charged or even charged at all, the process of opsonization is the first they will encounter when they enter the blood stream.

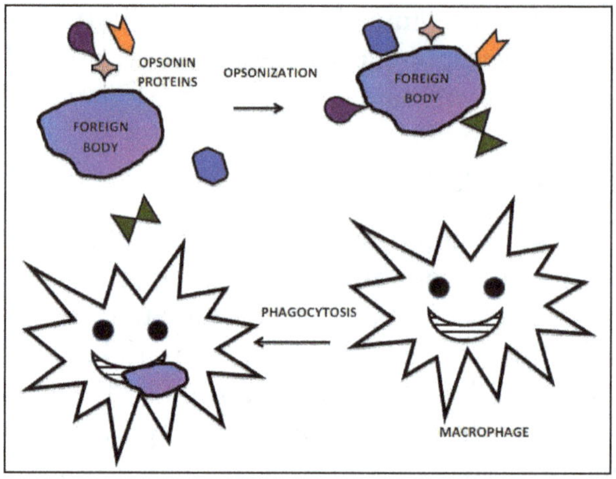

Cartoon showing the sequence of events during opsonisation.

What will happen if the nanocarrier could not be degraded by the MPS? In this event,the entire carrier molecule or part of it is anchored on the cell surface and 'presented' or exhibitedto the external environment by the MPS cells using major histocompatibility (MHC) proteins. This results in the activation of another component of the immune system, namely the T-lymphocytes, which in turn activate the B-lymphocytes leading to production of antibodies against the foreign body.

Among the range of opsonins that can adsorb on to a foreign body surface are a group of proteins called as the complement proteins. If the complement proteins adsorb on to the surface of the nanocarrier, then another cascade of events leading to the activation of the complement system is also triggered.

Opsonin

An opsonin is any molecule that enhances phagocytosis by marking an antigen for an immune response or marking dead cells for recycling (i.e., causes the phagocyte to "relish" the marked cell). *Opson* in ancient Greece referred to the delicious side-dish of any meal, versus the *sitos*, or the staple of the meal.

Opsonization (also, opsonisation) is the molecular mechanism whereby molecules, microbes, or apoptotic cells are chemically modified to have stronger interactions with - to be more "delicious" to - cell surface receptors on phagocytes and NK cells. With the antigen coated in opsonins, binding to immune cells is greatly enhanced. Opsonization also mediates phagocytosis via signal cascades from cell surface receptors.

Opsonins aid the immune system in a number of ways. In a healthy individual, they mark dead and dying self cells for clearance by macrophages and neutrophils, activate complement proteins, and target cells for destruction through the action of natural killer (NK) cells.

Mechanism

All cell membranes have negative charges (Zeta potential) which makes it difficult for two cells to come close together. When opsonins bind to their targets they boost the kinetics of phagocytosis by favoring interaction between the opsonin and cell surface receptors on immune cells. This overrides the negative charges from cell membranes. This principle holds true for clearance of pathogens as well as dead or dying *self* cells.

Varieties

Different opsonins perform different functions. Opsonin molecules include:

Antibodies

Antibodies are part of the adaptive immune response and are generated by B cells in response to antigen exposure. The Fab region of the antibody binds to the antigen, whereas the Fc region of the antibody binds to an Fc receptor on the phagocyte, facilitating phagocytosis. The antigen-antibody complex can also activate complement through the classical complement pathway. Phagocytic cells do not have an Fc receptor for immunoglobulin M (IgM), making IgM ineffective in assisting phagocytosis alone. However, IgM is extremely efficient at activating complement and is, therefore, considered an opsonin. IgG antibodies are also capable of binding immune effector cells via their Fc domain, triggering a release of lysis products from the bound immune effector cell (monocytes, neutrophils, eosinophils, and natural killer cells). This process, called antibody-dependent cellular cytotoxicity, can cause inflammation of surrounding tissues and damage to healthy cells.

Complement Proteins

The complement system is a part of the innate immune response. C3b, C4b, and C1q are important complement molecules that serve as opsonins. As a part of the alternative complement pathway, the spontaneous activation of a complement cascade converts C3 to C3b, a component that can serve as an opsonin when bound to an antigen's surface. Antibodies can also activate complement via the classical pathway, resulting in deposition of C3b and C4b onto the antigen surface. After C3b has bound to the surface of an antigen, it can be recognized by phagocyte receptors that signal for phagocytosis. Complement receptor 1 is expressed on all phagocytes and recognizes a number of complement opsonins, including C3b and C4b which are both parts of C3-convertase. C1q, a member of the C1 complex, is able to interact with the Fc region of antibodies.

Circulating Proteins

Pentraxins, collectins, and ficolins are all circulating proteins that are capable of serving as opsonins. They are secreted Pattern recognition receptors (PRRs). These molecules coat the microbes as opsonins and enhance neutrophil reactivity against them through a number of mechanisms.

Targets

Apoptotic Cells

Apoptosis is related to low tissue inflammation. A number of opsonins play a role in marking apoptotic cells for phagocytosis without a pro-inflammatory response.

Members of the pentraxin family can bind to apoptotic cell membrane components like phosphatidylcholine (PC) and phosphatidylethanolamine (PE). IgM antibodies also bind to PC. Collectin molecules such as mannose-binding lectin (MBL), surfactant protein A (SP-A), and SP-D interact with unknown ligands on apoptotic cell membranes. When bound to the appropriate ligand these molecules interact with phagocyte receptors, enhancing phagocytosis of the marked cell.

C1q is capable of binding directly to apoptotic cells. It can also indirectly bind to apoptotic cells via intermediates like IgM autoantibodies, MBL, and pentraxins. In both cases C1q activates complement, resulting in the cells being marked for phagocytosis by C3b and C4b. C1q is an important contributor to the clearance of apoptotic cells and debris. This process usually occurs in late apoptotic cells.

Opsonization of apoptotic cells occurs by different mechanisms in a tissue-dependent pattern. For example, while C1q is necessary for proper apoptotic cell clearance in the peritoneal cavity, it is not important in the lungs where SP-D plays an important role.

Pathogens

As part of the late stage adaptive immune response, pathogens and other particles are marked by IgG antibodies. These antibodies interact with Fc receptors on macrophages and neutrophils resulting in phagocytosis. The C1 complement complex can also interact with the Fc region of IgG and IgM immune complexes activating the classical complement pathway and marking the antigen with C3b. C3b can spontaneously bind to pathogen surfaces through the alternative complement pathway. Furthermore, pentraxins can directly bind to C1q from the C1 complex.

SP-A opsonizes a number of bacterial and viral pathogens for clearance by lung alveolar macrophages.

Phagocytosis

In cell biology, phagocytosis is the process by which a cell—often a phagocyte or a protist—engulfs a solid particle to form an internal compartment known as a phagosome. It is distinct from other forms of endocytosis like pinocytosis that involves the internalization of extracellular liquids. Phagocytosis is involved in the acquisition of nutrients for some cells. The process is homologous to eating at the level of single-celled organisms; in multicellular animals, the process has been adapted to eliminate debris and pathogens, as opposed to taking in fuel for cellular processes, except in the case of the animal *Trichoplax*.

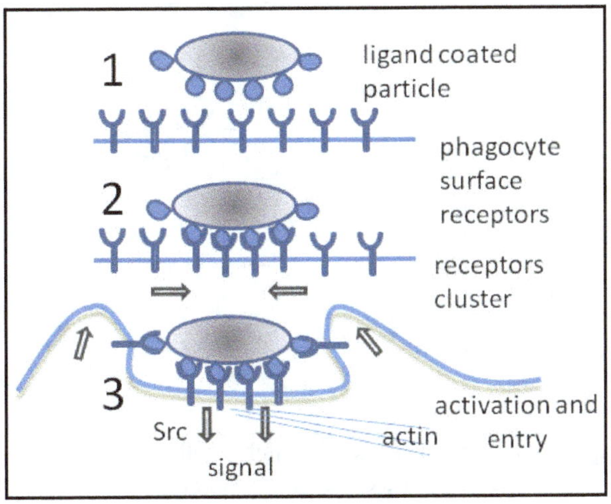

Phagocytosis in three steps:
1. Unbound phagocyte surface receptors do not trigger phagocytosis.
2. Binding of receptors causes them to cluster.
3. Phagocytosis is triggered and the particle is taken up by the phagocyte.

In an organism's immune system, phagocytosis is a major mechanism used to remove pathogens and cell debris. For example, when a macrophage ingests a pathogenic microorganism, the pathogen becomes trapped in a phagosome which then fuses with a lysosome to form a phagolysosome. Within the phagolysosome, enzymes and toxic peroxides digest the pathogen. Bacteria, dead tissue cells, and small mineral particles are all examples of objects that may be phagocytized. The process has triggered the name "Phagocytes" for the 1st line of defence in the immune system.

History

Phagocytosis was first noted by Canadian physician William Osler (1876), and later studied and named by Élie Metchnikoff (1880, 1883).

In Immune System

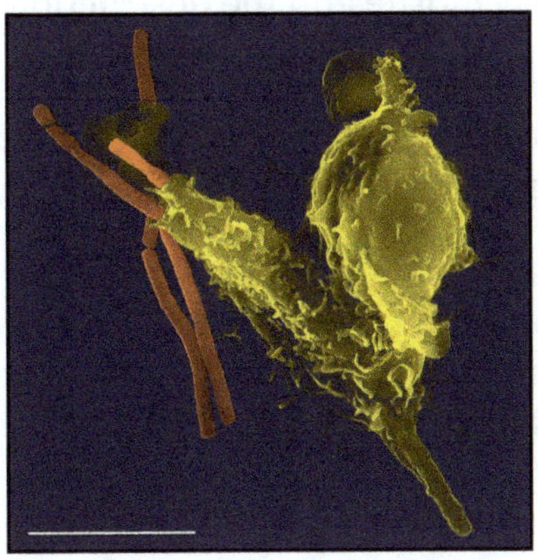

Scanning electron micrograph of a phagocyte (yellow, right) phagocytosing anthrax bacilli (orange, left).

Phagocytosis in mammalian immune cells is activated by attachment to pathogen-associated molecular patterns (PAMPS), which leads to NF-κB activation. Opsonins such as C3b and antibodies can act as attachment sites and aid phagocytosis of pathogens.

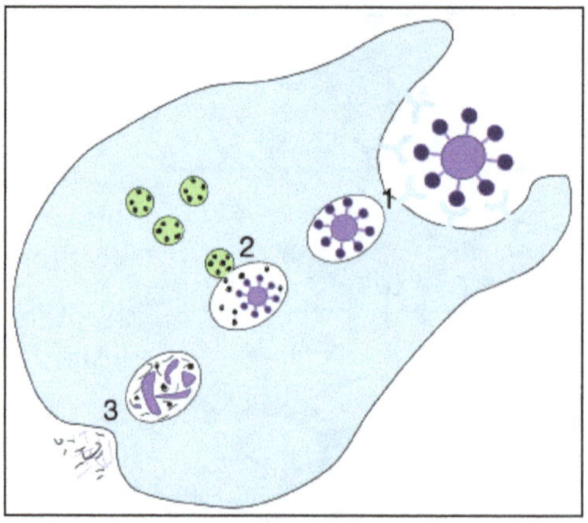

Phagocytosis is the process in which a cell engulfs a particle, digests it, and expels the waste products.

Process of phagocytosis:

1. A particle is ingested by a phagocyte after antigens are recognized which results in the formation of a phagosome.
2. The fusion of lysosomes with the phagosome creates a phagolysosome.
3. The particle is broken down by the digestive enzymes found in the lysosomes. The resulting waste material is discharged from the phagocyte by exocytosis.

Engulfment of material is facilitated by the actin-myosin contractile system. The phagosome of ingested material is then fused with the lysosome, forming a phagolysosome and leading to degradation.

Degradation can be oxygen-dependent or oxygen-independent:

- Oxygen-dependent degradation depends on NADPH and the production of reactive oxygen species. Hydrogen peroxide and myeloperoxidase activate a halogenating system, which leads to the creation of hypochlorite and the destruction of bacteria.
- Oxygen-independent degradation depends on the release of granules, containing proteolytic enzymes such as defensins, lysozyme, and cationic proteins. Other antimicrobial peptides are present in these granules, including lactoferrin, which sequesters iron to provide unfavourable growth conditions for bacteria.

It is possible for cells other than dedicated phagocytes (such as dendritic cells) to engage in phagocytosis. Some white blood cells in human immune system perform phagocytosis by gulping in some pathogenic and disease causing cells.

In Apoptosis

Following apoptosis, the dying cells need to be taken up into the surrounding tissues by macrophages in a process called efferocytosis. One of the features of an apoptotic cell is the presentation of a variety of intracellular molecules on the cell surface, such as calreticulin, phosphatidylserine (From the inner layer of the plasma membrane), annexin A1, oxidised LDL and altered glycans.

These molecules are recognised by receptors on the cell surface of the macrophage such as the phosphatidylserine receptor or by soluble (free-floating) receptors such as thrombospondin 1, GAS6, and MFGE8, which themselves then bind to other receptors on the macrophage such as CD36 and alpha-v beta-3 integrin. Defects in apoptotic cell clearance is usually associated with impaired phagocytosis of macrophages. Accumulation of apoptotic cell remnants often causes autoimmune disorders; thus pharmacological potentiation of phagocytosis has a medical potential in treatment of certain forms of autoimmune disorders.

In Protists

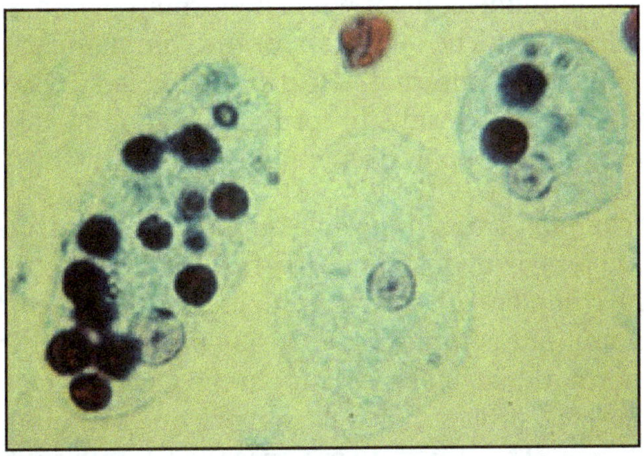

Trophozoites of *Entamoeba histolytica* with ingested erythrocytes.

In many protists, phagocytosis is used as a means of feeding, providing part or all of their nourishment. This is called phagotrophic nutrition, distinguished from osmotrophic nutrition which takes place by absorption.

- In some, such as amoeba, phagocytosis takes place by surrounding the target object with pseudopods, as in animal phagocytes. In humans, *entamoeba histolytica* can phagocytose red blood cells. Entamoeba histolytica is an anaerobic parasitic protozoan, part of the genus Entamoeba. Predominantly infecting humans and other primates, E. histolytica is estimated to infect about 50 million people worldwide. Previously, it was thought that 10% of the world population was infected, but these figures predate the recognition that at least 90% of these infections were due to a second species, E. dispar. Mammals such as dogs and cats can become infected transiently, but are not thought to contribute significantly to transmission.The word histolytic literally means "Tissue destroyer". This process is known as "erythrophagocytosis", and is considered the only reliable way to distinguish *Entamoeba histolytica* from noninvasive species such as *Entamoeba dispar*.
- Ciliates also engage in phagocytosis. In ciliates there is a specialized groove or chamber in the cell where phagocytosis takes place, called the cytostome or mouth.

As in phagocytic immune cells, the resulting phagosome may be merged with lysosomes containing digestive enzymes, forming a phagolysosome. The food particles will then be digested, and the released nutrients are diffused or transported into the cytosol for use in other metabolic processes.

Mixotrophy can involve phagotrophic nutrition and phototrophic nutrition.

Complement Activation

Complement is referred to a group of proteins that form part of the innate immune system, which does not change over the course of an individual's lifetime due to environmental factors. The complement proteins are synthesized in the liver. Some of the complement proteins possess enzymatic activity but under normal conditions, they exist as zymogens. Zymogens require a trigger such as lysis of a polypeptide sequence to transform into an active enzyme. The introduction of a foreign body can trigger this transformation resulting in a cascade of events leading to enhanced phagocytosis, recruitment of other immune components and formation of a membrane attack complex (MAC) to destruct the foreign body. The activation of the complement system can occur in three different modes – classical pathway, alternative pathway and lectin pathway. All three pathways converge to a common pathway that results in formation of the membrane attack complex that alters the membrane permeability of a pathogen resulting in its death.

The classical pathway for complement activation is initiated by the formation of an immune complex (antigen-antibody complex) involving the immunoglobulins IgM or IgG (antibodies). The immune complex binds to the complement protein complex C1. The C1 complex comprises one C1q, two C1r and two C1s. If the immune complex binds to C1q or if the C1q directly binds to the surface of the foreign body, a conformational change occurs in the C1 complex. This change activates the C1r sub-units. The C1r is a serine protease, i.e., it is an enzyme containing serine in its active site, and can cleave peptide linkages. The substrate of C1r is the C1s sub-unit, which on lysis gets activated. The C1s also is a serine protease. The C1 complex containing the active sub-units of C1r and C1s now act upon two other complement proteins C4 and C2 to form C2a, C2b, C4a and C4b fragments. The lysis of the proteins does not form symmetrical fragments (i.e., same sized fragments). The smaller fragment is denoted by the suffix 'a' and the larger fragment is denoted by the suffix 'b'. The C4b and C2a fragments bind to form another enzyme known as the C3 convertase.

The C4b2a acts on the complement protein C3 to form C3a and C3b. The C3b fragment binds with the C42b complex to from yet another new enzyme C4b2a3b known as the C5 convertase. The C5 convertase cleaves the complement protein C5 to form C5a and C5b. The C5b then associates with the complement proteins C6, C7, C8 and multiple C9 in a sequential manner to form the membrane attack complex (MAC). The MAC inserts into the microbial membrane to form a pore leading to drastic changes in the permeability resulting in the cell death.

What is the fate of the smaller fragments C4a, C3 and C5a that are formed during the complement activation? These are anaphylatoxins that promote inflammatory reactions in the body. They stimulate the cells of the mononuclear phagocyte system (MPS) and the mast cells (cells with granules containing histamine, heparin, serine proteases, leukotrienes, prostaglandins, cytokines etc.; it is found in many tissues and is thought to play a key role in wound healing) to degranulate and release their contents that contain many lytic enzymes as well as chemokines and cytokines. The chemokines serve to attract the other components of the immune system from remote locations in the body to the site of release leading to an increase in the immune activity at the site of release. This migration of the immune components occurs towards increasing concentration gradientand is referred to as 'chemotaxis'. Thus, the formation of the anaphylatoxins during the complement activation leads to an amplification of the immune response against the foreign body. It is interesting to note that the activation of one type of immune response also leads to the activation of other classes of immune response in the biological system.

In the case of the alternative pathway, the activation involves the binding of the C3 protein with the foreign body. The C3 protein under normal conditions undergoes spontaneous lysis to form a nascent active form denoted as C3b-like-C3 or C3b*. This fragment is unstable and is inactivated by hydrolysis under normal conditions. The C3b- like-C3 contains a thioester group using which it can bind to surfaces containing nucleophilic groups (containing electron rich centres) such as amino group ($-NH_2$) or hydroxyl group (-OH). Such binding is prevented in the normal cells due to the presence of inhibitory factors, namely Factor H and Factor I, on the cell surfaces. Such factors are however lacking in the foreign body and hence the binding of the C3b-like-C3 can occur on the foreign body. In the presence of a foreign body, the C3b-like-C3 binds to the surface by cleavage of the thioester link to form a covalent ester link with the surface of the foreign body. In the absence of reactive functional groups on the surface of the foreign body, this binding occurs through the opsonin proteins adsorbed on the surface of the foreign body. Figure shows the reaction that occurs when C3b-like-C3 encounters a foreign body.

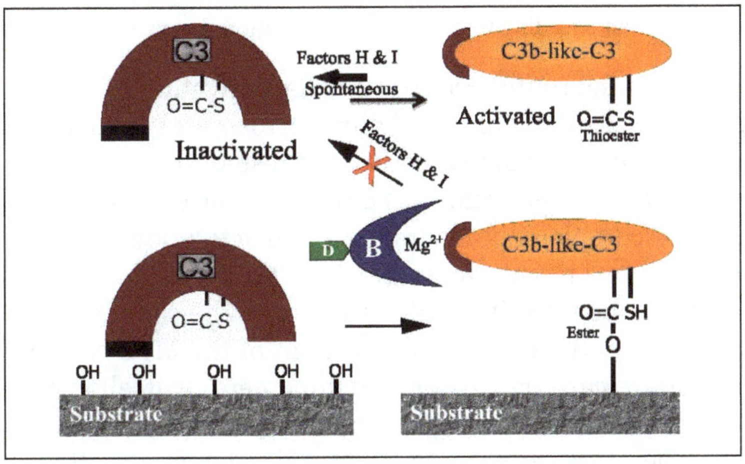

Binding of C3b-like-C3 to the surface of a foreign body.

Once bound to the surface, the Factors H and I cannot inhibit the C3b-like-C3 anymore. In the presence of divalent magnesium ions, the C3b-like-Cr in the presence of Factor D, cleavesthe protein Factor B to Ba and Bb fragments The C3b-like-C3 associates with the larger Bb fragment to form the C3bBb complex. This complex is stabilized by the protein properidin (also known as Factor P). The stabilized C3bBb complex is now called the C3 convertase. It cleaves C3 to form C3a and C3b, which then associates with C3bBb to from the C5 convertase C3bBb3b. The C5 convertase cleaves C5 to C5a and C5b. The C5b associates with C6, C7, C8 and multiple C9 proteins to form the MAC.

The lectin pathway is initiated by the binding of the mannose binding lectin (MBL) protein to the mannose moiety found in the surface glycoproteins of microorganisms. This binding brings about the activation of MBL-associated serine proteases 1 and 2 (MASP-1 and MASP-2), which then cleave the complement proteins C4 and C2 into C4a, C4b, C2a and C2b fragments. The C4b and C2a bind to form C4b2a complex that functions as the C3 convertase. The rest of the sequence is common as in the classical and alternative pathways. Figure summarizes the events leading to activation of complement through different routes.

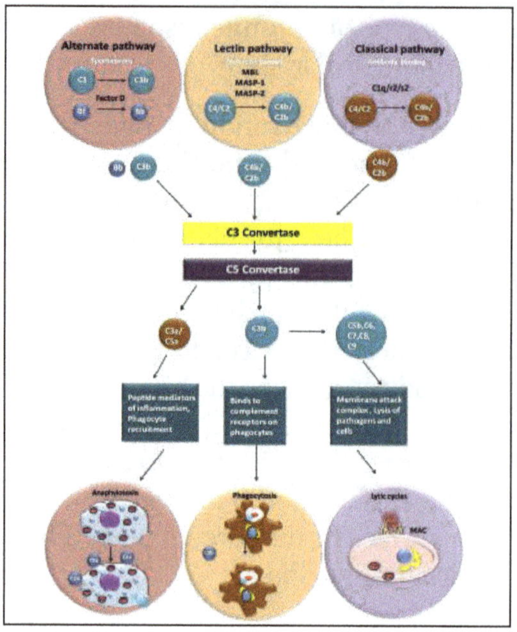

Summary of the events leading to complement activation through different pathways.

It is evident that regardless of the route of activation, the activation of the complement protein C3 remains the vital step that can lead to enhanced phagocytosis and destruction of the foreign body. The most common pathway that gets activated in the presence of a nanocarrier is the alternative pathway. The high surface-to-volume ratio of the nanocarrier enables more effective adsorption of the opsonin proteins and complement leading to rapid activation of the complement as well as phagocytosis. In the case of non- membranous carriers, the formation of MAC as a result of complement activation might not be effective in destruction of the carrier. But the generation of the anaphylatoxins and cytokines can aid in the recruitment of the MPS cells and the T-lymphocytes that can effectively destroy the carrier. Thus, the complement by itself is not effective in eliminating all foreign bodies, but it serves to alert and activate other powerful immune components that can eliminate the threat. Therefore suppression of opsonization might offer a good option to retard the complement activation and hence the immune recognition and response.

Coagulation Pathway

The blood also contains components of the coagulation pathway. What factors activate the coagulation pathway? The activation of the coagulation cascade could occur either due to intrinsic (direct contact pathway) or extrinsic(tissue contact pathway) pathways.The extrinsic pathway is activated with the release of thrombogenic tissue factor from the tissues as well as exposure of collagen from the extracellular matrix of injured tissue. The activation of the intrinsic pathway requires activation of many proteins in a sequential manner. Co-factors such as vitamin K and calcium ions are essential for all the coagulation processes. Figure depicts the various factors involved in the activation of the coagulation cascade by different types of nanoparticles.

Chemotaxis Assay

Chemotaxis assays are experimental tools for evaluation of chemotactic ability of prokaryotic or eukaryotic cells. A wide variety of techniques have been developed. Some techniques are qualitative - allowing an investigator to approximately determine a cell's chemotacic affinity for an analyte - while others are quantitative, allowing a precise measurement of this affinity.

Quality Control

In general, the most important requisite is to calibrate the incubation time of the assay both to the model cell and the ligand to be evaluated. Too short incubation time results no cells in the sample, while too long time perturbs the concentration gradients and measures more chemokinetic than chemotactic responses.

The most commonly used techniques are grouped into two main groups:

Agar-plate Techniques

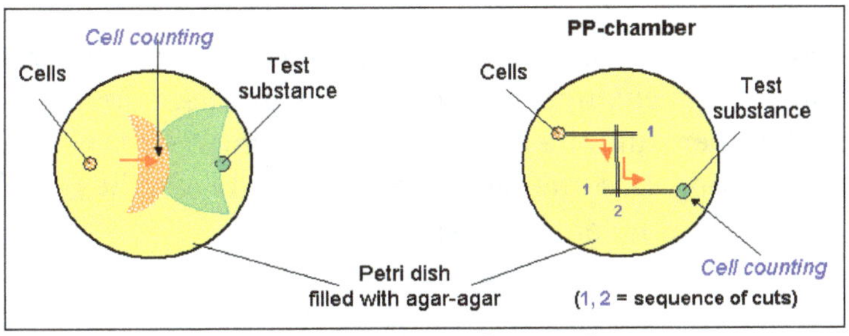

Chemotaxis assays with agar plates.

This way of evaluation deals with agar-agar or gelatine containing semi-solid layers made prior to the experiment. Small wells are cut into the layer and filled with cells and the test substance. Cells can migrate towards the chemical gradient in the semi solid layer or under the layer as well. Some variations of the technique deal also with wells and parallel channels connected by a cut at the start of the experiment (PP-technique). Radial arrangement of PP-technique (3 or more channels) provides the possibility to compare chemotactic activity of different cell populations or study preference between ligands.

Counting of cells: positive responder cells could be counted from the front of migrating cells, after staining or in native conditions in light microscope.

Two-chamber Techniques

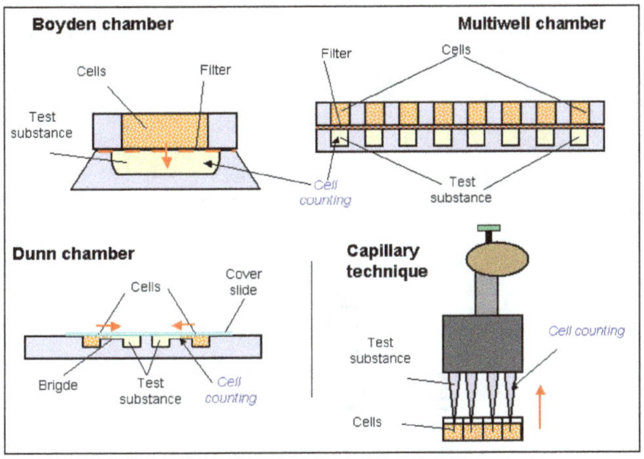

Chemotaxis chamber assays.

Boyden Chamber

Chambers isolated by filters are proper tools for accurate determination of chemotactic behavior. The pioneer type of these chambers was constructed by Boyden. The motile cells are placed into the upper chamber, while fluid containing the test substance is filled into the lower one. The size of the motile cells to be investigated determines the pore size of the filter; it is essential to choose a diameter which allows active transmigration. For modelling *in vivo* conditions, several protocols prefer coverage of filter with molecules of extracellular matrix (collagen, elastin etc.) Efficiency of the measurements was increased by development of multiwell chambers (e.g. NeuroProbe), where 24, 96, 384 samples are evaluated in parallel. Advantage of this variant is that several parallels are assayed in identical conditions.

Bridge Chambers

In another setting the chambers are connected side by side horizontally (Zigmond chamber) or as concentric rings on a slide (Dunn chamber) Concentration gradient develops on a narrow connecting bridge between the chambers and the number of migrating cells is also counted on the surface of the bridge by light microscope. In some cases the bridge between the two chambers is filled with agar and cells have to "glide" in this semisolid layer.

Capillary Techniques

Some capillary techniques provide also a chamber like arrangement, however, there is no filter between the cells and the test substance. Quantitative results are gained by the multiwell type of this probe using 4-8-12-channel pipettes. Accuracy of the pipette and increased number of the parallel running samples is the great advantage of this test.

Counting of cells: positive responder cells are count from the lower chamber (long incubation time) or from the filter (short incubation time). For detection of cells general staining techniques (e.g. trypan blue) or special probes (e.g. mt-dehydrogenase detection with MTT assay) are used. Labelled (e.g. fluorochromes) cells are also used, in some assays cells get labelled during transmigration the filter.

Other Techniques

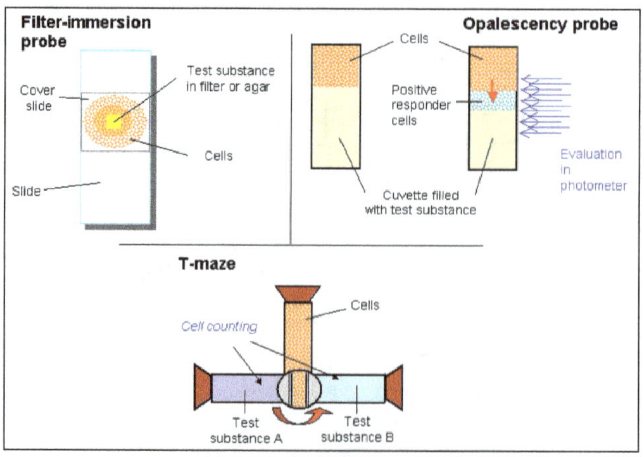

Other chemotaxis assay techniques.

Besides the above-mentioned two, most commonly used family of techniques a wide range of protocols were developed to measure chemotactic activity. Some of them are only qualitative, like aggregation tests, where small pieces of agar or filters are placed onto a slide and accumulation of cells around is measured.

In another semiquantitative technique cells are overlaid the test substance and changes in opalescence of the originally cell-free compartment is recorded during the incubation time.

The third very frequently used, however, qualitative technique is the T-maze and its adaptations for microplates. In the original version a container drilled in a peg is filled with cells. Then the peg is twisted and the cells get contact with two other containers filled with different substances. The incubation is stopped with resetting the peg, the cell number is counted from the containers.

Coagulation

Coagulation (also known as clotting) is the process by which blood changes from a liquid to a gel, forming a blood clot. It potentially results in hemostasis, the cessation of blood loss from a damaged vessel, followed by repair. The mechanism of coagulation involves activation, adhesion, and aggregation of platelets along with deposition and maturation of fibrin. Disorders of coagulation are disease states which can result in bleeding (hemorrhage or bruising) or obstructive clotting (thrombosis).

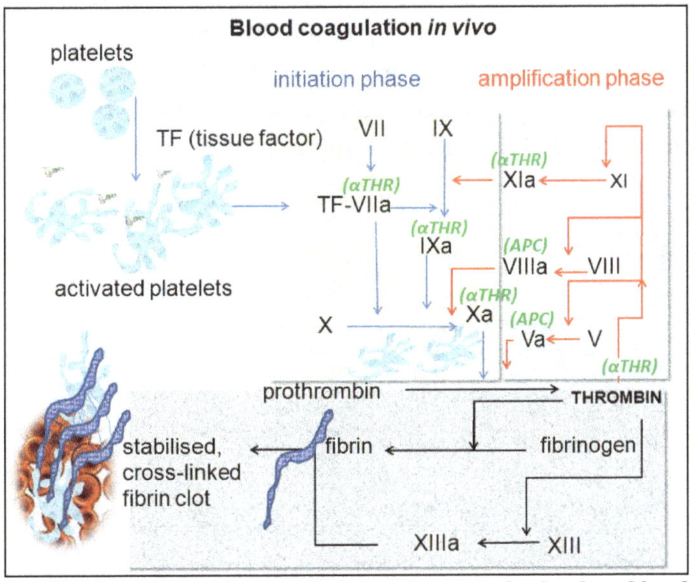

Blood coagulation pathways in vivo showing the central role played by thrombin.

Coagulation is highly conserved throughout biology; in all mammals, coagulation involves both a cellular (platelet) and a protein (coagulation factor) component. The system in humans has been the most extensively researched and is the best understood.

Coagulation begins almost instantly after an injury to the blood vessel has damaged the endothelium lining the vessel. Leaking of blood through the endothelium initiates two processes: changes in platelets, and the exposure of subendothilial tissue factor to plasma Factor VII, which ultimately leads to fibrin formation. Platelets immediately form a plug at the site of injury; this is called *primary hemostasis*. *Secondary hemostasis* occurs simultaneously: Additional coagulation factors or clotting factors beyond Factor VII (listed below) respond in a complex cascade to form fibrin strands, which strengthen the platelet plug.

Physiology

Platelet Activation

When the endothelium is damaged, the normally isolated, underlying collagen is exposed to circulating platelets, which bind directly to collagen with collagen-specific glycoprotein Ia/IIa surface receptors. This adhesion is strengthened further by von Willebrand factor (vWF), which is released from the endothelium and from platelets; vWF forms additional links between the platelets' glycoprotein Ib/IX/V and the collagen fibrils. This localization of platelets to the extracellular matrix promotes collagen interaction with platelet glycoprotein VI. Binding of collagen to glycoprotein VI triggers a signaling cascade that results in activation of platelet integrins. Activated integrins mediate tight binding of platelets to the extracellular matrix. This process adheres platelets to the site of injury.

Activated platelets will release the contents of stored granules into the blood plasma. The granules include ADP, serotonin, platelet-activating factor (PAF), vWF, platelet factor 4, and thromboxane A_2 (TXA_2), which, in turn, activate additional platelets. The granules' contents activate a G_q-linked protein receptor cascade, resulting in increased calcium concentration in the platelets' cytosol. The calcium activates protein kinase C, which, in turn, activates phospholipase A_2 (PLA_2). PLA_2 then modifies the integrin membrane glycoprotein IIb/IIIa, increasing its affinity to bind fibrinogen. The activated platelets change shape from spherical to stellate, and the fibrinogen cross-links with glycoprotein IIb/IIIa aid in aggregation of adjacent platelets (completing primary hemostasis).

Coagulation Cascade

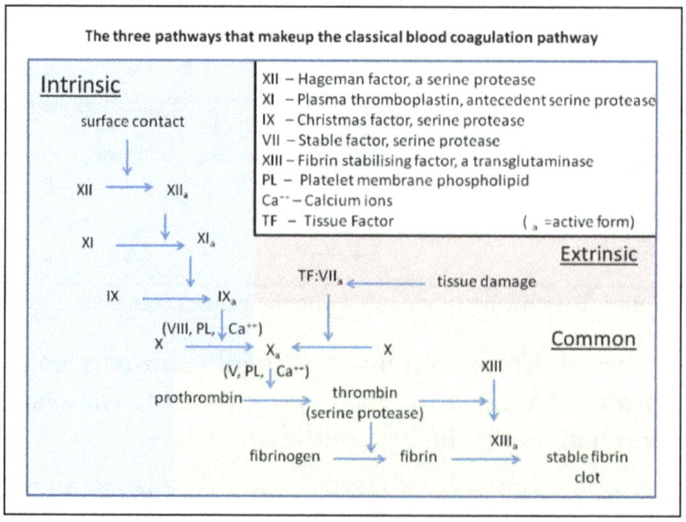

The classical blood coagulation pathway.

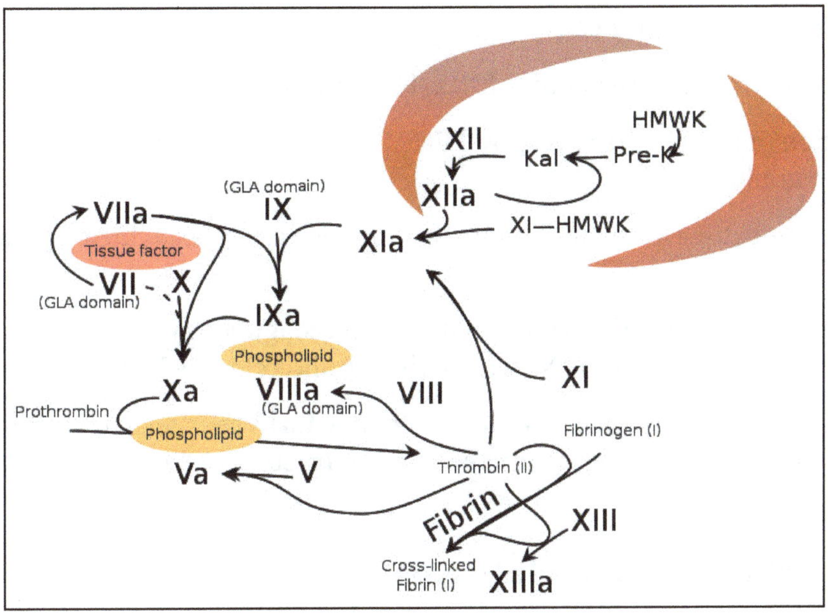

Modern coagulation pathway.

The coagulation cascade of secondary hemostasis has two initial pathways which lead to *fibrin* formation. These are the *contact activation pathway* (also known as the intrinsic pathway), and the *tissue factor pathway* (also known as the extrinsic pathway) which both lead to the same fundamental reactions that produce fibrin. It was previously thought that the two pathways of coagulation cascade were of equal importance, but it is now known that the primary pathway for the initiation of blood coagulation is the *tissue factor* (extrinsic) pathway. The pathways are a series of reactions, in which a zymogen (inactive enzyme precursor) of a serine protease and its glycoprotein co-factor are activated to become active components that then catalyze the next reaction in the cascade, ultimately resulting in cross-linked fibrin. Coagulation factors are generally indicated by Roman numerals, with a lowercase *a* appended to indicate an active form.

The coagulation factors are generally serine proteases (enzymes), which act by cleaving downstream proteins. There are some exceptions. For example, FVIII and FV are glycoproteins, and Factor XIII is a transglutaminase. The coagulation factors circulate as inactive zymogens. The coagulation cascade is therefore classically divided into three pathways. The *tissue factor* and *contact activation* pathways both activate the "final common pathway" of factor X, thrombin and fibrin.

Tissue Factor Pathway (Extrinsic)

The main role of the tissue factor pathway is to generate a "thrombin burst", a process by which thrombin, the most important constituent of the coagulation cascade in terms of its feedback activation roles, is released very rapidly. FVIIa circulates in a higher amount than any other activated coagulation factor. The process includes the following steps:

1. Following damage to the blood vessel, FVII leaves the circulation and comes into contact with tissue factor (TF) expressed on tissue-factor-bearing cells (stromal fibroblasts and leukocytes), forming an activated complex (TF-FVIIa).

2. TF-FVIIa activates FIX and FX.
3. FVII is itself activated by thrombin, FXIa, FXII and FXa.
4. The activation of FX (to form FXa) by TF-FVIIa is almost immediately inhibited by tissue factor pathway inhibitor (TFPI).
5. FXa and its co-factor FVa form the prothrombinase complex, which activates prothrombin to thrombin.
6. Thrombin then activates other components of the coagulation cascade, including FV and FVIII (which forms a complex with FIX), and activates and releases FVIII from being bound to vWF.
7. FVIIIa is the co-factor of FIXa, and together they form the "tenase" complex, which activates FX; and so the cycle continues. ("Tenase" is a contraction of "ten" and the suffix "-ase" used for enzymes.)

Contact Activation Pathway (Intrinsic)

The contact activation pathway begins with formation of the primary complex on collagen by high-molecular-weight kininogen (HMWK), prekallikrein, and FXII (Hageman factor). Prekallikrein is converted to kallikrein and FXII becomes FXIIa. FXIIa converts FXI into FXIa. Factor XIa activates FIX, which with its co-factor FVIIIa form the tenase complex, which activates FX to FXa. The minor role that the contact activation pathway has in initiating clot formation can be illustrated by the fact that patients with severe deficiencies of FXII, HMWK, and prekallikrein do not have a bleeding disorder. Instead, contact activation system seems to be more involved in inflammation, and innate immunity. Despite this, interference with the pathway may confer protection against thrombosis without a significant bleeding risk.

Final Common Pathway

The division of coagulation in two pathways is mainly artificial, it originates from laboratory tests in which clotting times were measured after the clotting was initiated by glass (intrinsic pathway) or by thromboplastin (a mix of tissue factor and phospholipids). In fact thrombin is present from the very beginning, already when platelets are making the plug. *Thrombin* has a large array of functions, not only the conversion of fibrinogen to fibrin, the building block of a hemostatic plug. In addition, it is the most important platelet activator and on top of that it activates Factors VIII and V and their inhibitor protein C (in the presence of thrombomodulin), and it activates Factor XIII, which forms covalent bonds that crosslink the fibrin polymers that form from activated monomers.

Following activation by the contact factor or tissue factor pathways, the coagulation cascade is maintained in a prothrombotic state by the continued activation of FVIII and FIX to form the tenase complex, until it is down-regulated by the anticoagulant pathways.

Cofactors

Various substances are required for the proper functioning of the coagulation cascade:

Calcium and Phospholipid

Calcium and phospholipid (a platelet membrane constituent) are required for the tenase and prothrombinase complexes to function. Calcium mediates the binding of the complexes via the terminal gamma-carboxy residues on FXa and FIXa to the phospholipid surfaces expressed by platelets, as well as procoagulant microparticles or microvesicles shed from them. Calcium is also required at other points in the coagulation cascade.

Vitamin K

Vitamin K is an essential factor to a hepatic gamma-glutamyl carboxylase that adds a carboxyl group to glutamic acid residues on factors II, VII, IX and X, as well as Protein S, Protein C and Protein Z. In adding the gamma-carboxyl group to glutamate residues on the immature clotting factors Vitamin K is itself oxidized. Another enzyme, *Vitamin K epoxide reductase*, (VKORC) reduces vitamin K back to its active form. Vitamin K epoxide reductase is pharmacologically important as a target of anticoagulant drugs warfarin and related coumarins such as acenocoumarol, phenprocoumon, and dicumarol. These drugs create a deficiency of reduced vitamin K by blocking VKORC, thereby inhibiting maturation of clotting factors. Vitamin K deficiency from other causes (e.g., in malabsorption) or impaired vitamin K metabolism in disease (e.g., in liver failure) lead to the formation of PIVKAs (proteins formed in vitamin K absence) which are partially or totally non-gamma carboxylated, affecting the coagulation factors' ability to bind to phospholipid.

Regulators

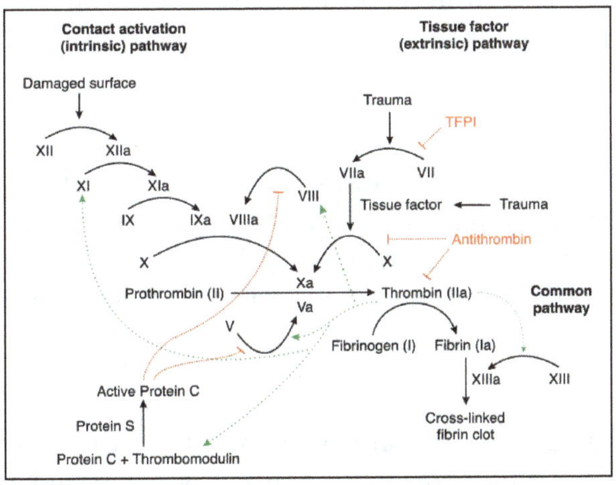

Coagulation with arrows for negative and positive feedback.

Five mechanisms keep platelet activation and the coagulation cascade in check. Abnormalities can lead to an increased tendency toward thrombosis:

Protein C

Protein C is a major physiological anticoagulant. It is a vitamin K-dependent serine protease enzyme that is activated by thrombin into activated protein C (APC). Protein C is activated in a sequence that starts with Protein C and thrombin binding to a cell surface protein thrombomodulin.

Thrombomodulin binds these proteins in such a way that it activates Protein C. The activated form, along with protein S and a phospholipid as cofactors, degrades FVa and FVIIIa. Quantitative or qualitative deficiency of either (protein C or protein S) may lead to thrombophilia (a tendency to develop thrombosis). Impaired action of Protein C (activated Protein C resistance), for example by having the "Leiden" variant of Factor V or high levels of FVIII also may lead to a thrombotic tendency.

Antithrombin

Antithrombin is a serine protease inhibitor (serpin) that degrades the serine proteases: thrombin, FIXa, FXa, FXIa, and FXIIa. It is constantly active, but its adhesion to these factors is increased by the presence of heparan sulfate (a glycosaminoglycan) or the administration of heparins (different heparinoids increase affinity to FXa, thrombin, or both). Quantitative or qualitative deficiency of antithrombin (inborn or acquired, e.g., in proteinuria) leads to thrombophilia.

Tissue Factor Pathway Inhibitor (TFPI)

Tissue factor pathway inhibitor (TFPI) limits the action of tissue factor (TF). It also inhibits excessive TF-mediated activation of FVII and FX.

Plasmin

Plasmin is generated by proteolytic cleavage of plasminogen, a plasma protein synthesized in the liver. This cleavage is catalyzed by tissue plasminogen activator (t-PA), which is synthesized and secreted by endothelium. Plasmin proteolytically cleaves fibrin into fibrin degradation products that inhibit excessive fibrin formation.

Prostacyclin

Prostacyclin (PGI_2) is released by endothelium and activates platelet G_s protein-linked receptors. This, in turn, activates adenylyl cyclase, which synthesizes cAMP. cAMP inhibits platelet activation by decreasing cytosolic levels of calcium and, by doing so, inhibits the release of granules that would lead to activation of additional platelets and the coagulation cascade.

Fibrinolysis

Eventually, blood clots are reorganised and resorbed by a process termed *fibrinolysis*. The main enzyme responsible for this process (plasmin) is regulated by various activators and inhibitors.

Role in Immune System

The coagulation system overlaps with the immune system. Coagulation can physically trap invading microbes in blood clots. Also, some products of the coagulation system can contribute to the innate immune system by their ability to increase vascular permeability and act as chemotactic agents for phagocytic cells. In addition, some of the products of the coagulation system are directly antimicrobial. For example, beta-lysine, an amino acid produced by platelets during coagulation, can cause lysis of many Gram-positive bacteria by acting as a cationic detergent. Many acute-phase proteins of inflammation are involved in the coagulation system. In addition, pathogenic bacteria may secrete agents that alter the coagulation system, e.g. coagulase and streptokinase.

Assessment

Numerous tests are used to assess the function of the coagulation system:

- Common: aPTT, PT (also used to determine INR), fibrinogen testing (often by the Clauss method), platelet count, platelet function testing (often by PFA-100), thrombodynamics test.
- Other: TCT, bleeding time, mixing test (whether an abnormality corrects if the patient's plasma is mixed with normal plasma), coagulation factor assays, antiphospholipid antibodies, D-dimer, genetic tests (e.g. factor V Leiden, prothrombin mutation G20210A), dilute Russell's viper venom time (dRVVT), miscellaneous platelet function tests, thromboelastography (TEG or Sonoclot), euglobulin lysis time (ELT).

The contact activation (intrinsic) pathway is initiated by activation of the "contact factors" of plasma, and can be measured by the activated partial thromboplastin time (aPTT) test.

The tissue factor (extrinsic) pathway is initiated by release of tissue factor (a specific cellular lipoprotein), and can be measured by the prothrombin time (PT) test. PT results are often reported as ratio (INR value) to monitor dosing of oral anticoagulants such as warfarin.

The quantitative and qualitative screening of fibrinogen is measured by the thrombin clotting time (TCT). Measurement of the exact amount of fibrinogen present in the blood is generally done using the Clauss method for fibrinogen testing. Many analysers are capable of measuring a "derived fibrinogen" level from the graph of the Prothrombin time clot.

If a coagulation factor is part of the contact activation or tissue factor pathway, a deficiency of that factor will affect only one of the tests: Thus hemophilia A, a deficiency of factor VIII, which is part of the contact activation pathway, results in an abnormally prolonged aPTT test but a normal PT test. The exceptions are prothrombin, fibrinogen, and some variants of FX that can be detected only by either aPTT or PT. If an abnormal PT or aPTT is present, additional testing will occur to determine which (if any) factor is present as aberrant concentrations.

Deficiencies of fibrinogen (quantitative or qualitative) will affect all screening tests.

Role in Disease

Coagulation defects may cause hemorrhage or thrombosis, and occasionally both, depending on the nature of the defect.

Platelet Disorders

Platelet conditions may be congenital or acquired. Some inborn platelet pathologies are

Glanzmann's thrombasthenia, Bernard-Soulier syndrome (abnormal glycoprotein Ib-IX-V complex), gray platelet syndrome (deficient alpha granules), and delta storage pool deficiency (deficient dense granules). Most are rare conditions. Most inborn platelet pathologies predispose to hemorrhage. Von Willebrand disease is due to deficiency or abnormal function of von Willebrand factor, and leads to a similar bleeding pattern; its milder forms are relatively common.

Decreased platelet numbers may be due to various causes, including insufficient production (e.g., in myelodysplastic syndrome or other bone marrow disorders), destruction by the immune system (immune thrombocytopenic purpura/ITP), and consumption due to various causes (thrombotic thrombocytopenic purpura/TTP, hemolytic-uremic syndrome/HUS, paroxysmal nocturnal hemoglobinuria/PNH, disseminated intravascular coagulation/DIC, heparin-induced thrombocytopenia/HIT). Most consumptive conditions lead to platelet activation, and some are associated with thrombosis.

Disease and Clinical Significance of Thrombosis

The best-known coagulation factor disorders are the hemophilias. The three main forms are hemophilia A (factor VIII deficiency), hemophilia B (factor IX deficiency or "Christmas disease") and hemophilia C (factor XI deficiency, mild bleeding tendency). Hemophilia A and B are X-linked recessive disorders, whereas Hemophilia C is a much more rare autosomal recessive disorder most commonly seen in Ashkenazi Jews.

Von Willebrand disease (which behaves more like a platelet disorder except in severe cases), is the most common hereditary bleeding disorder and is characterized as being inherited autosomal recessive or dominant. In this disease, there is a defect in von Willebrand factor (vWF), which mediates the binding of glycoprotein Ib (GPIb) to collagen. This binding helps mediate the activation of platelets and formation of primary hemostasis.

Bernard-Soulier syndrome is a defect or deficiency in GPIb. GPIb, the receptor for vWF, can be defective and lead to lack of primary clot formation (primary hemostasis) and increased bleeding tendency. This is an autosomal recessive inherited disorder.

Thrombasthenia of Glanzmann and Naegeli (Glanzmann thrombasthenia) is extremely rare. It is characterized by a defect in GPIIb/IIIa fibrinogen receptor complex. When GPIIb/IIIa receptor is dysfunctional, fibrinogen cannot cross-link platelets, which inhibits primary hemostasis. This is an autosomal recessive inherited disorder.

In liver failure (acute and chronic forms), there is insufficient production of coagulation factors by the liver; this may increase bleeding risk.

Deficiency of Vitamin K may also contribute to bleeding disorders because clotting factor maturation depends on Vitamin K.

Thrombosis is the pathological development of blood clots. These clots may break free and become mobile, forming an embolus or grow to such a size that occludes the vessel in which it developed. An embolism is said to occur when the thrombus (blood clot) becomes a mobile embolus and migrates to another part of the body, interfering with blood circulation and hence impairing organ function downstream of the occlusion. This causes ischemia and often leads to ischemic necrosis of tissue. Most cases of venous thrombosis are due to acquired states (older age, surgery, cancer, immobility) or inherited thrombophilias (e.g., antiphospholipid syndrome, factor V Leiden, and various other genetic deficiencies or variants).

Mutations in factor XII have been associated with an asymptomatic prolongation in the clotting time and possibly a tendency toward thrombophlebitis. Other mutations have been linked with a rare form of hereditary angioedema (type III) essentialism.

Pharmacology

Procoagulants

The use of adsorbent chemicals, such as zeolites, and other hemostatic agents are also used for sealing severe injuries quickly (such as in traumatic bleeding secondary to gunshot wounds). Thrombin and fibrin glue are used surgically to treat bleeding and to thrombose aneurysms.

Desmopressin is used to improve platelet function by activating arginine vasopressin receptor 1A.

Coagulation factor concentrates are used to treat hemophilia, to reverse the effects of anticoagulants, and to treat bleeding in patients with impaired coagulation factor synthesis or increased consumption. Prothrombin complex concentrate, cryoprecipitate and fresh frozen plasma are commonly used coagulation factor products. Recombinant activated human factor VII is increasingly popular in the treatment of major bleeding.

Tranexamic acid and aminocaproic acid inhibit fibrinolysis, and lead to a *de facto* reduced bleeding rate. Before its withdrawal, aprotinin was used in some forms of major surgery to decrease bleeding risk and need for blood products.

Anticoagulants

Anticoagulants and anti-platelet agents are amongst the most commonly used medications. Anti-platelet agents include aspirin, dipyridamole, ticlopidine, clopidogrel, ticagrelor and prasugrel; the parenteral glycoprotein IIb/IIIa inhibitors are used during angioplasty. Of the anticoagulants, warfarin (and related coumarins) and heparin are the most commonly used. Warfarin affects the vitamin K-dependent clotting factors (II, VII, IX,X) and protein C and protein S, whereas heparin and related compounds increase the action of antithrombin on thrombin and factor Xa. A newer class of drugs, the direct thrombin inhibitors, is under development; some members are already in clinical use (such as lepirudin). Also under development are other small molecular compounds that interfere directly with the enzymatic action of particular coagulation factors (e.g., rivaroxaban, dabigatran, apixaban).

Coagulation Factors

Coagulation factors and related substances		
Number and/or name	**Function**	**Associated genetic disorders**
I (fibrinogen)	Forms clot (fibrin)	Congenital afibrinogenemia, Familial renal amyloidosis
II (prothrombin)	Its active form (IIa) activates I, V, X, VII, VIII, XI, XIII, protein C, platelets	Prothrombin G20210A, Thrombophilia

III (tissue factor or tissue thromboplastin)	Co-factor of VIIa (formerly known as factor III)	
IV Calcium	Required for coagulation factors to bind to phospholipid (formerly known as factor IV)	
V (proaccelerin, labile factor)	Co-factor of X with which it forms the prothrombinase complex	Activated protein C resistance
VI	*Unassigned* – old name of Factor Va	
VII (stable factor, proconvertin)	Activates IX, X	congenital factor VII deficiency
VIII (Antihemophilic factor A)	Co-factor of IX with which it forms the tenase complex	Haemophilia A
IX (Antihemophilic factor B or Christmas factor)	Activates X: forms tenase complex with factor VIII	Haemophilia B
X (Stuart-Prower factor)	Activates II: forms prothrombinase complex with factor V	Congenital Factor X deficiency
XI (plasma thromboplastin antecedent)	Activates IX	Haemophilia C
XII (Hageman factor)	Activates factor XI, VII and prekallikrein	Hereditary angioedema type III
XIII (fibrin-stabilizing factor)	Crosslinks fibrin	Congenital Factor XIIIa/b deficiency
von Willebrand factor	Binds to VIII, mediates platelet adhesion	von Willebrand disease
prekallikrein (Fletcher factor)	Activates XII and prekallikrein; cleaves HMWK	Prekallikrein/Fletcher Factor deficiency
high-molecular-weight kininogen (HMWK) (Fitzgerald factor)	Supports reciprocal activation of XII, XI, and prekallikrein	Kininogen deficiency
fibronectin	Mediates cell adhesion	Glomerulopathy with fibronectin deposits
antithrombin III	Inhibits IIa, Xa, and other proteases	Antithrombin III deficiency
heparin cofactor II	Inhibits IIa, cofactor for heparin and dermatan sulfate ("minor antithrombin")	Heparin cofactor II deficiency
protein C	Inactivates Va and VIIIa	Protein C deficiency
protein S	Cofactor for activated protein C (APC, inactive when bound to C4b-binding protein)	Protein S deficiency
protein Z	Mediates thrombin adhesion to phospholipids and stimulates degradation of factor X by ZPI	Protein Z deficiency
Protein Z-related protease inhibitor (ZPI)	Degrades factors X (in presence of protein Z) and XI (independently)	
plasminogen	Converts to plasmin, lyses fibrin and other proteins	Plasminogen deficiency, type I (ligneous conjunctivitis)
alpha 2-antiplasmin	Inhibits plasmin	Antiplasmin deficiency
tissue plasminogen activator (tPA)	Activates plasminogen	Familial hyperfibrinolysis and thrombophilia
urokinase	Activates plasminogen	Quebec platelet disorder
plasminogen activator inhibitor-1 (PAI1)	Inactivates tPA & urokinase (endothelial PAI)	Plasminogen activator inhibitor-1 deficiency
plasminogen activator inhibitor-2 (PAI2)	Inactivates tPA & urokinase (placental PAI)	
cancer procoagulant	Pathological factor X activator linked to thrombosis in cancer	

History

Initial Discoveries

Theories on the coagulation of blood have existed since antiquity. Physiologist Johannes Müller (1801–1858) described fibrin, the substance of a thrombus. Its soluble precursor, fibrinogen, was thus named by Rudolf Virchow (1821–1902), and isolated chemically by Prosper Sylvain Denis (1799–1863). Alexander Schmidt suggested that the conversion from fibrinogen to fibrin is the result of an enzymatic process, and labeled the hypothetical enzyme "thrombin" and its precursor "prothrombin". Arthus discovered in 1890 that calcium was essential in coagulation. Platelets were identified in 1865, and their function was elucidated by Giulio Bizzozero in 1882.

The theory that thrombin is generated by the presence of tissue factor was consolidated by Paul Morawitz in 1905. At this stage, it was known that *thrombokinase/thromboplastin* (factor III) is released by damaged tissues, reacting with *prothrombin* (II), which, together with calcium (IV), forms *thrombin*, which converts fibrinogen into *fibrin* (I).

Coagulation Factors

The remainder of the biochemical factors in the process of coagulation were largely discovered in the 20th century.

A first clue as to the actual complexity of the system of coagulation was the discovery of *proaccelerin* (initially and later called Factor V) by Paul Owren (1905–1990) in 1947. He also postulated its function to be the generation of accelerin (Factor VI), which later turned out to be the activated form of V (or Va); hence, VI is not now in active use.

Factor VII (also known as *serum prothrombin conversion accelerator* or *proconvertin*, precipitated by barium sulfate) was discovered in a young female patient in 1949 and 1951 by different groups.

Factor VIII turned out to be deficient in the clinically recognised but etiologically elusive hemophilia A; it was identified in the 1950s and is alternatively called *antihemophilic globulin* due to its capability to correct hemophilia A.

Factor IX was discovered in 1952 in a young patient with hemophilia B named Stephen Christmas (1947–1993). His deficiency was described by Dr. Rosemary Biggs and Professor R.G. MacFarlane in Oxford, UK. The factor is, hence, called Christmas Factor. Christmas lived in Canada, and campaigned for blood transfusion safety until succumbing to transfusion-related AIDS at age 46. An alternative name for the factor is *plasma thromboplastin component*, given by an independent group in California.

Hageman factor, now known as factor XII, was identified in 1955 in an asymptomatic patient with a prolonged bleeding time named of John Hageman. Factor X, or Stuart-Prower factor, followed, in 1956. This protein was identified in a Ms. Audrey Prower of London, who had a lifelong bleeding tendency. In 1957, an American group identified the same factor in a Mr. Rufus Stuart. Factors XI and XIII were identified in 1953 and 1961, respectively.

The view that the coagulation process is a "cascade" or "waterfall" was enunciated almost simultaneously by MacFarlane in the UK and by Davie and Ratnoff in the USA, respectively.

Nomenclature

The usage of Roman numerals rather than eponyms or systematic names was agreed upon during annual conferences (starting in 1955) of hemostasis experts. In 1962, consensus was achieved on the numbering of factors I-XII. This committee evolved into the present-day International Committee on Thrombosis and Hemostasis (ICTH). Assignment of numerals ceased in 1963 after the naming of Factor XIII. The names Fletcher Factor and Fitzgerald Factor were given to further coagulation-related proteins, namely prekallikrein and high-molecular-weight kininogen, respectively.

Factors III and VI are unassigned, as thromboplastin was never identified, and actually turned out to consist of ten further factors, and accelerin was found to be activated Factor V.

Other Species

All mammals have an extremely closely related blood coagulation process, using a combined cellular and serine protease process. In fact, it is possible for any mammalian coagulation factor to "cleave" its equivalent target in any other mammal. The only non-mammalian animal known to use serine proteases for blood coagulation is the horseshoe crab.

Stealth Characteristics in Nanoparticles

It is evident from the various immune recognition processes that protein adsorption is one of the prime factors that results in activation of the immune response. Hence, any strategy to avoid immune response should be directed in interfering with the protein adsorption process.

Dysopsonins

Just as the adsorption of the opsonin proteins promote immune system activation, there is another category of proteins found in the blood known as dysopsonins that have the ability to promote the desorption of the opsonin proteins from the surface. If the dysopsonins adsorb on to the surface, further adsorption of the opsonins is inhibited. A few serum DNA binding proteins (SDBP) have been implicated to possess dysopsonin activity. However, the dysopsonins present in the blood are yet to positively identified thus limiting the scope of application of this strategy. Moreover, it remains to be confirmed whether the introduction of a dysopsonin protein from a particular species in another species can stimulate antibody production against it.

Role of Ions in Immune Recognition

In several trials involving nanocarriers of the same size, shape and chemical nature, the magnitude of opsonization on the nanocarriers was measured in the presence of citrate ions and in the absence of citrate ions. It was found that the levels of protein adsorption sharply decreased in the presence of citrate ions. What could be the reason? Citrate is a good chelator of ions, especially the divalent ions. It is therefore evident that when citrate was present in the medium, all divalent ions such as Ca^{2+} and Mg^{2+} were chelated. This implies that the opsonization, complement activation and coagulation process is dependent on the presence of divalent ions in the medium. Though this experiment provides excellent insight into the importance of divalent ions in the immune system activation, the feasibility of utilizing this property for imparting stealth characteristics is poor. This is because chelating divalent ions in vivo will lead to the disruption of the fragile ion balance that can cause undesirable effects on other metabolic functions of the system.

Role of Hydroxyl Functionalities in Immune Recognition

Zymosan, a polysaccharide from yeast, when introduced into the biological system generates a strong immune response through complement activation. Similarly, membranes made from regenerated cellulose have also induced high levels of complement activation. Sephadex or cross-linked dextran also has been found to activate the complement. What is common between these three examples apart from the fact that they all belong to the polysaccharide family? All of them contain a large number of hydroxyl groups (-OH). This suggests that presence of abundant surface –OH groups might serve to elicit a high degree of immune response through complement activation. Indeed, it was observed that hydroxymethyl polystyrene containing over 90% hydroxyl groups strongly activated the complement. On the other hand substitution of the hydroxyl groups in sephadex with diethylaminoethyl (DEAE) groups significantly reduced the magnitude of complement activation. This is on expected lines since the alternative pathway for activation of the complement involves lysis of the thioester bond in C3b- like-C3 and formation of an ester link with –OH groups on the surface of the foreign body.

Intriguingly, polystyrene, which does not contain any hydroxyl groups also activates the complement. This indicates that the presence of hydroxyl groups on the surface may not be the main reason for complement activation; rather it may serve to enhance the immune response once activated! This highlights the fact that the nature and extent of protein adsorption on the surface of the nanocarrier may have a major role in determining the magnitude of immune response. Therefore, the right strategy would be to inhibit the protein adsorption on the surface.

Inhibition of Opsonization by Steric Repulsion

The non-ionic surfactant Pluronics® (also known as Poloxamer) when coated on nanoparticles was found to reduce the protein adsorption and hence exhibit a marked decrease in the complement activation. Pluronics® is a tri-block copolymer of poly(ethylene oxide)-co-poly(propylene oxide)-co-poly(ethylene oxide). Figure shows the structure of this polymer:

$$HO(CH_2CH_2O)_n(CH_2\underset{\substack{|\\ CH_3}}{C}HO)_m(CH_2CH_2O)_pH$$

The high hydrophilicity and chain mobility of the poly(ethylene oxide) segments could be instrumental in conferring protein repulsion properties. Poly(ethylene oxide) has hydrophilic character that tends to attract water molecules around itself. Also, the poly(ethylene oxide) segments possess the ability to move about due to the small groups in their chain. A constantly moving segment with a layer of water molecules presents a surface on which protein adsorption becomes difficult.

When nanocarriers were coated with poly(ethylene glycol) that have the same structure as poly(ethylene oxide) but only differ in the molecular weight, protein adsorption was significantly reduced indicating the hydrophilicity and fast chain dynamics are both essential to provide protection against opsonization processes. Surfaces that possess only hydrophilic nature but lack chain mobility (as in the case of sephadex), exhibit very high immune response. The discovery of the protein-repelling property of poly(ethylene glycol) (PEG) has been widely exploited in developing drug delivery systems that exhibit considerably less degree of immune activation owing to the retardation in opsonization. The process of introducing poly(ethylene glycol) to a nanocarrier is termed as 'PEGylation'.

PEGylation

PEGylation (often styled pegylation) is the process of both covalent and non-covalent attachment or amalgamation of polyethylene glycol (PEG, in pharmacy called macrogol) polymer chains to molecules and macrostructures, such as a drug, therapeutic protein or vesicle, which is then described as PEGylated (pegylated). PEGylation is routinely achieved by incubation of a reactive derivative of PEG with the target molecule. The covalent attachment of PEG to a drug or therapeutic protein can "mask" the agent from the host's immune system (reduced immunogenicity and antigenicity), and increase the hydrodynamic size (size in solution) of the agent which prolongs its circulatory time by reducing renal clearance. PEGylation can also provide water solubility to hydrophobic drugs and proteins.

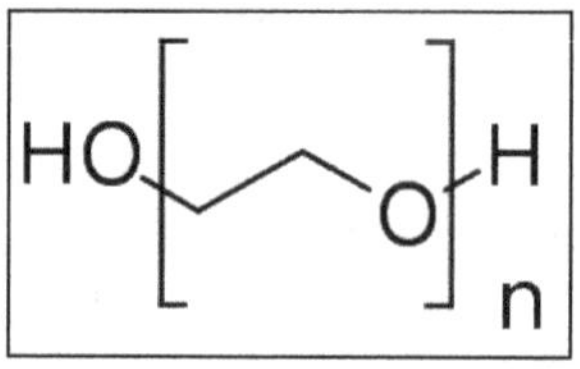

Polyethylene glycol (in pharmacy called macrogol).

History

Around 1970, Frank F. Davis, a professor of biochemistry at Rutgers University, became interested in developing a process to render usable bioactive proteins of potential medical value. After considerable study, he concluded that the attachment of an inert and hydrophilic polymer might extend blood life and control immunogenicity of the proteins. Polyethylene glycol was chosen as the polymer. A team of Davis, Theodorus Van Es and Nicholas C. Palczuk conducted animal studies and found that PEG attachment greatly extended blood life and controlled immunogenicity of the proteins. A patent application was filed in 1973 and patent issued in 1979. The inventors and Abraham Abuchowski conducted extensive additional PEGylation studies on various enzymes. In 1981 Davis and Abuchowski founded Enzon, Inc., which brought three PEGylated drugs to market. Abuchowski later founded and is CEO of Prolong Pharmaceuticals.

Overview

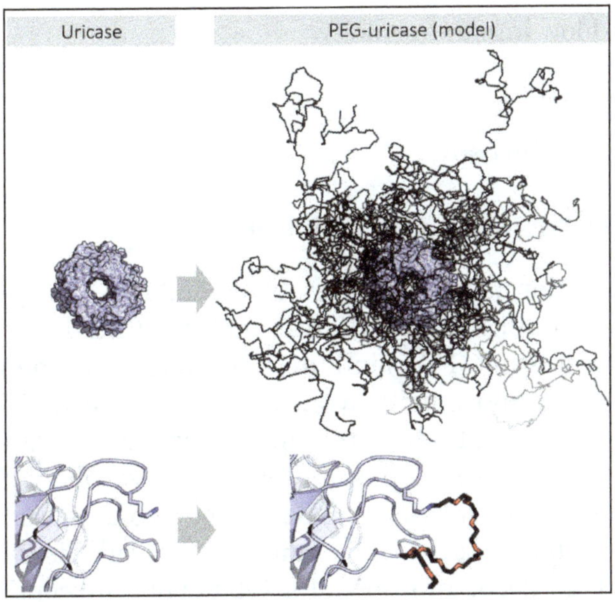

A comparison of uricase and PEG-uricase. PEG-uricase includes 40 polymers of 10kDa PEG. PEGylation improves its solubility at physiological pH, increases serum half-life and reduces immunogenicity without compromising activity. Upper images show the whole tetramer, lower images show one of the lysines that is PEGylated.

PEGylation is the process of attaching the strands of the polymer PEG to molecules, most typically peptides, proteins, and antibody fragments, that can improve the safety and efficiency of many therapeutics. It produces alterations in the physiochemical properties including changes in conformation, electrostatic binding, hydrophobicity etc. These physical and chemical changes increase systemic retention of the therapeutic agent. Also, it can influence the binding affinity of the therapeutic moiety to the cell receptors and can alter the absorption and distribution patterns.

PEGylation, by increasing the molecular weight of a molecule, can impart several significant pharmacological advantages over the unmodified form, such as:

- Improved drug solubility.
- Reduced dosage frequency, without diminished efficacy with potentially reduced toxicity.
- Extended circulating life.
- Increased drug stability.
- Enhanced protection from proteolytic degradation.

PEGylated drugs also have the following commercial advantages:

- Opportunities for new delivery formats and dosing regimens.
- Extended patent life of previously approved drugs.

PEG is a particularly attractive polymer for conjugation. The specific characteristics of PEG moieties relevant to pharmaceutical applications are:

- Water solubility.

- High mobility in solution.
- Lack of toxicity and low immunogenicity.
- Ready clearance from the body.
- Altered distribution in the body.

PEGylated Pharmaceuticals on the Market

The clinical value of PEGylation is now well established. ADAGEN (pegademase bovine) manufactured by Enzon Pharmaceuticals, Inc., US was the first PEGylated protein approved by the U.S. Food and Drug Administration (FDA) in March 1990, to enter the market. It is used to treat X-linked severe combined immunogenicity syndrome, as an alternative to bone marrow transplantation and enzyme replacement by gene therapy. Since the introduction of ADAGEN, a large number of PEGylated protein and peptide pharmaceuticals have followed and many others are under clinical trial or under development stages. Sales of the two most successful products, Pegasys and Neulasta, exceeded $5 billion in 2011. All commercially available PEGylated pharmaceuticals contain methoxypoly(ethylene glycol) or mPEG. PEGylated pharmaceuticals currently on the market (in reverse chronology by FDA approval year) include:

- Adynovate — PEGylated Antihemophilic Factor VIII for the treatment of patients with Hemophilia A. (Baxalta, 2015).
- Plegridy — PEGylated Interferon Beta-1a for the treatment of patients with relapsing forms of multiple sclerosis. (Biogen, 2014).
- Naloxegol (Movantik) — PEGylated naloxol for the treatment of opioid-induced constipation in adults patients with chronic non-cancer pain (un-pegylated methadone can cause adverse gastrointestinal reactions). (AstraZeneca, 2014).
- Peginesatide (Omontys) — once-monthly medication to treat anemia associated with chronic kidney disease in adult patients on dialysis (Affymax/Takeda Pharmaceuticals, 2012).
- Pegloticase (Krystexxa) — PEGylated uricase for the treatment of gout (Savient, 2010).
- Certolizumab pegol (Cimzia) — monoclonal antibody for treatment of moderate to severe rheumatoid arthritis and Crohn's disease, an inflammatory gastrointestinal disorder (Nektar/UCB Pharma, 2008).
- Methoxy polyethylene glycol-epoetin beta (Mircera) — PEGylated form of erythropoetin to combat anemia associated with chronic kidney disease (Roche, 2007).
- Pegaptanib (Macugen) — used to treat neovascular age-related macular degeneration (Pfizer, 2004).
- Pegfilgrastim (Neulasta) — PEGylated recombinant methionyl human granulocyte colony-stimulating factor for severe cancer chemotherapy-induced neutropenia (Amgen, 2002).

- Pegvisomant (Somavert) — PEG-human growth hormone mutein antagonist for treatment of Acromegaly (Pfizer, 2002).
- Peginterferon alfa-2a (Pegasys) — PEGylated interferon alpha for use in the treatment of chronic hepatitis C and hepatitis B (Hoffmann-La Roche, 2001).
- Doxorubicin HCl liposome (Doxil/Caelyx) — PEGylated liposome containing doxorubicin for the treatment of cancer (Ortho Biotech/Schering-Plough, 2001).
- Peginterferon alfa-2b (PegIntron) — PEGylated interferon alpha for use in the treatment of chronic hepatitis C and hepatitis B (Schering-Plough/Enzon, 2000).
- Pegaspargase (Oncaspar) — PEGylated L-asparaginase for the treatment of acute lymphoblastic leukemia in patients who are hypersensitive to the native unmodified form of L-asparaginase (Enzon, 1994). This drug was recently approved for front line use.
- Pegademase bovine (Adagen) — PEG-adenosine deaminase for the treatment of severe combined immunodeficiency disease (SCID) (Enzon, 1990).

PEGylation Process

The first step of the PEGylation is the suitable functionalization of the PEG polymer at one or both terminals. PEGs that are activated at each terminus with the same reactive moiety are known as "homobifunctional", whereas if the functional groups present are different, then the PEG derivative is referred as "heterobifunctional" or "heterofunctional." The chemically active or activated derivatives of the PEG polymer are prepared to attach the PEG to the desired molecule.

The overall PEGylation processes used to date for protein conjugation can be broadly classified into two types, namely a solution phase batch process and an on-column fed-batch process. The simple and commonly adopted batch process involves the mixing of reagents together in a suitable buffer solution, preferably at a temperature between 4 and 6 °C, followed by the separation and purification of the desired product using a suitable technique based on its physicochemical properties, including size exclusion chromatography (SEC), ion exchange chromatography (IEX), hydrophobic interaction chromatography (HIC) and membranes or aqueous two phase systems.

The choice of the suitable functional group for the PEG derivative is based on the type of available reactive group on the molecule that will be coupled to the PEG. For proteins, typical reactive amino acids include lysine, cysteine, histidine, arginine, aspartic acid, glutamic acid, serine, threonine, tyrosine. The N-terminal amino group and the C-terminal carboxylic acid can also be used as a site specific site by conjugation with aldehyde functional polymers.

The techniques used to form first generation PEG derivatives are generally reacting the PEG polymer with a group that is reactive with hydroxyl groups, typically anhydrides, acid chlorides, chloroformates and carbonates. In the second generation PEGylation chemistry more efficient functional groups such as aldehyde, esters, amides etc. made available for conjugation.

As applications of PEGylation have become more and more advanced and sophisticated, there has been an increase in need for heterobifunctional PEGs for conjugation. These heterobifunctional PEGs are very useful in linking two entities, where a hydrophilic, flexible and biocompatible spacer is needed. Preferred end groups for heterobifunctional PEGs are maleimide, vinyl sulfones, pyridyl disulfide, amine, carboxylic acids and NHS esters.

Third generation pegylation agents, where the shape of the polymer has been branched, Y shaped or comb shaped are available which show reduced viscosity and lack of organ accumulation.

Limitations of PEGylation

Unpredictability in clearance times for PEGylated compounds may lead to the accumulation of large molecular weight compounds in the liver leading to inclusion bodies with no known toxicologic consequences. Furthermore, alteration in the chain length may lead to unexpected clearance times *in vivo*.

Future Perspectives

Four decades of development in PEGylation technology have proven its pharmacological advantages and acceptability. As a multibillion-dollar annual business with growing interest from both emerging biotechnology and established multinational pharmaceutical companies, there is great scientific and commercial interest in improving present methodologies and in introducing innovative process variations.

Polyethylene Glycol

Polyethylene glycol (PEG) is a polyether compound with many applications from industrial manufacturing to medicine. PEG is also known as polyethylene oxide (PEO) or polyoxyethylene (POE), depending on its molecular weight. The structure of PEG is commonly expressed as $H-(O-CH_2-CH_2)_n-OH$.

Available Forms and Nomenclature

PEG, *PEO*, and *POE* refer to an oligomer or polymer of ethylene oxide. The three names are chemically synonymous, but historically *PEG* is preferred in the biomedical field, whereas *PEO* is more prevalent in the field of polymer chemistry. Because different applications require different polymer chain lengths, *PEG* has tended to refer to oligomers and polymers with a molecular mass below 20,000 g/mol, *PEO* to polymers with a molecular mass above 20,000 g/mol, and *POE* to a polymer of any molecular mass. PEG and PEO are liquids or low-melting solids, depending on their molecular weights. PEGs are prepared by polymerization of ethylene oxide and are commercially available over a wide range of molecular weights from 300 g/mol to 10,000,000 g/mol. While PEG and PEO with different molecular weights find use in different applications, and have different physical properties (e.g. viscosity) due to chain length effects, their chemical properties are nearly identical. Different forms of PEG are also available, depending on the initiator used for the polymerization process – the most common initiator is a monofunctional methyl ether PEG, or methoxypoly(ethylene glycol), abbreviated mPEG. Lower-molecular-weight PEGs are also available as purer oligomers, referred to as monodisperse, uniform, or discrete. Very high purity PEG has recently been shown to be crystalline, allowing determination of a crystal structure by x-ray diffraction. Since purification and separation of pure oligomers is difficult, the price for this type of quality is often 10–1000 fold that of polydisperse PEG.

PEGs are also available with different geometries.

- *Branched* PEGs have three to ten PEG chains emanating from a central core group.
- *Star* PEGs have 10 to 100 PEG chains emanating from a central core group.
- *Comb* PEGs have multiple PEG chains normally grafted onto a polymer backbone.

The numbers that are often included in the names of PEGs indicate their average molecular weights (e.g. a PEG with n = 9 would have an average molecular weight of approximately 400 daltons, and would be labeled PEG 400.) Most PEGs include molecules with a distribution of molecular weights (i.e. they are polydisperse). The size distribution can be characterized statistically by its weight average molecular weight (Mw) and its number average molecular weight (Mn), the ratio of which is called the polydispersity index (Mw/Mn). MW and Mn can be measured by mass spectrometry.

PEGylation is the act of covalently coupling a PEG structure to another larger molecule, for example, a therapeutic protein, which is then referred to as a *PEGylated* protein. PEGylated interferon alfa-2a or −2b are commonly used injectable treatments for hepatitis C infection.

PEG is soluble in water, methanol, ethanol, acetonitrile, benzene, and dichloromethane, and is insoluble in diethyl ether and hexane. It is coupled to hydrophobic molecules to produce non-ionic surfactants.

PEGs potentially contain toxic impurities, such as ethylene oxide and 1,4-dioxane. Ethylene Glycol and its ethers are nephrotoxic if applied to damaged skin.

Polyethylene oxide (PEO, M_w 4 kDa) nanometric crystallites (4 nm).

Polyethylene glycol (PEG) and related polymers (PEG phospholipid constructs) are often sonicated when used in biomedical applications. However, as reported by Murali et al., PEG is very sensitive to sonolytic degradation and PEG degradation products can be toxic to mammalian cells. It is, thus, imperative to assess potential PEG degradation to ensure that the final material does not contain undocumented contaminants that can introduce artifacts into experimental results.

PEGs and methoxypolyethylene glycols are manufactured by Dow Chemical under the tradename *Carbowax* for industrial use, and *Carbowax Sentry* for food and pharmaceutical use. They vary in consistency from liquid to solid, depending on the molecular weight, as indicated by a number following the name. They are used commercially in numerous applications, including as surfactants, in foods, in cosmetics, in pharmaceutics, in biomedicine, as dispersing agents, as solvents, in ointments, in suppository bases, as tablet excipients, and as laxatives. Some specific groups are lauromacrogols, nonoxynols, octoxynols, and poloxamers.

Macrogol, used as a laxative, is a form of polyethylene glycol. The name may be followed by a number which represents the average molecular weight (e.g. macrogol 3350, macrogol 4000 or macrogol 6000).

Production

Polyethylene glycol 400, pharmaceutical quality.

Polyethylene glycol 4000, pharmaceutical quality.

Polyethylene glycol is produced by the interaction of ethylene oxide with water, ethylene glycol, or ethylene glycol oligomers. The reaction is catalyzed by acidic or basic catalysts. Ethylene glycol and its oligomers are preferable as a starting material instead of water, because they allow the creation of polymers with a low polydispersity (narrow molecular weight distribution). Polymer chain length depends on the ratio of reactants.

$$HOCH_2CH_2OH + n(CH_2CH_2O) \rightarrow HO(CH_2CH_2O)_{n+1}H$$

Depending on the catalyst type, the mechanism of polymerization can be cationic or anionic. The anionic mechanism is preferable because it allows one to obtain PEG with a low polydispersity.

Polymerization of ethylene oxide is an exothermic process. Overheating or contaminating ethylene oxide with catalysts such as alkalis or metal oxides can lead to runaway polymerization, which can end in an explosion after a few hours.

Polyethylene oxide, or high-molecular weight polyethylene glycol, is synthesized by suspension polymerization. It is necessary to hold the growing polymer chain in solution in the course of the polycondensation process. The reaction is catalyzed by magnesium-, aluminium-, or calcium-organoelement compounds. To prevent coagulation of polymer chains from solution, chelating additives such as dimethylglyoxime are used.

Alkaline catalysts such as sodium hydroxide (NaOH), potassium hydroxide (KOH), or sodium carbonate (Na_2CO_3) are used to prepare low-molecular-weight polyethylene glycol.

Uses

Medical Uses

- PEG is the basis of a number of laxatives. Whole bowel irrigation with polyethylene glycol and added electrolytes is used for bowel preparation before surgery or colonoscopy.
- PEG is also used as an excipient in many pharmaceutical products.
- When attached to various protein medications, polyethylene glycol allows a slowed clearance of the carried protein from the blood.

The remains of the 16th century carrack *Mary Rose* undergoing conservation treatment with PEG in the 1980s.

Chemical Uses

- Polyethylene glycol has a low toxicity and is used in a variety of products. The polymer is used as a lubricating coating for various surfaces in aqueous and non-aqueous environments.
- Since PEG is a flexible, water-soluble polymer, it can be used to create very high osmotic

pressures (on the order of tens of atmospheres). It also is unlikely to have specific interactions with biological chemicals. These properties make PEG one of the most useful molecules for applying osmotic pressure in biochemistry and biomembranes experiments, in particular when using the osmotic stress technique.

- Polyethylene glycol is also commonly used as a polar stationary phase for gas chromatography, as well as a heat transfer fluid in electronic testers.
- PEG has also been used to preserve objects that have been salvaged from underwater, as was the case with the warship *Vasa* in Stockholm, the Mary Rose in England, the Ma'agan Michael Ship in Israel, and artifacts from the Steamboat Arabia in Kansas City, Missouri. It replaces water in wooden objects, making the wood dimensionally stable and preventing warping or shrinking of the wood when it dries. In addition, PEG is used when working with green wood as a stabilizer, and to prevent shrinkage.
- PEG has been used to preserve the painted colors on Terra-Cotta Warriors unearthed at a UNESCO World Heritage site in China. These painted artifacts were created during the Qin Shi Huang Di dynasty (first emperor of China). Within 15 seconds of the terra-cotta pieces being unearthed during excavations, the lacquer beneath the paint begins to curl after being exposed to the dry Xian air. The paint would subsequently flake off in about four minutes. The German Bavarian State Conservation Office developed a PEG preservative that when immediately applied to unearthed artifacts has aided in preserving the colors painted on the pieces of clay soldiers.
- PEG is often used (as an internal calibration compound) in mass spectrometry experiments, with its characteristic fragmentation pattern allowing accurate and reproducible tuning.
- PEG derivatives, such as narrow range ethoxylates, are used as surfactants˙
- PEG has been used as the hydrophilic block of amphiphilic block copolymers used to create some polymersomes.

Biological Uses

- PEG is commonly used as a precipitant for plasmid DNA isolation and protein crystallization. X-ray diffraction of protein crystals can reveal the atomic structure of the proteins.
- PEG is used to fuse two different types of cells, most often B-cells and myelomas in order to create hybridomas.
- Polymer segments derived from PEG polyols impart flexibility to polyurethanes for applications such as elastomeric fibers (spandex) and foam cushions.
- In microbiology, PEG precipitation is used to concentrate viruses. PEG is also used to induce complete fusion (mixing of both inner and outer leaflets) in liposomes reconstituted *in vitro*.
- Gene therapy vectors (such as viruses) can be PEG-coated to shield them from inactivation by the immune system and to de-target them from organs where they may build up and have a toxic effect. The size of the PEG polymer has been shown to be important, with larger polymers achieving the best immune protection.

- PEG is a component of stable nucleic acid lipid particles (SNALPs) used to package siRNA for use *in vivo*.
- In blood banking, PEG is used as a potentiator to enhance detection of antigens and antibodies.
- When working with phenol in a laboratory situation, PEG 300 can be used on phenol skin burns to deactivate any residual phenol.

Commercial Uses

- PEG is the basis of many skin creams (as *cetomacrogol*) and personal lubricants (frequently combined with glycerin).
- PEG is used in a number of toothpastes as a dispersant. In this application, it binds water and helps keep xanthan gum uniformly distributed throughout the toothpaste.
- PEG is also under investigation for use in body armor, and in tattoos to monitor diabetes.
- In low-molecular-weight formulations (i.e. PEG 400), it is used in Hewlett-Packard designjet printers as an ink solvent and lubricant for the print heads.
- PEG is also one of the main ingredients in paintball fills, because of its thickness and flexibility. However, as early as 2006, some Paintball manufacturers began substituting cheaper oil-based alternatives for PEG.
- PEG is also used as an anti-foaming agent in food – its INS number is 1521 or E1521 in the EU.

Industrial Uses

- Nitrate ester-plasticized polyethylene glycol is used in Trident II submarine-launched ballistic missile solid rocket fuel.
- Dimethyl ethers of PEG are the key ingredient of Selexol, a solvent used by coal-burning, integrated gasification combined cycle (IGCC) power plants to remove carbon dioxide and hydrogen sulfide from the gas waste stream.
- PEG has been used as the gate insulator in an electric double-layer transistor to induce superconductivity in an insulator.
- PEG is also used as a polymer host for solid polymer electrolytes. Although not yet in commercial production, many groups around the globe are engaged in research on solid polymer electrolytes involving PEG, with the aim of improving their properties, and in permitting their use in batteries, electro-chromic display systems, and other products in the future.
- PEG is injected into industrial processes to reduce foaming in separation equipment.
- PEG is used as a binder in the preparation of technical ceramics.

Health Effects

PEG is generally considered biologically inert and safe. However, a minority of people are allergic to it. Allergy to PEG is usually discovered after a person has been diagnosed with an allergy to an increasing number of seemingly unrelated products, including processed foods, cosmetics, drugs, and other substances that contain PEG or were manufactured with PEG.

Factors Influencing PEGylation

Mode of Attachment

The first important aspect in PEGylation is that the PEG coating should be able to withstand the shear forces that can be experienced by the carrier as it moves through the circulating system. Therefore, physical adsorption of PEG on the carrier surface might not be the best technique as there is a strong possibility that the coating might be desorbed in the biological system. Hence, introduction of PEG polymeric chains on the surface of the carrier through covalent binding is preferred and widely practiced, especially in the case of polymeric nanocarrier, liposomes and protein carriers. The covalent linking of PEG to different functional groups present in various carriers is simple as the terminal group in PEG is the reactive –OH group that is also amenable to modifications by other molecules so that the terminal group can be tailored to complement the functionality present in the carrier group. The PEG chains containing amine, carboxyl, methoxy, maleimide, N-hydroxysuccinimide etc. are available. The PEG can be introduced as a block with the polymer forming the carrier core. The block of PEG can be introduced at the end or in the middle depending upon whether a diblock or triblock copolymer is desired. In the case of di- and triblock copolymers containing PEG, the PEGylation is referred to as 'end-on' type, as the PEG groups will be present in the backbone of the polymer.Alternately, the PEG chains can be grafted to the main chain of the polymer. The number of PEG chains attached to a single polymer chain can be modified in this technique. This will influence the surface density as well as morphology of the PEG chains. This type of PEGylation is referred to as 'side-on' type, as PEG is present in the side chains and not in the backbone of the polymer. Figure depicts an end-on type and side-on type of PEGylation in polymers.

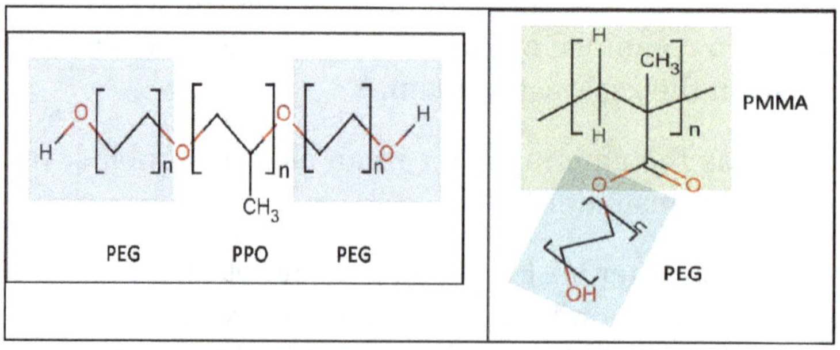

End-on and side-on type of PEGylation.

When a polymer carrier is modified with PEG, due to the highly hydrophilic nature of the PEG chains, they tend to project outwards into the aqueous environment so as to maximize their contact with water. The polymer forming the carrier molecule becomes the core. Experiments have revealed that the PEG segments always try to localize in the surface that is beneficial for the protein repulsion requirement. The nature of PEG linkage determines the configuration of the PEG chains in the carrier. This in turn can influence the protein repelling ability of the PEG chains. Four types of PEG configurations on polymeric carriers have been defined – brush, loop, bunch and star. Figure shows the different types of configurations of PEG on nanocarriers.

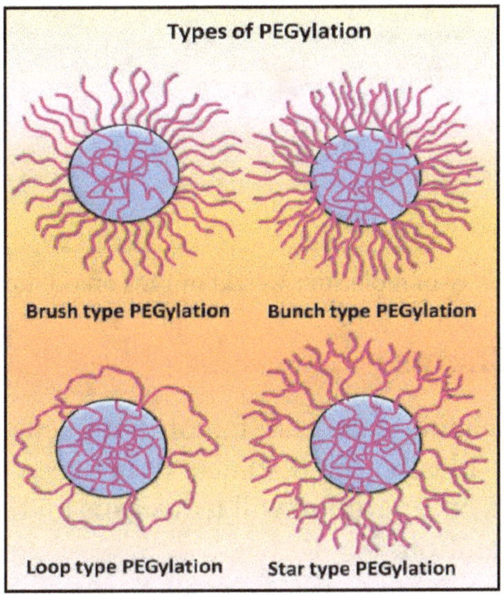

Different types of PEG configurations in nanocarriers.

The brush configuration can be achieved with diblock copolymers and is the most commonly employed. The triblock copolymers containing PEG in the central segment form structures with a loop configuration of PEG. The graft polymers containing multiple PEG chains in the side chains form bunch type of configuration. The star configuration is least explored and is obtained using special type of branched copolymer designs. The brush and loop configurations were found to be comparatively more effective than the other configurations in repelling proteins. This difference may mainly arise from the inter-chain steric hindrances that may be posed by the PEG chains in the bunch and star configurations. In other words, the 'end-on' configuration works best for PEGylated carriers!

Molecular Weight of PEG

Another important parameter that can influence the protein repelling characteristics of the PEG coating is the molecular weight of PEG. As PEG is a linear polymer, the molecular weight is directly responsible for its chain length. It has been found that the protein adsorption remains unchanged if the molecular weight of the surface PEG chains is less than 1000 Da. Hence the minimum molecular weight of PEG that needs to be used is 1000 Da. As the molecular weight increases, the protein adsorption decreases up to a molecular weight of 5000 Da, beyond which not much significant decrease in protein adsorption is observed. Increasing the molecular weight beyond 5000 Da will no doubt be able to decrease the opsonization and immune response enabling it to remain in the blood circulation for a longer time. However, it will also retard cell uptake of the nanocarrier as the same factors that retard opsonization and immune response will also contribute to reduction in cell uptake. Hence, it is essential that the molecular weight of the PEG chains be in the range 2000-5000 Da to ensure sufficient retardation of opsonization as well as not compromise too much on the cell uptake of the nanocarrier. Figure represents the influence of the molecular weight of PEG on cell uptake.

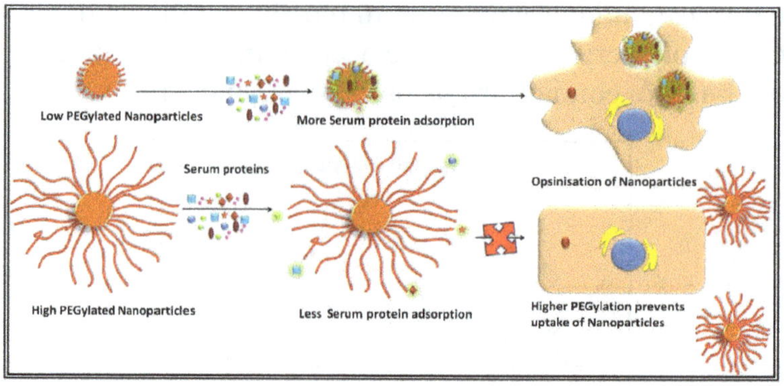

Influence of molecular weight of PEG on cell uptake.

Surface Density of PEG Chains

The surface density of the PEG chains is another factor that needs to be in the optimum range for effective stealth characteristics. Ideally, the distance 'd' between the PEG chains on the surface should be 2.2 nm. Higher values of separation will leave gaps on the surface where proteins could adsorb leading to immune recognition.

Attachment of Targeting Ligand

If the surface is totally covered with PEG, then the cell uptake will be reduced considerably. Hence, current approaches involve covalently linking a targeting ligand to the terminus of the PEG chains to ensure that the carrier can selectively enter a target cell using the concept of 'receptor-mediated endocytosis'. The choice of the targeting ligand is based on the expression of surface receptors on the cell surface that are unique to the target cell. To facilitate covalent linking of the targeting ligand, the terminal group of the PEG chain could be suitably chosen.

Why should the targeting ligand be linked to the terminus of the PEG chain and not directly on the surface of the nanocarrier itself? Well, if the targeting ligand were directly present on the surface of the nanocarrier itself, the PEG chains will mask the ligand from its target receptor and hence binding and subsequent entry of the nanocarrier will never occur. On the other hand, if the targeting ligand is on the terminus of the PEG chain, it can selectively bind with the target receptor.

It should be noted that the PEGylation cannot permanently avoid opsonization. The PEGylation only aids in increasing the blood circulation time of the nanocarrier thereby increasing its chances of reaching the target! Thus while PEGylation of a nanocarrier can enhance the residence time of the carrier in blood from a few minutes (less than 10 min in most cases) to a few hours, the ultimate destination of the PEGylated carrier still will be the MPS. Therefore, better strategies to completely inhibit opsonization and MPS activation need to be identified.

Other Alternatives to PEG

Other alternatives to PEGylation have been investigated and a few candidates have been identified to have the potential to impart stealth characteristics to the nanocarrier. Whether these could replace PEG? The answer is still NO, as they have not been demonstrated to be superior to PEG. The list of potential replacements for PEG includes dextran (the soluble form), pullulan, monosialoganglioside (GM1),heparin, poly(vinyl pyrrolidone), poly(acryl amide), poly(2-methyl2-oxazoline) and poly(2-ethyl oxazoline). Among these, dextran has been the most promising. Interestingly, the loop configuration for dextran is not effective in protein repulsion unlike PEG. In the case of dextran, a side- on type of linkage provides better results with the brush configuration and bunch configuration being more effective.The linking of heparin to a nanocarrier offers an additional advantage in that it does not allow platelet adhesion or aggregation on the surface. It also does not allow uptake of the nanocarriers by the macrophages and other members of the MPS. As the heparin chains on the surface also retard opsonization due to steric repulsion effects, its anti-complement, anti-MPS and anti-coagulant action makes it a better option than other candidates being explored for imparting stealth characteristics.

References

- Elzoghby A, Samy W, Elgindy N. Protein-based nanocarriers as promising drug and gene delivery systems. Journal of Controlled Release. 2012;161(1):38-49
- Sigel H, Sigel A, eds. (2009). Metallothioneins and Related Chelators (Metal Ions in Life Sciences). Metal Ions in Life Sciences. 5. Cambridge, England: Royal Society of Chemistry. ISBN 1-84755-899-2
- Nutrition, Center for Food Safety and Applied. “Potential Contaminants - 1,4-Dioxane A Manufacturing By-product”. www.fda.gov. Retrieved 26, May 2020
- Kahovec, J.; Fox, R. B.; Hatada, K. (2002). “Nomenclature of regular single-strand organic polymers”. Pure an Applied Chemistry. 74 (10): 1921–1956. doi:10.1351/pac200274101921
- Schmaier, Alvin H.; Lazarus, Hillard M. (2011). Concise guide to hematology. Chichester, West Sussex, UK: Wiley-Blackwell. p. 91. ISBN 978-1-4051-9666-6
- David Lillicrap; Nigel Key; Michael Makris; Denise O’Shaughnessy (2009). Practical Hemostasis and Thrombosis. Wiley-Blackwell. pp. 1–5. ISBN 1-4051-8460-4
- Iverson, Cheryl, et al. (eds) (2007). “12.1.1 Use of Italics”. AMA Manual of Style (10th ed.). Oxford, Oxfordshire: Oxford University Press. ISBN 978-0-19-517633-9
- Basu, A. (2010). “Cellular Responses to Cisplatin-Induced DNA Damage”. Journal of Nucleic Acids. 2010: 1–16. doi:10.4061/2010/201367
- Veronese, F. M. (2001). “Peptide and protein PEGylation: A review of problems and solutions”. Biomaterials. 22 (5): 405–17. doi:10.1016/s0142-9612(00)00193-9. PMID 11214751
- “Current EU approved additives and their E Numbers”. UK Government – Food Standards Agency. Retrieved 21, August 2020
- Fee, Conan J.; Damodaran, Vinod B. (2012). “Biopharmaceutical Production Technology”: 199. doi:10.1002/9783527653096.ch7. ISBN 9783527653096
- Vignais, Paulette M.; Pierre Vignais (2010). Discovering Life, Manufacturing Life: How the experimental method shaped life sciences. Berlin: Springer. ISBN 90-481-3766-7
- Yu M, Zhao J, Feng S. Vitamin E TPGS prodrug micelles for hydrophilic drug delivery with neuroprotective effects. International Journal of Pharmaceutics. 2012;438(1-2):98-106
- Rossi, J.J. (2006). “RNAi therapeutics: SNALPing siRNAs in vivo”. Gene Therapy. 13 (7): 583–584. PMID 17526070. doi:10.1038/sj.gt.3302661
- Alǻn D. Michelson (26 October 2006). Platelets. Academic Press. pp. 3–5. ISBN 978-0-12-369367-9. Retrieved 18, August 2020
- American Psychological Association (2010), “4.21 Use of Italics”, The Publication Manual of the American Psychological Association (6th ed.), Washington, DC, USA: APA, ISBN 978-1-4338-0562-2

- Tosi, Michael F. (2005-08-01). "Innate immune responses to infection". Journal of Allergy and Clinical Immunology. 116 (2): 241–249. doi:10.1016/j.jaci.2005.05.036
- Veronese, FM; Harris, JM (2002). "Introduction and overview of peptide and protein pegylation". Advanced drug delivery reviews. 54 (4): 453–6. doi:10.1016/S0169-409X(02)00020-0. PMID 12052707
- Barret, James (1980). Basic Immunology and its Medical Application (2 ed.). St.Louis: The C.V. Mosby Company. ISBN 0-8016-0495-8
- Rothman, S. S. (2002). Lessons from the living cell: the culture of science and the limits of reductionism. New York: McGraw-Hill. ISBN 0-07-137820-0

Permissions

All chapters in this book are published with permission under the Creative Commons Attribution Share Alike License or equivalent. Every chapter published in this book has been scrutinized by our experts. Their significance has been extensively debated. The topics covered herein carry significant information for a comprehensive understanding. They may even be implemented as practical applications or may be referred to as a beginning point for further studies.

We would like to thank the editorial team for lending their expertise to make the book truly unique. They have played a crucial role in the development of this book. Without their invaluable contributions this book wouldn't have been possible. They have made vital efforts to compile up to date information on the varied aspects of this subject to make this book a valuable addition to the collection of many professionals and students.

This book was conceptualized with the vision of imparting up-to-date and integrated information in this field. To ensure the same, a matchless editorial board was set up. Every individual on the board went through rigorous rounds of assessment to prove their worth. After which they invested a large part of their time researching and compiling the most relevant data for our readers.

The editorial board has been involved in producing this book since its inception. They have spent rigorous hours researching and exploring the diverse topics which have resulted in the successful publishing of this book. They have passed on their knowledge of decades through this book. To expedite this challenging task, the publisher supported the team at every step. A small team of assistant editors was also appointed to further simplify the editing procedure and attain best results for the readers.

Apart from the editorial board, the designing team has also invested a significant amount of their time in understanding the subject and creating the most relevant covers. They scrutinized every image to scout for the most suitable representation of the subject and create an appropriate cover for the book.

The publishing team has been an ardent support to the editorial, designing and production team. Their endless efforts to recruit the best for this project, has resulted in the accomplishment of this book. They are a veteran in the field of academics and their pool of knowledge is as vast as their experience in printing. Their expertise and guidance has proved useful at every step. Their uncompromising quality standards have made this book an exceptional effort. Their encouragement from time to time has been an inspiration for everyone.

The publisher and the editorial board hope that this book will prove to be a valuable piece of knowledge for students, practitioners and scholars across the globe.

Index

www.ingramcontent.com/pod-product-compliance
Lightning Source LLC
LaVergne TN
LVHW080725210926
848485LV00003B/522